Geriatric Bioscience

Geriatric Bioscience

The Link between Aging and Disease

DAVID HAMERMAN, M.D.
Distinguished University Professor of Medicine,
Division of Geriatrics
Albert Einstein College of Medicine
and Montefiore Medical Center
Bronx, New York

Foreword by Robert N. Butler, M.D.

The Johns Hopkins University Press
Baltimore

© 2007 The Johns Hopkins University Press
All rights reserved. Published 2007
Printed in the United States of America on acid-free paper
2 4 6 8 9 7 5 3 1

The Johns Hopkins University Press
2715 North Charles Street
Baltimore, Maryland 21218-4363
www.press.jhu.edu

Library of Congress Cataloging-in-Publication Data

Hamerman, David.
Geriatric bioscience : the link between aging and disease / David Hamerman ; foreword by
Robert N. Butler.
p. ; cm.
Includes bibliographical references and index.
ISBN 13: 978-0-8018-8692-8 (hardcover : alk. paper)
ISBN 10: 0-8018-8692-9 (hardcover : alk. paper)
1. Geriatrics. I. Title. [DNLM: 1. Aging—pathology. 2. Age Factors. 3. Disease Susceptibility—
physiopathology.
4. Geriatrics.   WT 104 H214g 2007]
RC952.H247 2007
618.97—dc22   2007013963

A catalog record for this book is available from the British Library.

Contents

Color plates follow page 168.

Foreword

In 1825, Benjamin Gompertz, a London actuary, observed a logarithmic increase in the "force of mortality" every seven years after maturity. This curve, known today as the Gompertz curve, is the external manifestation of the underlying basic biology of aging.

Today, with the contributions of genomics, regenerative medicine, and the growing knowledge of the neuroendocrine immune system, the possibilities exist of extending both the quality and the length of life through a better understanding of aging. Nonetheless, in 2005 no more than $100 to $200 million of the nearly $30 billion budget of the National Institutes of Health was devoted to the basic science of aging. Notwithstanding the successful disease and organ mission approach most characteristic of the National Institutes of Health, it is time for a shift in both our thinking and the funding of biomedical research to dramatically increase our attention to the basic biology of aging and the interrelationship of aging and disease.

One percent of annual Medicare expenditures would be an appropriate amount to set aside for research on aging. This would be approximately $3 billion for FY 2007, with $1 billion used to study the basic biology of aging. I would recommend another $1 billion for the study of the interaction of aging and disease through the creation of a trans-NIH initiative (a technique that has worked before). The remaining $1 billion should be devoted to social-behavioral and environmental factors as well as clinical trials and infrastructure to provide an understanding of the lifestyle factors in which genes express themselves, and to research on health services delivery.

Geriatric Bioscience: The Link between Aging and Disease is a wonderful book that builds on our growing understanding of aging and the interaction of aging and disease. A direct and appropriate descendent of Gompertz's work, it speaks to the strong possibilities of prevention and treatment.

Robert N. Butler, M.D.
President and CEO
International Longevity Center

Preface

Geriatrics has been for me both a personal evolution and a professional endeavor. I entered the field in 1983, when Louis Sherwood, a noted endocrinologist and chair of the Department of Medicine at Montefiore Medical Center, Albert Einstein College of Medicine, in New York (a position I had left three years earlier), asked me to start a division of geriatrics. Perhaps my background in rheumatology with a chronic disease orientation, or possibly because he did not quite know what to do with me, may have accounted for his choice.

Other colleagues' stories are similar. Knight Steel reflected on the limited "visibility and vitality" of geriatrics in the United States three decades ago: when asked about what he was going to do in Boston in 1977, he responded, "I am interested in geriatrics." His questioner, looking puzzled, asked, "Jerry who?" (Steel, 2004). T. Franklin Williams, in an academic environment at the University of North Carolina and specializing in diabetes as a key chronic disease "not well addressed in the aspect of people's lives," was invited to Rochester in 1968 to head the Monroe County Infirmary for chronically ill patients. He subsequently went on to direct the National Institute on Aging (NIA) (Williams, 2004). James Goodwin (2004) experienced the "intrinsic appeal of the cohort that is now very old," with thoughts about new approaches to their health care.

Many specialists or generalists in internal medicine embarked on a new career in starting a division of geriatrics. Our challenge was to learn about the content of this emerging field: exploring its unique diversity of health care sites—including the nursing home, into which we had never imagined entering, much less teaching or doing research; trying to define geriatrics as a specialty or as a primary care service; responding to the opportunity to recruit new faculty and fellows not yet decided on a career; seizing the opportunity to encourage and awaken an interest in medical students; and, perhaps most significantly and difficult, establishing a foothold for this discipline within a Department of Medicine where other divisions, much more visible, experienced, and traditional, already existed. Only a few schools would have the endowment and unique prestigious individual to create an independent Department of Geriatrics. Robert Butler, who was the first director of the National Institute on Aging, established the first Department of Geriatrics at Mt. Sinai School of Medicine in New York, and others followed at medical schools in Arkansas, Oklahoma, Hawaii, and Florida.

My academic career, now of considerable duration, evolved in segments measured in decades: the first was spent doing basic laboratory research on the biochemistry of components of human joints; the second was as chair of a Department of Medicine; and the third and fourth were involved in starting a Division of Geriatrics and then moving on to direct an endowed Gerontology Center. Now, in an emeritus capacity as director of that center, I have begun the fifth decade by writing this book. In it, I hope to convey aspects of the interrelations of the biology of aging and disease—what I call geriatric bioscience—to readers in the course of their education, training, academic career in geriatrics, or engagement in aging-related research.

For those who are contemplating a career in geriatrics or who have not been in the field long, I want first to share my experience of the emergence of geriatrics and convey some thoughts on its future. Students and trainees need a bit of historical perspective if they are to make an informed choice, especially in the pursuit of a full-time academic career in geriatrics. This path has become less frequently traveled for new trainees, not only because they perceive early-on the rewards and priorities of patient care in Departments of Medicine, but also because they are uninformed about the structure of academic life and its expectations.

In 1987, in an editorial on geriatric practice, Claire Maklan and I (Hamerman and Maklan, 1987) wrote about this emerging field within internal medicine, surveying the opportunities for geriatrics to address the perceived "primary care crisis"; exploring the diversity of research opportunities, both basic and clinical, with reference to aging and the inception of disease; and citing the added dimension beyond geriatrics itself that *gerontology* represented: a *team endeavor* with an array of additional health care professionals—including nurses, social workers, pharmacists, and rehabilitation specialists—among others, and engaging in an academic interchange with those concerned with *aging* more than with *disease*—bioethicists, lawyers, health policy planners, sociologists, and basic researchers, again among others.

Writing this book reawakened for me the excitement and interest I earlier experienced at approaching the care of patients from the perspective of the basic science underlying their disease. I would like readers at all levels of training—from medical students and fellows to seasoned clinicians and researchers—to share in this excitement about which they may not have been aware or may have missed the role models to inspire them in this way. The "science" of geriatrics that emerges in these chapters is a new way to look at the interactive path between the biology of aging and its relation to disease expression in older persons, with important implications for health care and for therapies as well.

Gaining familiarity with the basic biology of aging and its relevance to disease, as described in this book, may be a further challenge for heavily engaged medical practitioners, especially in the present environment, where geriatrics is seeking to define itself (Chapter 1), and uncertainty exists about the very "survival" of primary care, which is "facing a confluence of factors that could spell disaster" (Bodenheimer, 2006). But it is also evident that *medicine* is undergoing a transformation marked by dramatic advances in "genomic medicine" (Chapter 2), imposing added responsibilities on health care professionals (and lay persons) to be ready for this era by understanding its background and potential application to clinical practice (Rubenstein and Roy, 2005). Achieving this understanding will not be easy. Billings et al. (2005) describe intensive workshops on "genomic medicine" for professionals, executives, and physicians involved in health care organizations, in which the participants, sadly, reported their limited enlightenment as an outcome. There is now widespread interest in what is termed *conserved* genetic regulation of metabolic pathways—extending from yeast to humans—that sheds light on aging and longevity, which is extensively discussed in the Epilogue. Serlucca and Fishman (2006) note the relevance to human disease in a commentary in *Proceedings of the National Academy of Sciences* entitled "Big, bad hearts: from flies to man." Insights on genetic screens of heart function in the fly drosophila may shed light on human cardiomyopathy.

Thus, a challenge does exist—and further examples abound in this book—for the clinician to be aware of the "translation" of advances in the laboratory by basic scientists into possible applications at the bedside for diagnosis, therapy, and improved patient outcome (Hamerman and Zeleznik, 2001). The director of the National Institutes of Health (Zerhouni, 2005) called for a "new vision" for "advancing knowledge of human disorders in a translational context." The former director of the National Heart, Lung, and Blood Institute (Lenfant, 2003) urged that "clinical research to clinical practice not be lost in translation."

However, the "road from the laboratory to the bedside is seldom straight" (Marcum, 2005). To begin the journey, the clinician and the experimentalist have to develop a dialogue: each must understand what the other does. This process was eloquently expressed by Gerard Karsenty, a distinguished experimental bone biologist, aware of the power of clinical observation to engender experimental designs in molecular genetics. He wrote of the clinician and the experimentalist as a "not-so-odd couple" (Karsenty (2001), emphasizing the exceptionally hard work each line of endeavor entails—"for the clinician to care first and to understand second, and for the experimentalist to understand first and only second to cure, or 'rescue,' as one says in the world of animal experimentation." Yet Karsenty goes on to note that "the dialogue is necessary because their two goals are complementary, if not identical. Both practice biology only under a different form, and it is in this spirit that the dialogue between them can develop." I hope this book will enlighten both parties in ways useful and applicable to their needs and professional pursuits and will add incentives to their respective work by way of mutually greater understanding and interchange.

In the conception and writing of this book, I have been concerned about the depth of the presentation of the basic science background that is integral to the clinical expression of aging and related diseases. Part of the difficulty is the need to set limits on the inclusion of details, so the reader will not be overwhelmed or discouraged. Yet for me the conjunction of what the scientist understands about the molecular events and what the clinician appreciates about the disease is the key to creating the unity and dialogue I have referred to. To lay the groundwork for the scientific background, I asked E. Richard Stanley, professor and chair of the Department of Developmental and Molecular Biology at the Albert Einstein College of Medicine, to introduce concepts of molecular biology and to develop with me a brief glossary of abbreviations used in the text. I hope this will permit clinician readers to grasp more completely the key parts of the chapters on aging and disease even though some aspects may not have been previously familiar to them.

Another reason I engaged in writing this book was to continue assembling ideas that evolved for me about geriatrics and its wider ramifications. This text develops associations—what I call interconnectedness—in many areas, in particular, those related to cytokines, inflammation, and stress responses, which unify most of the conditions presented in Chapter 4.

Finally, I want to express my views about what new directions geriatric practice might take to enhance and broaden its scope. I hope this book will

promote greater awareness of the biological basis of aging and related diseases; introduce the relevance of early origins of disease as part of overall
development; and encourage geriatricians to be aware of assessing risk factors
to permit timely interventions. In this way, "preventive gerontology would really be on the way," as Fries (1997) wrote. These are themes that run through
this book.

My aim in this book is to present a series of well-documented essays, to emphasize first principles in the bioscience of aging and disease, with interactive
and overlapping discussions. I was inspired by what E. B. White wrote about
"the essayist as a self-liberated man, sustained by the childish belief that ev-
erything he thinks about is of general interest. He is a fellow who thoroughly
enjoys his work, just as people who take bird walks enjoy theirs" (White,
1977). In this book there is no intention of "completeness," but rather "selec-
tivity," based on my own interests and intensive reading. I found it useful to
quote, often at length, from the relevant reviews by experts.

In this approach I do not discuss evidence-based medicine or the interpretation of clinical data based on meta-analysis. These are beyond the subject
of this book although they form the content of what current clinical faculty
have been exposed to and understand. Although these aspects are important
and relevant for clinical practice, geriatric fellows and faculty are not equally
knowledgeable about or exposed to the bioscientific basis of aging and its
relation to disease. Yet, mechanisms of disease with a bioscience orientation
are increasingly part of postgraduate education and featured in most review
articles in clinical journals; witness the series on "Genomic Medicine" in the
New England Journal of Medicine and published collectively by the Johns
Hopkins University Press (Guttmacher et al., 2004).

Furthermore, recent articles in scientific journals focus on topics related to
aging and disease that might engage clinicians and scientists alike. An editorial in the February 2005 issue of *Cell* entitled "Aging research comes of age"
notes that "the science of aging has grown in depth, breadth, and molecular
detail," and eleven reviews in that issue "capture the current excitement in
the field." In this same issue, Chien and Karsenty (2005) highlight advances
in three organ systems: heart, skeletal muscle, and bone, "which are beginning
to form a molecular framework for the integrative physiology of human aging
and related diseases." In an introduction to a series of articles in a recent issue of *Science,* Martin et al. (2003) note "the end of the beginning" of research
on aging; they propose that we should increasingly "focus on physiological
mechanism underlying processes of aging rather than on a large array of de-

bilitating and costly disorders that so commonly emerge during the latter half of the life-spans of human beings." Two issues of the *New York Academy of Sciences* are devoted to "Molecular and Cellular Gerontology" (Toussaint et al., 2000) and to "Healthy Aging for Functional Longevity. Molecular and Cellular Interactions in Senescence" (Park et al., 2001).

Many fields in geriatric bioscience evolve rapidly. By the time this book appears, some topics discussed may need to be reevaluated in the light of later publications. Nevertheless, I hope that the interested reader will gain a sufficient exposure to and familiarity with subjects that would not have come to his or her attention otherwise.

At the start of my academic career, an outstanding neurochemist named Maurice Rapport related to me, with mirth, a basis for the temerity to write for scientific colleagues of varying expertise. A child of five returns home from school and announces to his mother that he has won a prize. "What for?," his mother exclaims. "For guessing how many legs the rhinoceros has," replied the child. "Oh, how many?," asks the mother. "Five," says the child. "But, dear," says the mother, "the rhinoceros has four legs." "I know," says the child, "but all the other kids guessed six." In the broad aspects of biology pertaining to aging and to medical illnesses of older persons, I have endeavored to present an accurate account with four legs of the many complex and interactive issues. I hope that readers will find ideas and associations that will give them new insights about aging biology that relate to the human condition; these readers may then discover added attraction in the pursuit of an academic career in geriatric practice, in which caring for older patients will be one of the dominant health and social issues in the twenty-first century.

My wife, Laura Lowy Hamerman, provided continuing encouragement, especially in tolerating my working many nights and weekends. My children, Alan, Jean, and Frederick, were pleased their father was busily engaged, and they followed my progress with enthusiasm.

The preparation of the manuscript, the accession of articles, and the computer generation of successive versions—all of which were essential to the book's evolution—were accomplished by Dawn Bowen-Jenkins, my outstanding assistant for more than 15 years.

Special thanks are due to two faculty members at the Albert Einstein College of Medicine who wrote sections. E. Richard Stanley, Ph.D., the Renée and Robert Belfer Chair and Professor, Department of Developmental and Molecular Biology, has been a long-time mentor for me and a collaborator in research studies. His expertise and knowledge made Chapter 2 and its illustrations—containing an introduction to molecular biology—an essential part of understanding the text. Peter Davies, Ph.D., the Judith and Burton P. Resnick Chair in Alzheimer's Disease Research and professor of pathology, has been a valued colleague on whom I have called for almost 30 years for guidance on the status of Alzheimer disease; he wrote section 8 of Chapter 4 and supplied the related figures. I am deeply grateful to Drs. Stanley and Davies for their contributions.

Robert Butler, president and CEO of the International Longevity Center is, in my view, one of the founders of gerontology in this country: present at the creation as the first director of the National Institute on Aging at the National Institutes of Health; as a Pulitzer Prize recipient for his book *Why Survive? Being Old in America;* for his founding and chairing the first Department of Geriatrics at Mt. Sinai Hospital in New York; and for his ongoing spokesmanship for gerontology and promotion of the concepts of longevity. It is a privilege, then, for me to convey my admiration for what he has—and continues to—accomplish, and to acknowledge my gratitude that he wrote the foreword to this book.

The wise council of Dr. Richard Marks provided inspiration and guidance during the conception, preparation, and completion of this book.

Robin Barr, Ph.D., acting director for extramural activities at the National Institute on Aging, provided a statement on funding programs for young investigators embarking on an academic career.

Many colleagues read parts of the manuscript and made additions and corrections. At the Albert Einstein College of Medicine these were Nir Barzilai, Anne Etgen, Meredith Hawkins, and Luciano Rossetti. Richard Kitsis and Philipp Scherer shared with me aspects of their exciting new work. At Montefiore Medical Center Laurie Jacobs and Joel Neugarten made helpful suggestions.

Ronald Adelman, Geriatrics Division, Department of Medicine at Weill-Cornell School of Medicine, read an early version of the text and made helpful suggestions.

I am most grateful to Judith Campisi, Senior Scientist at the Lawrence Berkeley National Laboratory, Berkeley, California, and Professor, Buck Institute for Age Research, Novato, California, for her friendship over many years and for her advice and guidance concerning the biology of cell senescence; to Ronald DePinho, Director, Center for Applied Cancer Science, at the Dana Farber Cancer Institute, and Professor of Medicine (Genetics) at Harvard Medical

School, Boston, Massachusetts, for discussions and references on the biology of cancer; and to Cynthia Kenyon, director of the Hillblom Center for the Biology of Aging, University of California at San Francisco, for helping me with longevity genes and for reprints of distinguished lectures she presented, in press.

Library staffs were gracious and helpful in obtaining references and texts: at the Albert Einstein College of Medicine—Karen Sorensen; at Montefiore Medical Center—Josie Lim, Sheigla Smalling, and Heather Barnabas.

Members of the Graphic Arts Department at the Albert Einstein College of Medicine were most helpful with the preparation of figures and tables: Donna Bruno, Tatyana Harris, and Camille Anastasio.

My thanks to the staff of the Johns Hopkins University Press for all their help. My first association with Wendy Harris, medical editor, occurred in 1996/97 on the preparation of a book I edited on osteoarthritis. In the solo venture that marked the preparation of *Geriatric Bioscience,* Ms. Harris provided valuable editorial assistance and continuing support and guidance. Nancy Wachter thoughtfully edited the manuscript.

CHAPTER ONE

The Emergence and Future of Geriatrics

Although the history of geriatrics can be traced back to "the beginning of time," according to a review by John Morley, its birth was the invention of the word *geriatrics* by Ignatz Leo Nascher in the United States around the turn of the twentieth century (Morley, 2004). "Modern geriatrics" originated in Great Britain, where, in 1935, Marjory Warren introduced a comprehensive rehabilitation program for older patients in a West Middlesex Hospital. Geriatric medicine became a specialty in the United Kingdom in 1948, the year after the founding of the British Geriatrics Society (Brockelhurst, 1997; Barton and Mulley, 2003). In the United States, the American Geriatrics Society was organized in 1942, and subsequently key figures interested in aging research emerged, perhaps most notably Nathan Shock and Reuben Andres, who established the Longitudinal Study on Aging in Baltimore in 1958. In 1972 Leslie Libow set up the first residency and geriatric fellowship training program in the United States at the Mt. Sinai City Hospital Center, in Elmhurst, New York. Yet even in 1974 Smits and Draper (1974) could write that, compared with "its specialty status advancing rapidly in Great Britain, geriatrics barely exists in the United States."

A new orientation toward aging programs in medical schools was emerging with funding from the Administration on Aging, and the underpinning of geriatrics in the United States was the creation in 1974 of the National Institute on Aging (NIA) (Butler, 2004). This meant there would be multiple initiatives for grant funding that permitted medical schools to recruit faculty interested in aging research within basic laboratories and to conduct longitudinal studies on aging persons in the community and, uniquely, in nursing homes. Recognition of new and established academic scholars created the leaders and future leaders that implanted geriatrics faculty in the medical establishment. This grounding in the science of the field led to the publication of many comprehensive textbooks (although those reviewed in 1999 were deemed "not yet of the essence"; Finucane and Loo, 1999); new philanthropies that funded aging research initiatives and personal awards for leadership; and the emergence of the Institute of Medicine with an interest in aging, among other areas. Publications from the institute's National Academy Press included *Our Future Selves* (National Advisory Council on Aging, 1978), *Toward an Independent Old Age: A National Plan for Research on Aging* (National Research on Aging Planning Panel, 1982), and *Extending Life, Enhancing Life: A National Research Agenda on Aging* (Institute of Medicine, 1991). About the time the NIA was created, the Veterans Administration linked programs to medical schools by way of Geriatric Research, Education and Clinical Centers (GRECC) for competitive funding.

While the exciting momentum of the early years brought geriatrics and with it the care of older persons into virtually all medical schools, teaching hospitals, and departments of internal medicine in the United States, there never was any intention or capability for geriatrics to take on the provision of medical health care for the collective population of older persons. Rather, geriatrics defined guiding practice principles for the health care of older persons and provided certification with added qualifications for those practitioners. Geriatrics classified multiple aging-related conditions and diseases in terms of functional impairment in both well and ill older persons residing in the community and facilities, and it introduced principles of comprehensive care and multidisciplinary assessment.

GERIATRICS AT PRESENT

Geriatrics now seems to be seeking to define or redefine itself, or, as strikingly put by Robert Kane (2002), "deciding what it wants to be when it grows up." Other divisions within departments of internal medicine may not be experiencing such a debate, soul-searching, or sense of "being at the crossroads" (Kane, 2002). In fact, many of these divisions, especially cardiology and gastroenterology, and perhaps hematology and nephrology, have become more firmly grounded in new technologies and therapeutic innovations. As a result, those divisions have promoted and practiced assessment of health risks and earlier interventions for patients, often with remedial therapies that may prevent the inception of disease or actually reduce the manifestations of organ damage or failure.

One of the greatest challenges facing geriatrics has to do with "generational succession"—the declining numbers of certified geriatricians and reduced applicants for fellowships, especially among graduates of U.S. medical schools. Among those who do complete fellowships, fewer stay within academic programs and pursue research. There is a further issue of identity: whether geriatrics should serve largely in a consultative role or in a more diversified one in primary care. Perhaps seeking to expand its image, geriatrics has uniquely and intensively taken the initiative (supported by the John A. Hartford Foundation) to disseminate its precepts—sometimes called gerontologizing—to internal medicine itself and to some of its specialty areas, to family medicine, and even to the surgical specialties (Hazzard et al., 1997). Geriatrics taking a role in enhancing the knowledge base of other professionals is desirable, but also raises a prevailing issue of the rationale, subsequently, for internists to refer patients to the geriatrician. There is a more serious issue, namely, that program leaders in general internal medicine see their *own role* in directing their training programs for the care of older persons. Landefeld, Callahan, and Woolard, in a supplement to the *Annals of Internal Medicine* (2003), found "many causes for concern about epistemologic boundaries and social standing of geriatrics: the field is not defined by a focus on an organ or a disease, as other subspecialties in internal medicine are" (indeed, in my view this is a *strength* of geriatrics; see the discussion on frailty, Chapter 4, section 6). A further presentation by Simon, Fabiny, and Kotch (2003) discusses "barriers": that "geriatrics may not be a scholarly discipline," and a perception exists that "geriatric fellows are not of the same intellectual caliber as general internal medicine fellows." But there may be hope: "Together (the two disciplines) can develop and promote a *science and culture* that can serve the needs of every elderly patient" (my italics).

Perhaps recognizing the need for a reappraisal at this time, a core working group drawn from a large and diversified task force of the American Geriatrics Society (2005) published a paper in the *Journal of the American Geriatrics Society,* entitled "Caring for Older Americans: The Future of Geriatric Medicine." This report reflects much that has been accomplished, but acknowledges much that remains to be done to care for a "burgeoning number of older persons" and to mobilize and effectively organize the "physician work force." As set forth in the paper, one of the primary missions of geriatrics—to care for and conduct research on "the needs of frail older persons and those with multiple illnesses"—seems to have been well met by the emergence of functional status assessment and disability measures; indeed, these activities have "flourished" in terms of clinical care and epidemiologic and clinical research, according to a thoughtful editorial by Jack Guralnik of the NIA. But, he goes on to say, "much remains to be learned and the effort to reduce disability in older persons remains an overriding goal of geriatric medicine and public health" (Guralnik, 2005). A paper by Leveille and colleagues (2004), "Advancing the Taxonomy of Disability in Older Adults," is an important addition to this endeavor.

An admirable summing up of the perceived present status of geriatrics in this country appears in an article entitled "Geriatrics lags in an age of high-tech medicine" on page 1 of the *New York Times* of October 18, 2006, by the reporter Jane Gross. This account portrays what geriatrics—perhaps uniquely—does so well: care for the very elderly patient comprehensively to limit invasive tests, unnecessary surgery, and overuse of medications, all to maintain the patient's well-being and functional capacity. Against these positive features is the account of the widespread negative perception of geriatrics by the medical community: the medical resident interested in pursuing a geriatric fellowship and whose attending mutters "waste of a mind;" the account of long hours, complex patient issues, limited reimbursement; the "lowly status of geriatrics in most American medical schools" contrasted with its high regard in Great Britain, where "geriatrics is the most popular specialty." Indeed, we can strive to bring greater recognition and "dignity" to geriatrics by many means—including the exposition of bioscience integral to clinical medicine—but placing geriatrics in the mainstream of medical practice will likely require major modifications in our culture, in medical education, and in the health care system in the United States—a tall order!

FUTURE PROSPECTS FOR GERIATRICS

Geriatricians might consider adopting a still broader agenda for the future. The responsibility of geriatrics for the care of older persons with chronic diseases needs to be better defined. A major change in direction has been called for to develop new organizational models and systems of delivering and paying for care to "address the growing crisis in financing and delivering chronic disease care." The societal organizations in geriatrics and gerontology can be strong advocates for this—"rearranging the deck chairs," as Lawrence (2005) put it—but this seems in the realm of public policy and politics rather than the *practice and content* of the provision of health care for those with chronic diseases. On a more global scale, geriatrics could be responsive to the call by the World Health Organization (WHO) and other international centers to "overcome impediments for the prevention and control of chronic diseases" in developing countries (Yach et al., 2004). Cardiovascular diseases and diabetes are in ascendance in those countries, and intervention, with the potential to extend life, would be very much in the province of geriatrics.

The task force report also calls for "expansion of knowledge related to health and aging" that would include "research on the basic sciences of aging and age-related diseases." This does not capture the necessary *affinity* between the "not-so-odd couple" of the experimentalist and the clinician I wrote about in the preface.

Geriatrics in its agenda for the future needs to move vigorously toward the precepts of preventive gerontology. This is taking a step into the future and also retaining the past. Those younger generations of patients need to be engaged when they are still at midlife, a time when preventive gerontology has to begin. Two decades later, this constituency will present to the geriatrician, potentially with established morbidities and disabilities. How to introduce this earlier affiliation that may go against the tradition of geriatrics, where practice is dominated by those over 65 years, poses a challenge. Creating an alliance with primary care practitioners might ensure a wider role for geriatricians in tracking the person's *evolution* of health before overt disease, a more longitudinal appraisal in which the assessment of risk factors becomes a major consideration for decisions about at-risk populations (Hamerman, 2002a). Such appraisal may promote primary intervention—strongly advocated in this text, especially for cardiovascular diseases (Chapter 4) and for dealing with lifestyle choices (Chapter 6), also advocated by William Hazzard in his "strategy for healthy aging" (1983).

Geriatrics also needs to look more intensively at early developmental history, as this provides continuity with the present and prospects for the future. Westendorp and Wimmer (2005) link *development* with *aging* to provide a new research and clinical agenda and state that "the process of healthy aging cannot be understood without considering the entire human life history." Alwin and Wray (2005) describe a life-span developmental perspective on the linkages between social status and health—"a conceptual framework to understand why people are differentially exposed to risks of disease or protective factors" within the social environment. Gans (2006) writes of "various biologic, psychologic, and social factors that act independently, cumulatively, and interactively throughout life" involving health and disease in the adult. Ben-Shlomo and Kuh (2002) describe a "life-course" epidemiologic approach in which the risk of disease is related to physical and social exposure during gestation, childhood, and adolescence that affect later adult life. Settersten (2005) notes the "emerging interdisciplinary field of childhood studies extending to the field of gerontology" as a framework for developing research on old people around their rights, experiences, and the interdependence of generations. It is good to see the social and psychological sciences join the medical and biological sciences in exploring transitional issues from youth to age. However, it is unlikely that, without further effort, training programs in medicine in general, or geriatrics in particular, will present such a broad retrospective and prospective orientation to trainees.

In keeping with defining future prospects for geriatrics, program directors individually and collectively need to look at their program's purpose, effectiveness, and scope. If indeed a "crisis in geriatrics" exists, it should be met by innovations for the decades ahead that will infuse a renewed vigor in redefining geriatric practice. It does not necessarily follow that speakers and leaders in the field can provide a single model to unify geriatrics and make it vital for these changing times. Such efforts may be better achieved within individual divisions of geriatrics led by directors who also engage their chairs of departments of medicine and other key institutional figures, identifying ways to enhance local strengths. Indeed, program diversity among divisions is likely to be the case already with variations in recognition and achievements: witness the John A. Hartford Centers of Excellence, the Paul Beeson Awards, the Donald Reynolds Centers, the Claude Pepper Centers. Yet vigorous geriatrics programs exist without such recognition. Spirited discussions within a geriatrics division can be unifying and strengthening, providing a major basis for innovations and promoting recruitment and a range of career choices for faculty development.

The American Geriatrics Society might take heart to learn that self-examination on the scientific aspects of aging has also been carried out in Great Britain by a committee of the House of Lords, no less. The report found "little evidence that policy has been sufficiently informed by scientific understanding of the ageing process." John Grimley Evans (2005), calling this a "lordly report," made note of aspects that are of interest to American readers in a comparative sense. He wrote, "Geriatrics has quietly revolutionized British medicine but has been a disappointment academically." I would like to think that at least the revolutionary aspect is comparable in U.S. geriatrics. Grimley Evans goes on: "the biological study of ageing in Britain almost died in the 1960s but now flourishes anew in the warm glow of molecular biology." This molecular glow has occurred in aging research in the United States as well—witness much of the contents of this book. Grimley Evans states that "for practical reasons geriatricians have been more aware of developments in social than in biological gerontology," which for American readers I would amend to read *clinical* for *social.* I was encouraged by this report and the comments of Grimley Evans; there seems to be a trans-Atlantic (if not universal) perception of the need to implement a bioscientific dialogue to mutually enhance aging research and geriatric practice. This interchange would "improve people's prospects of healthy and active life expectancy," which the report seeks to encourage.

GERIATRICS AS A CAREER FOR STUDENTS AND TRAINEES

Increasingly in medical schools in the United States, students experience throughout their curriculum, and in diverse ways, an exposure to issues of aging and to older persons. "Geriatrics" as a designation may not formally begin until a required or optional clinical rotation by that name in the third or fourth year. Until then, the students' background would have been heavily weighted in the basic sciences. The application of geriatric bioscience to the clinical conditions to which they are now exposed didactically or on the wards can be relevant, timely, and exciting.

In terms of students' subsequent choice of a professional career, geriatrics needs to be "highlighted" for them in its *academic mode.* In my experience, it has been unusual for a student at one of our teaching sessions to profess an interest in geriatrics, even tentatively, as a career choice. Other options appear to emerge more defined—ophthalmology, pediatrics, orthopedics—and selecting internal medicine itself is becoming more rare, especially primary care (Bodenheimer, 2006), and indeed may be actively discouraged (Woo, 2006). It seems important to encourage and enlighten students about geriatrics as a career choice, for this is essential to its future growth, stability, and recognition.

Many types of experience may kindle a student's interest in geriatrics: a caring interaction with a vigorous grandparent; volunteer work in an extended care facility; the inspiration of a mentor—encouraging the introduction to laboratory or clinical research, and with the advice of a dean of students, a grant from the American Federation for Aging Research (www.AFAR.org) for a Summer Research Training Program in Aging. Divisions of geriatrics usually designate a "point" faculty member to actively engage students and seek their participation in the American Geriatrics Society student organization.

During the early postgraduate years, the student-now-resident, and other house staff members, need to be exposed to the activities of a Division of Geriatrics. Enthusiasm for geriatrics is likely to wane when the assurance and technological innovations of other divisions of medicine become apparent, or when frustration occurs in attempting to communicate with a frail patient transferred from a nursing home, or when peer pressures and unfavorable attitudes about these patients prevail. Yet geriatrics-trained faculty in teaching hospitals can be especially important and visible role models on attending ward rounds, morning report, clinical conferences, and grand rounds, and can be keen and persuasive advocates to excite and recruit house staff. There are many unique and appealing activities—not necessarily part of the more traditional medical instruction to house staff—in which geriatrics faculty can capture the interests of students and house staff on the wards. Some examples in my own experience include (1) so-called skin rounds for decubitus ulcer care (Odierna and Zeleznik, 2003); (2) "ombudsman rounds," with nurses and psychiatrists-in-liaison to address, among other wrenching issues, conflicts among next-of-kin and staff members about the extent of interventions in futility cases or in end-of-life decisions (Strain and Hamerman, 1978); (3) rounds to intervene constructively with frail, disoriented patients admitted from a nursing home; (4) accessibility to emergency department and orthopedics staff to help with preoperative assessment of patients with a hip fracture; (5) teaching interactions with ward nurses to create a milieu of collegiality focused on geriatric patient care. The geriatrician can be a valuable resource in advising house staff about team interactions with ethicists, lawyers, and clergy to assist patients to prepare advanced directives; with nurses and neurologists to manage delirium; with the surgical specialists in pre- and postoperative care to salvage limbs threatened by amputation; with pharmacists to forestall or address the adverse effects of drug polypharmacy. One of the most effective ways to excite house staff about the dimensions of geriatric practice is to bring interdisciplinary staff—as noted above—to case-solving clinical conferences that aim to resolve complex issues that invariably arise in the care of very old patients (Hamerman et al., 1986; Hamerman and Fox, 1992). Practical and focused recommendations in these complex cases are extremely important to time-obsessed house staff members who deal with multiple patients and their attending physician constituencies. Word gets around among house staff about these broader ramifications of geriatrics and the detailed, pertinent, and coordinated chart notes that result.

Yet even the perception of interest in geriatrics by house staff members rests on a fragile branch. Such an interest goes against the tide of peer pressure that denigrates taking care of older persons on the medical service, or by a youth-oriented society that expresses a more negative view globally. It is worth recounting an impassioned determination nevertheless to go forward in an academic career in geriatrics by a noted younger role model—Maria Fiatarone Singh—who made important contributions on the effects of muscle strengthening (resistance) exercises among very elderly nursing home residents, as I'll discuss later. This is what she wrote (Singh, 2003):

> For a university or medical student to choose geriatric medicine requires an inherent sense of purpose and level of determination that is rarely praised by academic faculty or even fellow students. At medical school graduation in 1981, I remember my teachers actively trying to dissuade me from "wasting" my potential in a field with such a low level of prestige (and economic rewards). Although geriatric fellowships have since been established nationwide, and a geriatric curriculum has been added to many training programs, attitudinal shifts have lagged far behind the structural changes established. In addition to a basic problem of ageism in society, there are other factors that limit attractiveness of the field, such as fewer technological procedures, stressful or depressing working conditions, patients who do not recover or improve with care, the frustration and complexity of managing multiple systems disorders and social problems at the same time, and income levels perhaps not commensurate with the workload required to provide excellent care.

And, I might add, good *quality* of care judged by a scoring system for vulnerable older adults was associated with their better survival (Higashi et al., 2005).

In the course of the trajectory from student to medical resident, a mutually gratifying moment is the interview by faculty of the candidate they have nurtured for a geriatrics fellowship. To make this transition successful, the environment and conditions for fellowship training must be carefully presented. It is not just a question of recruiting and retaining a fellow who has become acquainted with the division; this is obviously of great importance, a source of satisfaction, and a widely visible message for the peers of the fellow-to-be. But beyond all these things is the opportunity of the division head and the faculty to foster the trainee's talent in the most advantageous manner for his or her professional career. I don't think this is an overstatement considering the opportunities for personal and professional growth and development that geriatrics offers in its academic mode. New trainees are unfamiliar with the breadth of these opportunities for they have not been exposed to them, and guidance on the part of a knowledgeable division head is essential.

For the aspiring fellow, some engagement with the scientific background heightens interest and understanding of the interconnections of research, teaching, and clinical practice. Awareness of the broader aspects of aging and the inception of disease enhances the year of clinical fellowship, helps define the goals and career choices in subsequent years, and for those trainees so inclined, prepares them for a formal research experience. Trainee identification with a mentor is key to encouraging the motivation and readiness for research. With appropriate preparation and evidence of a plan, research may begin with an application for an AFAR grant; if successful, application for NIA funding may follow. (The following write-up was contributed by Robin Barr, Ph.D., acting director for extramural activities, NIA.)

NIA has multiple routes for aspiring clinician-researchers to follow that are mapped to different amounts of research experience and differing needs for research training. For junior clinicians with limited research experience or training, appointment to an institutional training grant can give a thorough grounding in research. For a clinician with a year or two of research experience, including a publication history, an independent postdoctoral fellowship (F32) may be appropriate. The fellowship pairs the candidate with a particular sponsor or cosponsor and provides an intense apprentice-like training experience. Some NIA program career development awards (the K12 series) also allow appointment of junior clinician-researchers and represent substantial research training opportunities in targeted areas and selected institutions.

Benefits on these award mechanisms are not always sufficient to cover clinical loan repayment as well as the cost of living. NIA does also offer loan repayment contracts that may be combined with appointments on training grants or career development awards and individual fellowships. The programs target particular areas of research—including clinical research—and particular kinds of researchers. The combination of the two awards does offer reasonably competitive funding for many clinicians struggling to advance a research career while faced with high expenses.

A new program offered by the NIA-the "Pathway to Independence Award" (the K99-R00)—is designed to attract those who have caught the research bug, are still in research training, and find early success during their research fellowship or training years. The first part (the K99) provides additional support in the current training environment of the candidate. It provides both salary support above fellowship levels, and research expenses above fellowship levels that permit the candidate to begin an independent research project. The R00 portion begins once the trainee has moved to a faculty-type appointment in an environment that is supportive of research. Three additional years of research funding are allowed where the candidate can expand on and complete the project begun during the K99 phase. NIA strongly encourages awardees to seek independent (R01-like) funding following the R00 award.

An NIH individual mentored career development award (K08, K23) is a common follow-up award to a research training grant appointment (on a T32) or an individual fellowship award. These awards are designed to advance clinician-researchers to research independence. An aspiring clinician-researcher now with several years of research training and experience, often needs additional mentoring and skill development targeted to the particular research area that the clinician-researcher has now begun to focus on. The mentored career awards allow up to five years of protected time for research and research training to acquire these additional skills and develop a sufficient research track record that the candidate becomes competitive for and seeks independent research funding.

If the unusual and gifted trainee has pursued any part of these paths, application for a Paul Beeson Career Development Award, which is sponsored jointly with the NIA, or an Ellison Medical Foundation Award, both described on the AFAR Web site, will be a unique way to recognize those with great promise for future leadership in the field. Ultimately, other NIA initiatives, such as a FIRST or R-01, will complete the path to investigative independence. This, then, is a rather idealized journey toward academic development pursued by a limited number of trainees, yet one that the division head must encourage.

Alternatively, following the completion of a fellowship, new faculty appointed to divisions of geriatrics may become engaged full time in clinical care and teaching in an academic mode, highly valued and indispensable for the future of the field. The issue is the breadth of knowledge attained and passed on to future generations of students, house staff, and fellows. The professional satisfaction of faculty members will be based in large part on the scope of their learning, demonstrated clinical and teaching capabilities, productivity, and local and national recognition on presentation of their clinical studies at poster sessions or in published papers. These individuals must be intensively nurtured by the division head and department chair. Depending on the new faculty member's effectiveness, he or she will subsequently have a wide influence in fostering awareness of geriatrics and gerontologic precepts within the medical center and teaching hospital. Such faculty will be leaders in raising awareness of geriatric principles within other medical specialties, will be the strongest advocates for the vitality and validity of geriatrics, and will set the standards for the care of older patients as part of a division within a Department of Medicine. They will be the key faculty to promote geriatrics among medical students and house staff, and the major force for recruiting fellows. Thus, they are indispensable.

How will those planning a career in geriatrics, or those in academic geriatric practice, gain familiarity with aspects of the science of aging and its relationship to the inception and progression of disease? The process needs to begin in the medical school curriculum and be continued by geriatric faculty teaching in the hospital setting. The director and staff members at the NIA are interested in geriatricians acquiring a background in bioscience in the context of improving health care for an older population. In writing "The future of aging therapies" in *Cell*, staff members of the NIA point out that the thrust of research in aging bioscience is to understand cellular aging in models applicable to the human condition, reduce or delay the inception of disease and ensuing pathologies, and promote healthy longevity (Hadley et al., 2005). Their article provides examples of what the authors call "therapeutic foci" that represent new thoughts on those aging changes once considered to be "normal aging" but now associated with adverse consequences: specifically—*sarcopenia,* the reduction of muscle mass and strength with attendant weakness; *osteoporosis* in the postmenopausal state, with bone loss that may result in fragility fractures; and *vascular stiffness,* one of the risk factors for vascular disease. Thus, means to identify a variety of risk factors again become an important consideration. Of course, those in clinical practice cannot overlook the existing *disease* manifestations that constitute the bulk of their older patients' presentations; but geriatricians in academic settings who see older persons increasingly in manifest good health can gain greater understanding of those risk factors that may modify well-being and thereby apply diagnostic and therapeutic interventions earlier, aimed at limiting or delaying the inception of disease. Geriatric bioscience as it is emerging provides the foundation for new clinical insights and therapeutic initiatives (hence translation) that may well mark an era in which geriatric practice becomes reinvigorated and fulfills the promise that seemed to be its destiny earlier.

CHAPTER TWO

An Introduction to Concepts of Molecular Biology

E. RICHARD STANLEY, PH.D.

This chapter introduces concepts of cell and molecular biology concerning the complex regulation of mammalian cells within the organism, some knowledge of which is essential to comprehend the aging-related conditions and current therapies discussed in this book. It represents a "bare bones" introduction to intercellular signaling, signal transduction, and the regulation of gene expression coupled with a brief discussion of animal model systems.

INTERCELLULAR REGULATION BY CYTOKINES, GROWTH FACTORS, AND HORMONES

Coordination of the survival, proliferation, differentiation, and functions of cells as well as the homeostatic regulation of organ systems is in large part regulated by extracellular *signal molecules,* including cytokines, growth factors, and hormones. These signals are produced by one type of cell and regulate another type, the target cell, through target cell expression of specific signal *receptors.* Extracellular signals act at very low concentrations ($\leq 10^{-8}$ M) by high-affinity binding to their cognate receptors, which become activated to initiate intracellular signaling pathways that regulate the function of the target cell. Signals may be hydrophilic (e.g., peptides, or glycoprotein cytokines, growth factors, or hormones) acting via cell surface receptors; or hydrophobic (e.g., steroid hormones), freely diffusing across the plasma membrane to bind intracellular receptors. Thus, intercellular communication for any particular signal depends on the expression of the cognate receptor for the signal on the target cell.

Intracellular signaling may take several forms (Fig. 2.1). *Juxtacrine signaling* involves direct cell–cell contact mediated by the cell surface signal on the signaling cell and a cell surface receptor on the target cell (Fig. 2.1a); it is important in immunity and development. Signals that do not diffuse far and affect only target cells in the immediate environment of the signaling cells in a process called *paracrine signaling* (Fig. 2.1b) are often called *local mediators*. A particular example of this type of signaling is synaptic signaling. Signaling at a distance by signals that enter the bloodstream to act on a distant organ or tissue is termed *endocrine signaling* (Fig. 2.1c); it is not only typical of endocrine hormones but also is mediated by several growth factors and cytokines. A fourth form of signaling differs from the others in that the signaling cell is also the target cell, producing a signal that binds its cognate receptor expressed on the signaling cell. Known as *autocrine signaling* (Fig. 2.1d), this type of signaling can reinforce developmental decisions made by a cell, but is more common in pathologic situations such as inflammation and cancer, allowing cells to survive and proliferate autonomously in situations where normal cells cannot. Some signals, particularly growth factors, may engage in all four of the forms of signaling illustrated in Figure 2.1. For example, colony-stimulating factor-1 (CSF-1), the primary regulator of macrophages and osteoclasts, is expressed as a cell surface variant (isoform) that engages in juxtacrine signaling and, by cleavage of this isoform from the cell surface, in paracrine signaling. In addition, it has secreted isoforms that are involved in paracrine and endocrine signaling. During inflammation, CSF-1 is produced by macrophages and engages in autocrine signaling.

Any particular cell in a multicellular organism is programmed to respond to a variety of signals, different combinations of which activate the cell to differentiate, proliferate, survive, or even die (the last by a process known as programmed cell death, or *apoptosis*). The use of combinations of signaling molecules allows the organism to regulate many different kinds of cell behavior with relatively few signals.

SIGNAL TRANSDUCTION

How cells respond to their environment is determined not only by their expression of receptors for signals but also by how the intracellular machinery interprets the binding of signal to receptor, a process known as *signal trans-*

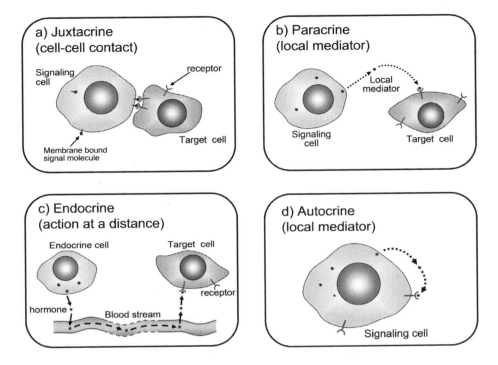

Figure 2.1. Modes of regulation by cellular signals, including cytokines, growth factors, and hormones. See text for details. (See color gallery.)

duction. Cell types expressing the same receptor may differ in their responses to the same signal because of differences in the downstream *signaling pathways* coupled to the receptor in different types of cell. Hydrophobic signal molecules that diffuse directly across the plasma membrane of target cells, including the steroid and thyroid hormones, prostaglandins, leukotrienes, retinoids, and vitamin D, bind receptors belonging to the nuclear receptor superfamily that possess separate ligand (signal)-binding, DNA-binding, and transcription-activating domains (Fig. 2.2a). These receptors all bind DNA sequences adjacent to genes the signal regulates, known as receptor-binding elements. However, some receptors are located primarily in the cytosol and enter the nucleus only to bind DNA after signal binding. In both cases, the binding of signal alters the receptor conformation, causing it to dissociate from an inhibitory complex that keeps it in an inactive state.

Two major classes of cell surface receptor (Fig. 2.2b) are the *G-protein-coupled* and *enzyme-linked* receptors. Following extracellular signal binding, G-

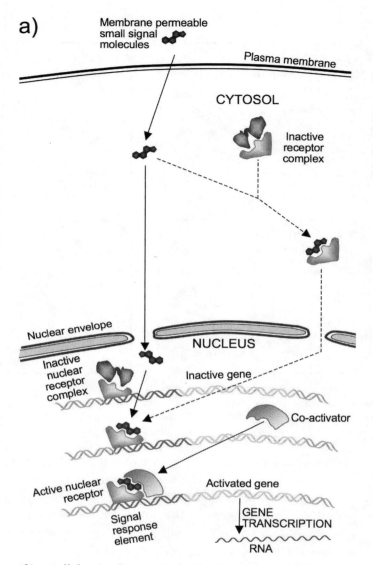

a)

Membrane permeable
small signal
molecules

Plasma membrane

CYTOSOL

Inactive
receptor
complex

Nuclear envelope

NUCLEUS

Inactive
nuclear
receptor
complex

Inactive gene

Co-activator

Active nuclear
receptor

Activated gene

Signal
response
element

GENE
TRANSCRIPTION

RNA

Figure 2.2. Types of intracellular signaling proteins acting downstream of cell surface and intracellular receptors. (a) The nuclear receptor subfamily. Binding of signal to a receptor leads to conformation change in the receptor, disassociation of an inhibitory complex, binding of coactivator proteins, and activation of target gene transcription. All nuclear receptors bind to DNA as homodimers or heterodimers. Monomeric forms are illustrated here for simplicity. Apart from the signals mentioned in the text, the prostaglandins and leukotrienes act via nuclear receptors termed PPARs that heterodimerize with RXR (see Chapter 5, section 4). (b) Cell surface receptors, showing how the signal of receptor activation is amplified, transduced, and distributed to regulate gene transcription. The functions of the labeled proteins are as follows: *Relay proteins* pass the message to the next signaling component. *Messenger proteins* carry the signal from one part of the cell to another without directly

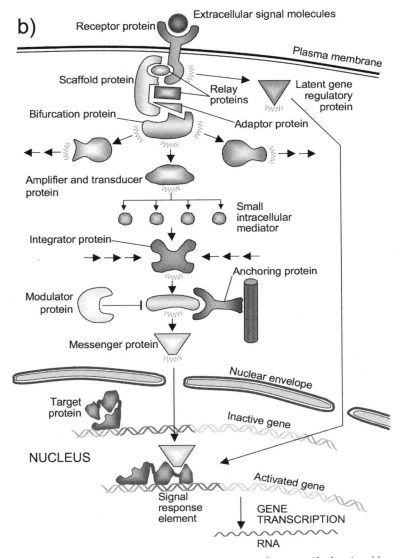

b)

Extracellular signal molecules

Receptor protein

Plasma membrane

Scaffold protein

Relay proteins

Latent gene regulatory protein

Bifurcation protein

Adaptor protein

Amplifier and transducer protein

Small intracellular mediator

Integrator protein

Anchoring protein

Modulator protein

Messenger protein

Nuclear envelope

Target protein

Inactive gene

NUCLEUS

Activated gene

Signal response element

GENE TRANSCRIPTION

RNA

conveying a signal. *Amplifier proteins* are enzymes that magnify the signal by producing large numbers of intracellular mediators (e.g., cyclic AMP) or by activating large numbers of downstream signaling proteins. *Transducer proteins* convert the signal to a different form. *Bifurcation proteins* distribute the signal to multiple signaling pathways. *Integrator proteins* receive signals from other signaling pathways and integrate them. *Latent gene regulatory proteins* are activated at the cell surface and translocated to the nucleus to regulate gene transcription. Other proteins with important roles in signaling include *scaffold proteins* that bind multiple signaling molecules in a functional complex, often at a specific location, *modulator proteins* that modify the activity of signaling proteins, and *anchoring proteins* that keep signaling molecules at specific cellular locations through tethers to a membrane or cytoskeleton. Adapted from Alberts et al. (2002). (See color gallery.)

protein-coupled receptor intracellular domains activate either plasma membrane-bound enzymes (e.g., adenylate cyclase) or an ion channel in a process
that is mediated by trimeric GTP-binding proteins (*G proteins*). Signal binding
to enzyme-linked receptors leads either to activation of enzymes that are an
intrinsic part of their cytoplasmic domains or, in receptors without such intrinsic catalytic activity, to activation of enzymes with which the receptor cytoplasmic domains are directly associated. Most of the enzyme-linked receptors are *protein kinases,* or associated with protein kinases that phosphorylate
the receptor and specific downstream signaling molecules. Phosphorylation
of these proteins can result in their activation and/or create docking sites for
other signaling proteins. The different types of signaling protein that have
been described to function downstream of both cell surface and intracellular
receptors are summarized in Figure 2.2. Some of the specific signaling pathways mentioned in this book are illustrated in Figure 2.3.

GENE EXPRESSION

A major effect of the actions of cellular signals and the activation of their
downstream receptors and signaling pathways is to regulate gene expression.
For appropriate cellular function, some genes are activated, while others need
to be repressed. The basic principle of DNA as the storehouse of genetic information, of *messenger RNAs (mRNAs)*, as transcripts of the DNA that are
translated into protein by the ribosome, and of proteins as the structural components and effector molecules, is firmly established. Recent studies demonstrated, however, that the *transcriptome,* the set of RNAs transcribed from the
genome, is significantly more diverse and complex than previously thought.
For example, a particularly exciting recent discovery is the existence of genes
encoding micro-RNAs (miRNAs) of approximately 22 nucleotides in length
that do not encode protein, but bind to mRNAs to inhibit their translation or
mediate their destruction. There are dozens to thousands of miRNAs in cells
and, via the inhibition of several mRNA targets, an individual miRNA can
regulate complex multicomponent processes, including signaling and differentiation. Furthermore, RNA silencing like this, using short inhibitory RNAs
(siRNAs), allows one to "knock down" expression of genes in a sequence-spe-
cific fashion and is used to silence genes implicated in diseases and to determine the function of various genes. In addition, recent studies also indicate
an increasingly important role of *epigenetics,* that is, changes in gene function that are often heritable, that occur without a change in the sequence of

the DNA (e.g., DNA methylation), in gene regulation, in development, and in disease. A simple illustration of our current understanding of gene expression indicating the points at which it may be regulated is shown in Figure 2.4.

With the sequencing of the human genome, the application of high-through-put technologies to understand the regulation of gene expression at the level of the entire genome promises to provide important new information concerning both genetic and epigenomic regulation of gene expression that should greatly contribute to our understanding of the aging process.

ANIMAL MODEL STUDIES

Animal genetics has long been important in understanding development, physiology, and disease. Apart from matching particular genes to disease states, genetics is a powerful tool in the analysis of mechanisms in biology. For example, loss-of-function mutations in genes encoding the components of signaling pathways are used to determine whether components are in the same pathway or whether a particular component is upstream or downstream in a pathway (*epistatic analysis*). Two advances in mammalian genetics have been critical to the explosion in our understanding of pathways and disease states during the past twenty years. One of these has been the ability to carry out targeted mutation of specific genes of interest in mice. This has meant that, instead of obtaining mutations by chance, one can now create valuable, total loss-of-function mutants in genes of choice. The creation of such mice has often led to a direct connection between loss of gene function and a particular disease state. Using a similar approach, one can "knock-in" particular mutations into specified genes (e.g., point mutations that confer disease in humans can be introduced into the homologous mouse gene). The other advance was the creation of *transgenic mice* that *express* any gene of interest (from mouse or from any other species) under the control of gene regulatory regions that confer expression in selected types of cell. This enables the effect of expression or overexpression of genes of interest to be examined in specific cell lineages or tissues. Furthermore, in conjunction with more sophisticated approaches, one can combine both these technologies to "knock-out" particular genes solely in a lineage- or tissue-specific manner. This approach is particularly useful for loss-of-function mutations that are lethal. When knocked out in specific lineages or tissues, to generate mutant cells in an otherwise wild-type mouse, the progeny of the mice are often viable, allowing the effect in that lineage/tissue to be examined. In addition, besides these "reverse

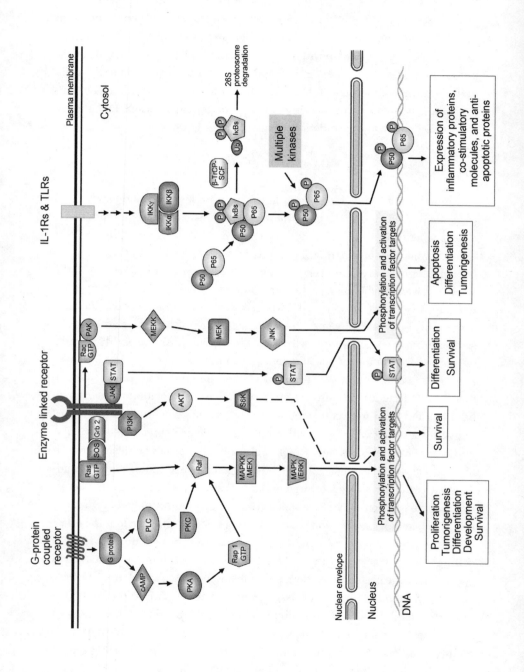

Figure 2.3. Some specific signaling pathways and the cellular functions they regulate. The pathways shown (left to right) are the G-protein-coupled pathways, the Ras/MAPK pathway, the PI3K pathway, the Jak/Stat pathway, the Jun kinase pathway, and the NF-kB pathway. Abbreviations: cAMP, cyclic AMP; PKA, protein kinase A; PLC, phospholipase C; PKC, protein kinase C; Grb2, Grb2 adapter; SOS, SOS guanine nucleotide exchange factor; RasGTP, activated small GTPase Ras; Raf, Raf serine/threonine kinase; MAPKK, mitogen-activated protein kinase kinase, also known as MEK, MAPK/extracellular signal-regulated kinase; MAPK, mitogen-activated protein kinase, also known as ERK, extracellular signal-regulated kinase; PI3K, phosphatidylinositol-3-kinase; AKT, Akt kinase, also known as protein kinase B; S6K, p70/ribosomal protein S6kinase; JAK, Janus kinase; STAT, signal transducer and activator of transcription; P, phosphorylation (not indicated for most kinase cascades such as the Raf/MAPKK/MAPK cascade); PAK, p21-activated kinase; MEKK, MAPK/ERK kinase; JNK, Jun N-terminal kinase; IL-1R, interleukin-1 receptor; TLR, toll-like receptor (mediate innate immune responses); IKK, IκB kinase; Ub, ubiquitin. (See color gallery.)

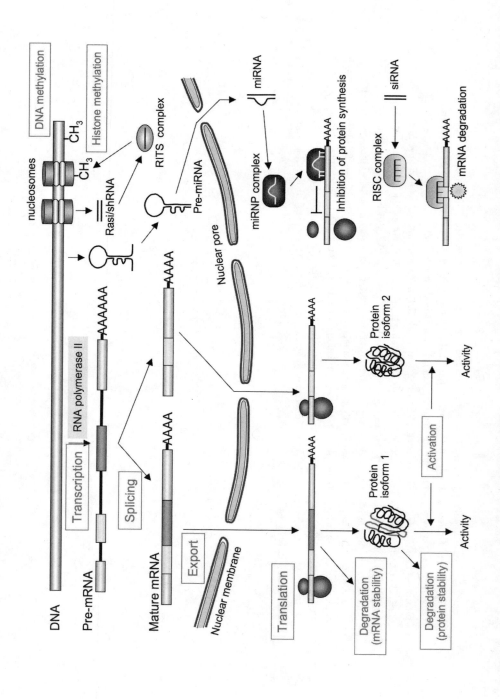

Figure 2.4. Regulation of gene expression. Transcription of the DNA by RNA polymerase II yields a primary transcript or mRNA precursor that contains *exons* (rectangles), including regions of the mature mRNA important for its regulation; and coding regions and noncoding regions known as *introns* (lines). Exonic sequences can also be removed as part of an intron to generate *alternative mRNAs* that can direct the synthesis of distinct *protein isoforms*. However, RNAs without coding capacity may also be similarly generated, including those that function as scaffolds for ribonuclear proteins that regulate processing of ribosomal RNA and other transcripts. An important example, represented here, is precursors of micro RNAs (miRNAs). Cleavage of these precursors by Drosha-type RNases in the nucleus and by Dicer-type enzymes in the cytoplasm generates 20–28 double-stranded miRNAs that assemble into RNP complexes that repress translation by complementary binding to the 3′-untranslated regions of mRNAs and/or cause mRNA degradation. Also shown is short inhibitory RNA (siRNA) that acts similarly and short hairpin RNA (shRNA), generated by bidirectional transcription and cleavage and often derived from repetitive DNA (rasiRNA), that can induce transcriptional silencing through histone and DNA methylation. Apart from this recently discovered and very important regulation by small double-stranded RNAs, gene expression may be regulated by many other mechanisms both normally and as a result of mutation. The steps at which regulation may occur include transcription, splicing, export of mRNA, mRNA stability, mRNA translation, and protein stability; these are indicated in red. Adapted from Mendes Soares and Varcárcel (2006). (See color gallery.)

genetic" approaches and because of the advances in genome mapping and sequencing, it is now possible to do "forward genetics" in vertebrates such as zebrafish and mice. In this approach, exposure of the organism to mutagens, followed by screens for mutations and their subsequent mapping, enables the genes mutated in disease resulting from the mutagenesis to be identified and the molecular basis of disease development determined. Experiments using these approaches or results obtained with them are referred to throughout this book.

SELECTED REFERENCES

Alberts B, Johnson A, Lewis J, Raff M, Roberts K, Walter P. 2002. *Molecular biology of the cell,* 4th ed. New York: Garland Science.

Guttmacher AE, Collins FS, Drazen JM (eds.). 2004. *Genomic medicine: articles from the New England Journal of Medicine.* Baltimore: Johns Hopkins University Press.

Kendrew J. (editor-in-chief). 1995. *The encyclopedia of molecular biology.* Cambridge, Mass.: Blackwell Science.

Mendes Soares LM, Valcárcel L. 2006. The expanding transcriptome: the genome as the 'Book of Sand.' *EMBO J* 25: 923–31.

Rodenhiser D, Mann M. 2006. Epigenetics and human disease: translating basic biology into clinical applications. *CMAJ* 174: 341–48.

Silver LM. 1995. *Mouse genetics: concepts and applications.* New York: Oxford University Press.

CHAPTER THREE

Aging

1. TRENDS IN THE AGING OF THE POPULATION

In the developed nations, the progressive increase in the average life span of human beings, which now extends into the 80s, is perhaps the dominant rationale for the medical practice of geriatrics, indeed sometimes called a demographic "revolution" or "an imperative" (Hamerman and Fox, 1992). Currently, persons 65 years and older make up about 15 percent of the population, and this number is expected to exceed 20 percent within three decades. Persons living longer are especially represented in those older than 85 years—sometimes referred to as the oldest-old (Suzman et al., 1992)—and constitute the most rapidly growing segment of the over-65 group (Fig. 3.1). The trade-off for society is that a century ago causes of death were primarily infectious diseases, and people recovered or died; survival, with extended morbidities, was uncommon. Now, death from infectious diseases is rare, and treatment for the major causes of death—cardiovascular diseases and cancer—and other conditions not fatal per se—Alzheimer disease, Parkinson disease—may extend life for years but with sustained disabilities. The sites for geriatric health care of an older population have therefore expanded from traditional ambulatory practice and the hospital to a unique added dimension in the home and, in what is now a growth industry, the nursing home and assisted living facilities. Although the maximum life span (about 120 years) does not seem to have changed, the increase in centenarians poses a new challenge for geriatrics to identify the social, ethical, psychological, and family interactions and health care costs as well as biological determinants of a very long life. Thus, *longevity* has emerged as a new issue for health care and research at the commencement of the twenty-first century (Butler et al., 2004).

Perls (2006) based an optimistic view of the health status of centenarians on his study of them in Boston. He notes that "90% of centenarians in a population-based study were functionally independent at the average age of 92 years."

The situation may not be as favorable for the age group exceeding the centenarians, *supercentenarians*—those who have lived to be 110 years or older. L. Stephen Coles (2004), in a demographic review, writes that "the number of worldwide members in this special group is much more manageable" for study: as of December 31, 2003, 45 living, 50 had died, compared with centenarians (about 180,000 worldwide and by 2050 projected to grow to 3.2 million!). From a personal interview of just seven persons, Coles makes note of a unifying feature of a very long life: "early health status was excellent, as was the case for their long-lived siblings, and most of them virtually never needed to see a doctor before age 90. This tells us immediately that our modern medical establishment cannot claim credit for those who exhibit extreme longevity," which seems to be the case. About 30 years ago, Aaron Wildavsky (1977), then the dean of the Graduate School of Public Policy at the University of California in Berkeley, expressed this aspect dramatically: "According to the Great Equation, Medical Care equals Health. But the Great Equation is wrong. More available medical care does not equal better health. The best estimates are that the medical system affects about 10% of the usual indices for measuring health. The remaining 90% are determined by factors over which doctors have little or no control." More recently, Berenson (2006), in an editorial, reported "no evidence that greater spending, more resource inputs, and more frequent use of hospitals or physician services are associated with improvements in survival, functional status, patient satisfaction with care, or better performance on technical measures of care processes." Surely, the argument for improved lifestyle with its demonstrated benefits—preventive gerontology, as discussed in Chapter 6—gains credibility as one alternative.

It is also apparent from Coles's interview that his subjects did not exhibit glowing health: often they were deaf, were blind, had limited taste and smell, had poor dentition, were frail, were weak (sarcopenia), and were poorly oriented. Thus, longevity comes at a functional price, although it has always been a cherished desire of mankind (Vijg and Suh, 2005). Yet this inevitable (so far) functional decline is often overlooked or denied by "hucksters, swindlers, and snake oil salesmen," in the words of Olshansky et al. (2004) about

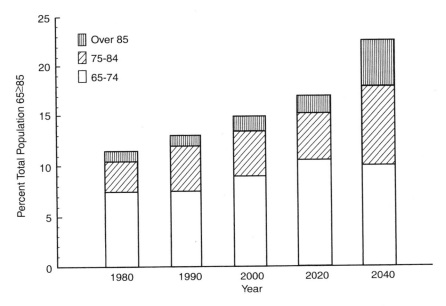

Figure 3.1. Graphic representation of population projections involving elderly persons. Those older than 85 years, sometimes designated the oldest-old or extreme aged (Suzman et al., 1992), will represent the fastest growing segment of the elderly population in the early decades of the twenty-first century. Adapted from McGinnis (1988) and Hamerman (1995), © British Medical Journal Publishing Group, with permission.

those who "hype" antiaging medicines. This entire subject, termed "antiaging medicine: the hype and the reality," is presented in an interesting issue of the *Journal of Gerontology,* to which readers are referred. But as bioethicist Arthur Caplan (2005) points out in an extended essay, "ageing is a chance phenomenon, and this makes ageing unnatural and in no way an intrinsic part of human nature. As such, there is no reason why it is intrinsically wrong to try to reverse or cure ageing." His view that there is no intrinsic ethical reason why we should not try to extend our lives seems contrary to those of Callahan, Kass, Fukuyama, and many others, who consider such activities *unethical* (emphasis added). There is much more to say about this, and interested readers can consult a Symposium on Science and Society in the *EMBO Reports,* volume 6, supplement 1 (2005). Later, I will discuss this subject in relation to frailty (Chapter 4, section 6).

2. WHAT IS AGING? CAN WE DISTINGUISH BETWEEN AGING AND DISEASE?

Aging in a medical context has rarely had an optimistic definition. Witness: "aging is a process of intrinsic deterioration that is reflected at the population level as an increase in the likelihood of death and a decline in the production of offspring" (Partridge and Gems, 2002); or, a "time-dependent loss of fitness" (Vijg and Suh, 2005). Kirkwood (2005a), who has contributed so much to the theoretical underpinnings and evolutionary biology of aging theories, notes that "aging is arguably the most familiar yet least well-understood aspect of human biology." When we use the word *aging,* we tend to think of old age (for example, "he's aging well") or a more persuasive medical link so widely used, "age-related diseases," which implies almost a synonymous association. My point in this section is to incline to separate them, although there are strong advocates for and against their linkages. A persuasive case can be made on either side by experts, as witness an extended, forceful, but cordial discussion between two leading investigators in biogerontology, Robin Holliday (2004) and Leonard Hayflick (2004a). I prefer separation but association.

First, aging is inevitable; disease is not. Aging is a continuum, measured by the passage of time from conception to birth, through development, maturity, and terminated by death. Butler et al. (2004) propose, as a "good overall working definition," that aging is a "nondescript colloquialism that can mean any change over time, whether during development, young adult life, or senescence." Masoro (2006) defines aging as a synonym for "senescence—deteriorative changes during the adult period of life, which underlie an increasing vulnerability to challenges, thereby decreasing the ability of the organism to survive." I have problems with the term *senescence,* which can refer widely, on the one hand, to fundamental molecular changes in cells, or on the other hand, to a more selective decline in mental acuity. Indeed, Butler et al. (2004) reinforce this latter sense in asserting that their "biomarkers of aging are the processes that progressively convert physiologically and cognitively fit healthy adults into less fit individuals with increasing vulnerability to injury, illness, and death." Kirkwood (2005a) also emphasizes "increasing vulnerability to environmental challenges and a growing *risk* (my emphasis) of disease and death." Slaets (2006) describes vulnerability in elderly persons in the context of frailty, which is discussed in Chapter 4, section 6. Leonard Hayflick (2004a), who has made fundamental contributions to the cellular basis of aging (which I will discuss in a moment as "replicative senescence"), draws a sharp demarcation when he states that "the aging process simply increases the vulnerability to disease or pathology."

It is understandable that the focus on aging is not on the evolving years of youth and early maturity, but some later time, when aging refers to "old age." The initial impression about older age is the inevitable but variable decline in an array of physiologic functions. Martin (2006), in his keynote lecture at the Halle Conference in Germany, states that he prefers the British spelling of *ageing* because this makes a "nicer match" with *sageing,* a term he has used to "refer to adaptive alterations in structure and function during the course of the life span. Sageing is called upon to compensate for the slow, insidious declines in structure and function that begin to emerge soon after the establishment of sexual maturity and the adult phenotype, i.e., the beginning, of what most of us regard as biological ageing."

Duthie (2004) drew decrements schematically to include, among a variety of responses, diminution in nerve-conduction velocity, resting cardiac index, vital capacity, renal blood flow, and creatinine clearance. Much attention has been given to changes in body composition, with loss of lean body mass—predominantly muscle—with increased abdominal fat mass and diminished total body water. These changes in body composition have importance for altering the distribution of drugs (wider in the case of fat-soluble ones, smaller in the case of water-soluble ones). Perhaps even more consequential is the accumulation of fat, in particular, in the abdomen (visceral fat), about which there is now great interest because of the profound implications with respect to altered glucose and lipid homeostasis arising from the associated metabolic syndrome, which will be discussed in Chapter, 4, section 3. Hormonal decline is also a feature of aging, with respect to growth hormone (GH) and insulin-like growth factor (IGF) (somatopause), testosterone (andropause), estrogen (menopause), and adrenal dehydroepiandrosterone (DHEA) (adrenopause) (Lamberts et al., 1997; Siiteri, 2005). The *inflammatory* underpinning of much of the biology of aging and disease is noteworthy with respect to cytokine interactions with the central nervous system, resulting in various physiologic, neuroendocrine, and behavioral responses (Sternberg, 1997). As I will discuss, "stress responses" involving interleukin-6 (IL-6) activation of the hypothalamic-pituitary-adrenal axis (Mobbs, 1996) release glucocorticoids in a manner tailored to stimulus intensity (Jacobson, 2005) (see Chapter 4, section 1). Franceschi et al. in an issue of the *Annals of the New York Academy of Science* (2000) proposes that with aging there is a "global reduction in the capacity to cope with a variety of stressors and a progressive increase in pro-inflammatory status" to which they refer as "inflamm-aging." Indeed, in their emphasis on inflammation and their proposal to "return the macrophage to its rightful place as a central actor in stress and in inflammation," the authors strike a recurring theme and emphasis of this book.

So far, I have not mentioned a disease, although diseases are incipient: osteoporosis, which may lead to fragility fractures; and obesity, contributing to the inception of the metabolic syndrome, diabetes, atherosclerosis, and cardiovascular diseases, as well as knee osteoarthritis. These are some of the links to be discussed in Chapter 4. Although many aging-related functions may be inevitable (e.g., bone loss), considerable individual variability nevertheless exists, and good personal health practices and attentive assessment by a health care provider may delay or prevent the inception of disease (e.g., fragility fractures). What appears to be generally applicable to aging is *vulnerability* to stress (i.e., organs or tissues that cannot adapt as well as when they were in a more youthful condition). The kidney is a good example. It is often assumed that renal function in older persons declines with aging. Yet this may not be universal. In a group of apparently "normal" persons enrolled in the Baltimore Longitudinal Study, Lindeman et al. (1985) reported a *mean* decrease in creatinine clearance of 0.75 ml/min/year. Yet, as the authors found, in accord with a Gaussian distribution, one-third of all subjects had no decrease in renal function. Nevertheless, a decline in renal function may not be *clinically* apparent. Blood tests may show normal blood urea nitrogen (BUN) and serum creatinine levels. The serum creatinine alone is not a reliable guide to renal function as people age; serum creatinine may remain within the normal range because of a concomitant reduction in muscle mass. One of several formulas that have been suggested (Levey et al., 2006) to estimate the glomerular filtration rate—the best overall index of renal function—uses serum creatinine, age, weight, and sex:

$$\frac{140 - \text{age (years)}}{\text{serum creatinine}}$$

The result is multiplied by weight (kilograms) times 72 (for women, multiply this result by 0.85).

Some pitfalls are associated with this formula. Recent papers discuss some of these, as well as the emergence of cystatin C (a protein that is a member of a family of competitive inhibitors of lysosomal cysteine proteases) as a marker possibly superior to creatinine—albeit still experimental—in "preclinical" chronic kidney disease patients (Coresh and Astor, 2006; Rule et al., 2006; Shlipak et al., 2006). Nevertheless, the formula presented here may be useful to assess *vulnerability* of the kidney in those older persons for whom use of nonsteroidal anti-inflammatory drugs (NSAIDs) is anticipated—and there are indeed many such persons! NSAIDs diminish clearance to an extent based on the type of cyclooxygenase inhibitor (1 or 2) used (see Chapter 5, section 1), the particular medication, dose, and duration of its use, the age of the person, and the preexisting renal status. Impairment of renal function may occur, with a rise in BUN and creatinine, water and salt retention, edema, and hypertension, which may lead to renal insufficiency. This is not a disease, but a vulnerability arising from existing aging changes. Drugs are a major contributor of untoward events in aging organ systems. The interested reader may wish to pursue a new classification and staging system for chronic kidney disease (from early to late) as presented by Perazella and Reilly (2003).

McEwen and coworkers (1997), using biological markers of physiologic responses, attempted to quantify vulnerability ("strain") of the body from stress. These investigators developed the "allosteric load" index, identifying blood pressure, waist-hip circumference, high-density lipoprotein (HDL) levels, hemoglobin Alc, DHEA, and other parameters, where higher values (except HDL, which was good) predicted poorer response to stress. I will go into the concept and ramifications of "stress" as a central feature of the impact of aging and disease in Chapter 4, section 1.

Lipsitz and Goldberger (1992) and Lipsitz (2002) looked at the physiologic basis of aging as a loss of complexity measured as "fractals" or oscillations. Complexity loss *within* elderly individuals could result in increased variability of a given response *between* individuals. "Thus compared with healthy young individuals, healthy elderly people may look more alike under steady state conditions, but less alike when challenged by internal or external perturbations." Once again, variability is as much a part of aging as vulnerability.

3. MECHANISMS OF AGING

One of the aims of promoting an interchange between basic scientists and academic geriatricians, or at least a greater understanding of basic mechanisms of aging on the part of the latter group, is to *translate* scientific advances to the bedside. This need surely must spur both groups toward a dialogue, in which "studying basic age-related molecular processes at the cellular level will allow us to understand the development of age-related disease and devise new diagnostic, preventive, and therapeutic approaches" (Wick and Xu, 1999). Even in this section we can begin to observe two trends toward translation, in which the understanding of basic research could lead to clinical application and profoundly modify aging trends. One example relates to experimentally extending the length of telomeres and thereby maintaining the replication of potentially senescent cells; a second example pertains to applying the knowledge gained from studies of human genetic polymorphisms (variations) to clinical testing, which may predict outcomes favoring longevity or, conversely, those at greater risk for disease inception. Barton Childs (1999, p. 112), in his perceptive text on a logic of disease, wrote, "What are the genes that contribute to the specificity of usual aging and to the diseases of old age? Gerontology has moved from a backwater to the forefront of science in a remarkably short time. Indeed, it is now possessed of that clutter of molecular information that at once enthralls the molecular geneticists and tries their patience as they struggle to fit new data to old hypotheses or to devise new ones."

Perhaps it is worthwhile to look back a few years to perceive how far insight into the molecular biology of aging has advanced. In 1997, Richard Miller, a distinguished investigator in this field, was prompted to ask when the biology of aging would become "useful": "Biogerontologists—members of the small cadre of researchers obsessed with the basic biology of aging and its putative links to disease—cannot yet look you straight in the eye and state honestly that they are doing useful research, research that is bound to reveal the secrets of aging, and thereby lead to improvements in human well-being" (Miller, 1997).

It would be interesting to know what Miller would say about the advances, many of which he had the foresight to project even then: comparative and evolutionary biology; mutants (induced and natural); invertebrate genetics; biomarker research; caloric restriction; and, above all, cellular senescence. All are presented in varying depths in this chapter and in relevant disease associations later. These aspects represent a virtual groundswell of new information, which has come to mean *translation,* a word not used much ten years ago. I think Miller and other biogerontological researchers would take pride in a steady and forthright gaze!

For a discussion of senescence as one aspect of aging mechanisms, I need first to introduce a diagram of the cell cycle and the factors that influence progression of the normally *replicating* (dividing) cell from G_1 to mitosis (M), as shown in Figure 3.2. The molecular-cellular manifestation of physiologic aging in humans may be cellular *senescence.* Cellular senescence was first

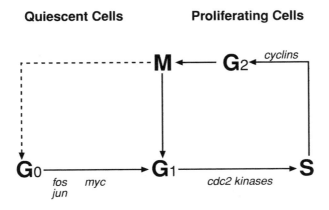

Quiescent Cells   **Proliferating Cells**

Figure 3.2. Diagram of the cell cycle. In a quiescent phase (G$_0$), cells are competent, and, in response to added growth factors that induce expression of some proto-oncogenes (c-*fos*, c-*jun*, c-*myc*), cells proliferate and enter the first part of the cycle (G$_1$). Then, in response to a complex of cdc2-like protein kinases and cyclins, the cells move into and out of S, G$_2$, and M (mitosis). A senescent cell, on the other hand, is essentially permanently growth arrested in G$_1$, and growth factors will not advance this cell further into the cell cycle. Reprinted from Hamerman (1993), with permission from Blackwell Publishing.

described as the process that limits the replication of normal human cells in a favorable culture system—hence replicative senescence. This finite capacity of normal cells to replicate in culture was the seminal discovery by Hayflick (1965) decades ago. Later studies showed that cells as they senesce lose replicative capacity and become unresponsive to many proliferative and apoptotic stimuli, including endogenous reactive oxygen species, growth factors, and exogenous carcinogens. But although senescent cells are no longer able to replicate, they are not dead. In fact, their viability is the cornerstone for relating the accumulation of senescent cells in an organ or tissue (i.e., not in a culture dish) to the in vivo basis for diminished organ function or tissue plasticity. Senescent cells could contribute to aging in at least two ways: first, senescent progenitor or stem cells, without the capability of renewal could impair tissue maintenance; second, senescent cells appear to secrete matrix-degrading proteases, cytokines, and other inflammatory factors that disrupt tissue function (Park et al., 2004; Pelicci, 2004; Gilley et al., 2005; Lombard et al., 2005). An enzyme called senescence-associated β-galactosidase has been identified as a cell marker of senescence, and cells with such a marker are present at sites of certain age-related pathologies, including atherosclerotic lesions, skin ulcers, and arthritic joints. These cells are "bad neighbors" that seem to alter tissue structure and function (Campisi, 2005).

An interesting and important suggestion is that senescent cells are the trade-off for preventing cancer. Cancer cells display uncontrolled and infinite replicative capacity. Thus, as part of aging, senescence in cells is a tumor suppressor mechanism, a "biological tradeoff," a "double-edged sword" (Campisi, 1997). This contradictory duality, as it were, in the example of cells in their youth (an era of growth) and cells in their old age (an era of senescence), has been termed "antagonistic pleiotropy" (i.e., what proves beneficial to the organism in early life may be deleterious at old age). There are other examples of antagonistic pleiotropy. Unger (2005) notes that overnutrition appears beneficial during the years of reproductive life yet harmful in later life because of surplus calories ingested—part of the obesity epidemic so prevalent now. Finch and Crimmins (2004) propose that adaptive responses to short-term infections early in life become mal-adaptive in the long term when chronic inflammatory mechanisms that arise (see Chapter 4) predispose to morbidity and mortality. This is also, in their view, a "double-edged sword" and a further example of antagonistic pleiotropy. Kirkwood (2005a) discusses the earlier literature on the existence of "pleiotropic genes—which have opposite effects at different ages." As a hypothetical example, a mutation may have a favorable effect on bone calcification during development but subsequently, in a different environment, is expressed as calcification in the arterial wall and atherosclerosis. In fact, this is a subject I have discussed elsewhere (Hamerman, 2005b) at some length in terms of the underlying biological linkages between bone and the arterial wall.

If geriatricians are to be aware of the extraordinary advances in the molecular biology of cell senescence, I should summarize this field. (See also Fig. 3.3.)

Activation of two *tumor suppressor* proteins called p53 and Rb is a key event in the induction of cell senescence in human cells (Fig. 3.3) (Helmbold et al., 2006). Multiple factors contribute to the activation of tumor suppressor protein, but most important is their critical responses to DNA damage—a kind of "molecular stress," compared to stresses that make older individuals vulnerable at the physiologic level. As Lombard and colleagues (2005) put it, "the imperfect maintenance of nuclear DNA likely represents a critical contributor to aging." Upon its activation, p53 induces expression of the cyclin-dependent kinase inhibitor p21, which promotes cell cycle *arrest* at the G_1/S

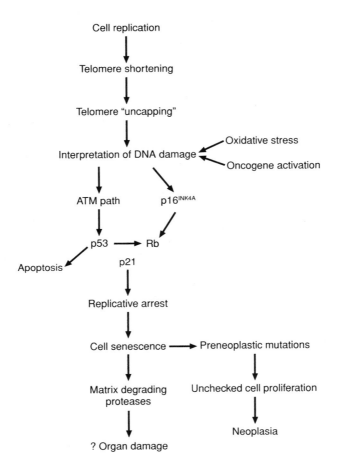

Cell replication

Telomere shortening

Telomere "uncapping"

Oxidative stress

Interpretation of DNA damage

Oncogene activation

ATM path   p16^{INK4A}

p53 ⟶ Rb

Apoptosis   p21

Replicative arrest

Cell senescence ⟶ Preneoplastic mutations

Matrix degrading   Unchecked cell proliferation
proteases

? Organ damage   Neoplasia

Figure 3.3. An overall survey of current ideas concerning a molecular basis for cell senescence. This diagram and legend aim to recapitulate in brief the discussion in the text. Findings from cell culture systems are thought to be applicable to the human condition of aging. With each cell replication, telomeres shorten because of the "end replication problem." Without telomerase to extend telomere length, telomeres reach a critical point of structural change and/or loss of protective proteins (referred to as "uncapping"). Although diverse stimuli can drive the senescence response, essentially two pathways are governed by the gatekeeper tumor suppressor proteins p53 and Rb. The p53 pathway is activated by dysfunctional telomeres and other events, as shown, including ataxia telangiectasia mutated (ATM) kinase—all interpreted by the cell as severe DNA damage. Oxidative stress, oncogene activation, cell cycle inhibitors, especially p16, activate Rb. The outcomes include programmed cell death (apoptosis); senescence of cells, which may contribute to organ damage; and unchecked proliferation and neoplasia in the "permissive environment" of senescent cells. Data from Balaban et al. (2005), Ben-Porath and Weinberg (2005), Campisi (2005), and Lombard et al. (2005).

checkpoint (Fig. 3.2) (Giono and Manfredi, 2006). The spectrum of cellular events brought about by tumor suppressor proteins p53 and Rb in response to DNA damage includes DNA repair, as well as growth arrest (for senescence this is permanent) and apoptosis (Fig. 3.3). Genes that code for p53 and Rb are frequently mutated (inactivated) in cancer cells: *loss* of checkpoint control supports the progression of cancer through different stages of tumorigenesis (Genovese et al., 2006)—alternatives that are key aspects of the biology of aging.

The activation of p53 in particular occurs in response to telomere shortening with each replication. Telomeres are specialized DNA that protect the ends of chromosomes ("capping"), and telomeres shorten with each replicative cycle of the cell due to "an end replication problem" (i.e., failure to fully replicate the entire length). Artandi (2006) presents a brief account of telomeres in "cancers and other disorders associated with aging," citing the 2006 Albert Lasker Award for Basic Medical Research to the three persons who made "seminal observations"—Elizabeth Blackburn, Carol Greider, and Jack Szostak. De Lange (2006), who has also made major contributions to the field, writes of "Lasker Laurels for Telomerase," citing the three recipients and tracing the historical identification of the enzyme telomerase: a reverse transcriptase copying part of the associated RNA template. De Lange also notes the signal experiment I cited above, in which the human counterpart of this reverse transcriptase (hTERT) introduced into primary human cells "reconstituted robust telomerase activity and thus abrogated telomere shortening. As predicted, such cells do not undergo replicative senescence; they are immortal." And—in line with a theme of this book—"hopefully, the story of telomerase will not come to an end (i.e., with these Awards) before its tremendous promise has paid off in the clinic." More detailed accounts of telomeres in aging and cancer appear in reviews published in 2004 by Sharpless and DePinho and in 2005 by Blackburn and by Blasco.

When telomeres undergo structural modification and/or shorten to a critical, defining length called "uncapping" (Gilley et al., 2005), the message the cell receives is one of profound damage to DNA integrity. Somatic cells (i.e., non-stem cells) lack the enzymatic capacity by way of *telomerase* that almost all cancer cells possess to extend telomere length and so maintain replication. The critically shortened telomere is recognized in the cell as a double-stranded DNA break, which activates the DNA damage machinery and p53 tumor suppressor protein, and contributes to replicative arrest. Telomere shortening is only part of the multifaceted basis for the onset of cell senescence. Yet the tight relationship between replicative senescence and telomere shortening in cultured human cells supports a molecular explanation for cellular senescence in aged tissues. In a mammalian model, workers in the laboratory of Ronald DePinho (Rudolph et al., 1999) generated telomerase-deficient mice, and despite the customary long telomeres of other mice species, these mice (designated mTR$^{-/-}$) showed progressive shortening of telomeres with each successive generation. In a beautifully illustrated paper, late-generation mice showed a "hallmark feature of aging—a reduced capacity to tolerate acute stress," with a shortened life span, poor wound healing, limited response to blood cell depletion, and increase in the rate of spontaneous tumors, suggesting genetic instability. These mice did not exhibit cataracts, osteoporosis, glucose intolerance, or vascular disease.

As I noted, support for a role of telomere integrity in cell senescence derives from basic experiments with important clinical implications, in which researchers added telomerase to cells *in culture* at the end of the cell's replicative life and thereby prolonged the cells' replicative capacity. Maintaining cell functions by extending telomere length, while still entirely experimental, has profound significance. Short telomeres may already be present in human diseases that are associated with aging, such as heart disease, atherosclerosis, and premature aging syndromes, discussed below (Blasco, 2005). Indeed, assessment of telomere length is approaching clinical studies (Wong and Collins, 2004). Valdes et al. (2005) found that in women judged obese by measures of body mass index (BMI) and who smoked, "telomere length decreased steadily with age, emphasizing the pro-ageing effects of obesity and cigarette smoking." Of course, it is still a wide stretch to equate telomere shortening with the health-adverse affects of obesity and smoking.

As Figure 3.3 shows, an additional activation input line feeds directly to Rb—the other tumor suppressor protein—and that is p16[INK4A], a cell inhibitor of cyclin-dependent kinases CDK4 and CDK6 (Kim and Sharpless, 2006). These kinases can stimulate progression of the cell cycle by phosphorylating Rb; but as a result of the action of p16[INK4a], Rb remains unphosphorylated and tightly bound to E2F transcription factors, preventing them from stimulating replication of DNA. In this way, p16[INK4a] blocks entry of the cell into the G_1-S phase of the cell cycle (Fig. 3.2) (Knudsen and Knudsen, 2006; Mooi and Peeper, 2006); p16[INK4a] is induced by many types of stress and highly expressed in senescent cells (Dimri, 2004). Senescence can thus be activated by either the p53→p21 arm or by the p16[INK4a] arm, or by both in parallel (Lombard et al., 2005), and there is "crosstalk" between the p53 and p16[INK4a] pathways. As Ben-Porath and Weinberg (2005) put it, "p53 and Rb function as a complex stress-signal integration and processing unit, whose input is the total level of stress and damage a cell is experiencing, and whose output is the activation of the appropriate cellular response." But keep in mind the "double-edged sword" as the alternative to cell senescence—namely, neoplastic transformation: p21 has been implicated in tumor-promoting signaling effects that senescent fibroblasts may impose on epithelial cells in their environment (Gilley et al., 2005). I will have more to say about the "permissive environment" that senescent cells confer on epithelial cells transformation in Chapter 4, section 7.

The alternatives of the genetic link between "life span and cancer" made national news in the *New York Times* of September 7, 2006. Citing three articles online in *Nature,* the reporter Nicholas Wade wrote of the "gene with the unmemorable name of p16^{INK4a}" from which the protein acts as tumor suppressor (noted above), accumulating in aging cells and directing cells toward senescence. To cite one example (Janzen et al., 2006) of the articles in *Nature:* "crucial functions of human stem cells relevant to their persistence and survival under stress were noted in older mice in the absence of p16^{INK4a}, with improvement in tissue regeneration and animal survival after injury." Indeed, the results of deletion of p16^{INK4a} do suggest a molecular basis for *less* vulnerability in the course of aging. Yet in an interesting essay on mouse models developed for cancer drug testing, Sharpless and DePinho (2006) remind us how a combined *deficiency* in p16^{INK4a} and *activation* of the tumor-producing oncogene RAS can induce cancer in these mice. Thus, the presence of p16^{INK4a} may limit the proliferative potential of cells. Further, in the presence of active tumor suppressor proteins that limit uncontrolled proliferation, the balance may be tipped toward senescence by the actions of certain oncogenes (Serrano et al., 1997; Mooi and Peeper, 2006). It almost seems that beyond the "double-edged sword" of senescence versus neoplasia cited by Campisi (1997) (and see Fig. 3.3), there is a "molecular struggle," in which so many factors may drive cells either way, crucially moving toward a virtually irreversible choice that ultimately affects the clinical course of later life. Indeed, Kim and Sharpless (2006) have conveyed the "drama" of the struggle between proliferation and senescence. They note that senescence requires at least two things: signaling induced by strong cellular stress (such as telomere dysfunction or oxygen radicals) and some coincident period of signaling to promote proliferation—citing the work in this regard of Satyanarayana et al. (2004). Whereas growth arrest is transient, senescence requires sustained expression of p16^{INK4a}. "Senescence can be considered a response to cellular desperation after a prolonged period of mixed signals, including signals for both stress and growth."

Oncogenes can also drive the p53 pathway toward cell senescence when reactive oxygen species (ROS) induce DNA damage, an aspect that will figure prominently in this book (Fig. 3.3). The reactive oxygen species are part of the free-radical theory of aging, proposed more than 50 years ago by Harman (1956). ROS have high chemical reactivity and include nonradical species such as hydrogen peroxide (H_2O_2); and free radicals with one or more unpaired electrons—such as superoxide (O_2^-), and hydroxyl radicals ($OH^.$), which are unstable and highly reactive when trying to capture the needed electron from other compounds to gain stability (Packer, 2005; Cave et al., 2006). In health, ROS generation is counteracted by the activity of enzymatic and nonenzymatic antioxidant systems that scavenge or reduce ROS levels, thereby maintaining an appropriate redox balance in cells and tissues. Perturbation of this normal balance due to increased ROS production and/or reduced antioxidant reserve leads to a state of *oxidative stress,* with damage to biological macromolecules, especially DNA, but also lipids and proteins (Frisard and Ravussin, 2006; Kajstura et al., 2006). Thus, a rise in intracellular oxidant levels has two potentially important effects: damage to various cell components, as noted, and triggering of the activation of specific signaling pathways—some of which are discussed in Chapter 2: signal-regulated kinase (ERK), c-Jun amino-terminal kinase (JNK), mitogen-activated protein kinase (MAPK), the phosphatidylinositol-3-kinase (PI3K)/Akt pathway, the nuclear factor (NF-kB) signaling system, and p53 activation, as noted above (Finkel and Holbrook, 2000; Chung et al., 2005).

Oxidative stress has come to be regarded as one of the fundamental mechanisms of aging (Lombard et al., 2005; Packer, 2005), associated with many age-related clinical conditions—especially oxidative modification of low-density lipoprotein (LDL) in the vascular intima—a key part of the pathogenesis of atherosclerosis (Chapter 4, section 2) (Balaban et al., 2005; Chung et al., 2005; Cave, 2006; Kaliora et al., 2006). The mitochondria are a major source of ROS, using oxidative phosphorylation to convert calories into usable energy and generating ROS as a toxic by-product. Despite glutathione and other systems to maintain redox (antioxidant) balance (Kaliora et al., 2006; Koehler et al., 2006), mitochondrial DNA is susceptible to ROS-generated damage (Wallace, 2005) due to proximity to oxidant generation and limited DNA repair (Finkel and Holbrook, 2000; Chan, 2006). Indeed Wallace (2005) proposed that the delayed onset and progressive course of age-related diseases result from accumulation of somatic mutations in mitochondrial DNA.

The free-radical theory of aging gained further credence from the numerous natural protective mechanisms to prevent oxyradical-induced cellular damage—intrinsic antioxidants that include superoxide dismutase, catalase, glutathione peroxidase, and glutathione reductase—as well as extrinsic antioxidants, vitamins C, E, carotenoids, flavinoids, etc. (Knight, 2000). Bruce Ames (2005) has long been an advocate for micronutrients to counter the adverse oxidative effects that may accompany their deficiency: "multivitamin and mineral supplementation is good health insurance and would markedly protect us against heart disease, cancer, immune deficiencies and cataracts"; but he cautions that "Mae West's dictum about sex—'too much of a good thing is wonderful,' does not apply to micronutrients"—where restraint may be wise. Finkel and Holbrook (2000) echo this view—that "given the role of ROS as mediators of normal signaling processes, determination of the optimal dosage of supplements may require fine tuning to avoid overshooting the desired effects to the point of perturbing the delicate redox balance required for the maintenance of normal cell functions." Kaliora et al. (2006) suggested the application of "genomic tools to study the integrated effects of nutrients on gene regulation, namely, nutrigenomics"—to improve our understanding of how nutrients affect the "whole organism in health and disease." Along the lines of nutrition and "healthy aging" (see Chapter 6, section 3), Duff et al. (2006) note that "available studies do not support a nutritional remedy for aging, but dietary intervention may delay certain deleterious effects of time." I will discuss later neuroendocrine signaling and metabolic pathways involved in nutritional sensing—"especially important because they converge with crucial regulators of energy homeostasis" (Picard and Guarente, 2005).

One further word on another family of tumor suppressor proteins that link cancer, senescence, and apoptosis, which is a central theme here. Russell et al. (2006) describe the "founding member" of this family as *in*hibitor of *g*rowth, or ING 1, which, when overexpressed, promotes G_1 arrest of the cell cycle (Fig. 3.2) and yet when inhibited encourages transformation in vitro and tumor formation in vivo. Thus, ING impinges on many aspects of p53 control of cell function having to do with apoptosis, DNA damage repair, cell cycle regulation, senescence, and tumorigenesis.

Before I consider basic studies on genetic expression that may provide clinical application for either longevity or accelerated aging, I should mention genetic modifications that produce *"premature aging"* or progeroid syndromes that resemble—but only *in part*—physiologic aging, hence the added word *segmental* progeroid syndromes (Martin, 2005). Of most relevance to geriatricians is the syndrome described by Werner in 1904. About 1,300 people worldwide have this condition, and hence, although it is quite rare, it is the most common of the progeroid syndromes and a "living expression" of the acceleration of many manifestations of aging. Werner syndrome begins early. The first sign of the disease is absence of the pubertal growth spurt by the third decade of life. Patients have short stature, premature graying, hair loss, skin atrophy, fat deposition in the trunk, diabetes mellitus, osteoporosis of long bones, cataracts, and atherosclerosis. Cancers are frequent, an interesting possible association with cell senescence, as noted above. The genetic hallmark of Werner syndrome is genome instability. Persons with Werner syndrome have mutations in the WRN gene, which encodes a helicase needed for DNA repair. There is also a model of a short-lived mouse with "progeroid features" (Bartke, 2006) that is instructive for early onset of aging—the Klotho mouse. Mice with deletion of the Klotho gene develop infertility, ectopic calcifications, lipodystrophy, skin and muscle atrophy, osteoporosis, and atherosclerosis and die at about two months of age (Katic and Kahn, 2005). Overexpression of the Klotho gene appears to increase mouse life span by inhibiting the insulin/IGF-1 signaling pathway (Kurosu et al., 2005; Bartke, 2006).

Another progeroid syndrome is ataxia telangiectasia; persons with perturbations in the ataxia-telangiectasia mutated (ATM) kinase required for detecting DNA damage and initiating repair responses age prematurely. These persons are immunodeficient, manifest cerebellar degeneration, are prone to cancer, and also die young (Lombard et al., 2005). The *senescence* response is tied to ATM by the fact that ATM kinase phosphorylates and thus activates p53, which we have seen coordinates aspects of cell growth, survival, and senescence in response to DNA damage.

4. LONGEVITY: GENETICS OF EXTENDED LIFE SPAN OR RISK FOR ADVERSE EVENTS

Human longevity, manifest particularly in centenarians, is perhaps the biological opposite of those with progeroid syndromes, and also exceeds by a generation the current traditional extension of life span of those persons in their 80s. Again, in the decades ahead, there are profound personal, medical, and societal implications as life span for many more persons begins to approach 100 years. One can imagine a true survival of the fittest among centenarians in whom possibly the earlier decades are much more free of disabling conditions; indeed, medical costs of those 60–70 year olds dying of disease were about threefold greater than those in persons dying after the age of 100 (Barzilai and Shuldiner, 2001). Leaving aside personal health (environmental) factors, there is a growing understanding of the genetic aspects that promote or limit longevity in humans.

Genes that enhance or diminish survival (i.e., risk factors for diseases of aging), exist in a variety of genetic variants—called alleles—or polymorphisms. Examples of such genes that have been intensively studied are shown in Table 3.1 (Hamerman and Zeleznik, 2001). As might be expected, most population studies focus on genetic variants that relate to cardiovascular disease, especially lipoprotein metabolism. An extensive review by Ordovas and Mooser (2005) examines many of these gene relationships to aging and longevity, emphasizing "the marked inter-ethnic differences . . . and the impact of gene-environment, gene-gender, and/or gene-gene interactions in aging." The predictive value of genetic information "is also pertinent to ethical concerns" (Finch and Kirkwood, 2000). The ε4 allele of apolipoprotein E (apoE4) is a susceptibility factor for cardiovascular disease and a high risk for Alzheimer disease, and thus "may be among the most common susceptibility factors for disease-specific mortality after 60 years. In current thinking the apoE4 allele modulates the time of onset but not the course of the disease" (Finch and Kirkwood, 2000, p. 197).

Many of the genetic variants may not only have effects on the cardiovascular system but also involve lipid homeostasis that is important in insulin sensitivity and diabetes (Ordovas and Mooser, 2005). Thus, traditionally *reduced* risk involves low levels of LDL and higher levels of HDL, but Barzilai et al. (2003) have shown that larger lipoprotein particle size also confers a healthy aging phenotype. Peroxisome proliferator-activated receptor γ (PPARγ) is an important regulator of adipose tissue metabolism and insulin sensitivity, as well as lipid metabolism, and will be discussed much more in the sections on the metabolic syndrome (Chapter 4, section 3) and therapies (Chapter 5, section 4). Interest in polymorphism and regulation of PPARγ and its coactivator PGG1 highlights the link between insulin sensitivity, lipid metabolism, and inflammation, which are factors promoting atherogenesis and clearly central to issues of longevity or, conversely, high risk for disease inception.

While cell senescence in culture systems appears to capture the essence of the molecular biology of *cellular aging,* the biology of *longevity* seems to be

TABLE 3.1
Examples of genes that display polymorphisms associated with longevity or early morbidity

Gene	Designation	Function
APOB	Apolipoprotein B	Lipid metabolism
REN	Renin	Vascular tone
SOD1	Superoxide dismutase	Scavenger of superoxide radicals
SOD2	Superoxide dismutase	Scavenger of superoxide radicals
THO	Tyrosine hydroxylase	Catecholamines
mtDNA	Mitochondrial locus	Oxidative phosphorylation
HLA-DR	Human leukocyte antigen	Immune
ACE	Angiotensin-converting enzyme	Vascular tone
APOE	Apolipoprotein E	Lipid metabolism
PAI-1	Plasminogen activator inhibitor	Clotting

Source: Reprinted from Hamerman and Zeleznik (2001), © 2001, with permission from Elsevier.

expressed in endocrine signaling pathways across many species, especially in evidence that mutations in the insulin/IGF pathway influence life span and interact with caloric restriction (CR). CR refers to a dietary regimen designed for mice in a laboratory setting in which a reduced amount of energy ingested with constant micronutrients extends the life span—indeed, "the most effective and reproducible intervention in a variety of animal species, including mammals" (Hursting et al., 2003). This is not malnutrition, but rather a regimen in which the calories ingested are 20–60 percent below those of mice fed ad libitum (Koubova and Guarente, 2003); mice with a surfeit of food "can be considered models for aging, sedentary humans at risk for obesity and associated diseases" (Corton and Brown-Borg, 2005). Extension of life span can approach 50 percent in these CR rodents who are spared a usual array of cancers, autoimmune diseases, kidney disease, and diabetes (Koubova and Guarente, 2003). In a recent review, Masoro (2006), one of the leaders in the field of CR, reevaluated many long-held concepts. He discussed food restriction (FR) that may involve *specific nutrients,* versus caloric restriction (CR) that implies decreased intake of *energy,* and proposed CR as the "major, if not the sole, dietary factor responsible for the life-extending action of long-term FR, at least in the rat." Caution is needed because the mechanism by which FR (his term) extends the life of rats, mice, and the fly *Drosophila* may differ. Sinclair (2005) traced the history of CR and longevity regulation from the past to the present—a synthesis entailing the "hormesis hypothesis" (Fig. 3.4). The term *hormesis* refers to *beneficial* actions resulting from the response of an organism to a low-intensity stressor. Thus, "CR imposes a low-intensity biological stress on the organism that elicits a defense response to protect it against the causes of aging." CR induces *active* defenses in the organism, including lon-gevity/survival signaling proteins, increased cell defenses, attenuation of cell death, altered metabolism, reduced ROS, fat mobilization, insulin sensitiza-tion, as illustrated in Figure 3.4.

It is remarkable how researchers have traced an *interactive* path between genetic manipulations (enhanced or reduced) and actual CR, not only in the traditional laboratory model—the mouse—but also in species that the geriatri-cian, dealing with humans, may be unfamiliar with, yet must come to under-stand because of the "homologies between the determinants of longevity in budding yeast (*Saccharomyces cerevisiae*), nematode worms (*Caenorhabditis elegans*), fruit flies (*Drosophila melanogaster*), and *Homo sapiens*" (Grimley Evans, 2005). The diversity of these models and their individual and collec-tive responses to CR enhances the interest and excitement of this approach to life span extension; but also "highlights awareness of ignorance," as Partridge and Brand (2005) put it in a special issue of *Mechanisms of Aging and Devel-opment.* In what follows, I discuss many of the articles in this special issue. Note also that the 2003 Kleemeier Award from the Gerontological Society of America was given to Thomas E. Johnson, of the University of Colorado at Boulder, who, following earlier studies by Michael Klass, first applied genetic analyses to the study of the aging process in the nematode, so this may be fa-miliar to some readers (Johnson, 2005). These multispecies models are ideal in the laboratory because they are simple and inexpensive to culture and can be manipulated over days to measure size and survival time, and even rep-licative capacity, thus opening up new horizons for the study of *longevity* (Butler et al., 2004). Yet, as Partridge and Gems (2002) put it, "for work on these organisms to be relevant to research on humans, their mechanisms of aging need to be common with those in mammals. We need to know which mechanisms are 'public'—those shared across distantly related evolutionary lineages—and which are 'private'—those peculiar to particular evolutionary lineages." Kirkwood and Shanley (2005) echo the need to know if common pathways affecting aging and longevity exist in the different species. Although the nomenclature changes across species, there indeed seems to be a basis for public lineage. Longo and Finch (2003) note that "the similarities in longev-ity regulatory pathways between yeast, worm, and fly suggest that portions of these pathways have evolved from common ancestors," or to put it an-other way, have been partially conserved throughout evolution (Fabrizio et al., 2005). In yeast, CR is a highly regulated response to food deprivation with a sensing step and activation of a genetic program to extend life span. One pos-

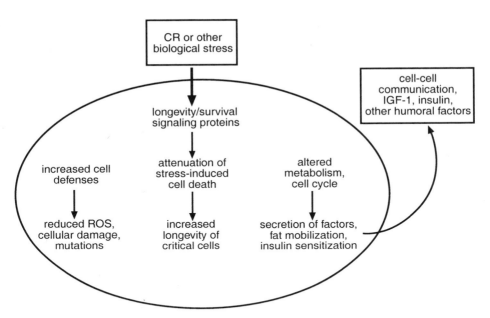

Figure 3.4. The hormesis hypothesis of caloric restriction (CR). The theory states that CR is a mild stress that provokes a survival response in the organism, which boosts resistance to stress and counteracts the causes of aging. The theory unites previously disparate observations about reactive oxygen species (ROS) defenses, apoptosis, metabolic changes, stress resistance, and hormonal changes. Reprinted from Sinclair (2005), © 2005, with permission from the author and Elsevier.

sible regulatory gene is *silent information regulator* protein 2 (SIR 2), which stimulates replicative capacity in yeast, and extends life span in worms, flies, and possibly higher species in response to food scarcity (Longo and Kennedy, 2006). The SIR 2 gene encodes a nuclear nicotinamide adenine dinucleotide (NAD)–dependent histone deacetylase, enabling the gene to monitor cellular metabolism (Picard and Guarente, 2005). The corresponding mammalian SIR 2 gene is sirtuin 1 (SIRT 1), also down-regulating stress-responsive inducers of cell death (i.e., decreasing apoptosis), and also deacetylating and inactivating p53, thereby promoting escape from senescence and allowing cell proliferation to resume (Argmann et al., 2005; Guarente, 2005; Guarente and Picard, 2005; Liu and Chen, 2006; Longo and Kennedy, 2006).

Also in *C. elegans,* the ability of SIR2 genes to extend life span depends on the gene daf16 (Haigis and Guarente, 2006). The mammalian homolog (counterpart) of daf16 is the transcription factor family known as FOXO (*f*orkhead box—containing protein, *O* subfamily), named after the original forkhead gene in *Drosophila,* indicating the conservation of this pathway, from yeast to humans (Wijchers et al., 2006). FOXO occupies an important place in the molecular links between aging, longevity, and cancer (Greer and Brunet, 2005), about which I will comment here, and again in the epilogue along with insulin and leptin and their roles in energy (food) intake and energy expenditure. Engelman et al. (2006) present an impressive account of the evolution and actions of the family of phosphatidylinositol-3-kinases (PI3Ks) as regulators of growth and metabolism. PI3K, which will be discussed in Chapter 4, section 3, in relation to insulin action and resistance, is also highly conserved from *C. elegans* to mammals. Figure 2 in the Engelman et al. article illustrates the overall "canonical" pathway crossing species' lines: insulin activation of its receptor → insulin receptor substrate adaptor proteins → class I PI3K expression → protein kinase B/Akt → FOXO. Protein kinase B/Akt triggers the phosphorylation of FOXO, its translocation from the nucleus to the cytoplasm, and its inactivation. This degradation of FOXO proteins often accompanies cell transformation and "may be a critical initiation step towards tumorigenesis" (Greer and Brunet, 2005) and, in a sense, longevity extension. "Daf-16 (worms) and FOXO (flies, mammalian species) appear to induce a program of genes that coordinately regulate longevity by promoting resistance to oxidative stress" (Greer and Brunet, 2005). Conversely, the expression of active forms of FOXO transcription factors prevents transcriptional activation of an array of cell cycle entry inducers (Wijchers et al., 2006), thereby fulfilling a role comparable to tumor suppressors in aging (Greer and Brunet, 2005).

In the fly *Drosophila,* mutations of either the insulin receptor (dIR) or receptor substrate (termed Chico) have led to enhanced longevity. "These changes link the invertebrate insulin/IGF-like signalling mutants with CR model of life extension in mammals" (Richardson et al., 2004). Fat-specific insulin receptor knockout (FIRKO) mice maintain normal glucose tolerance, have insulin sensitivity, and live longer than control mice. In at least three dwarf mice mutant strains deficiencies occur in several pituitary hormones, but the evidence suggests that reduced growth hormone is the basis for the longevity of these mice (Partridge and Brand, 2005). These are just "samples" of a large, complex literature and nomenclature, which the interested reader can consult for details.

There are other longevity-enhancing examples in worms. A long-lived *C. elegans* carried a mutation in age-1 (PI3K), a longevity mutant described by Johnson (2005) that sends signals to a gene in the insulin/IGF-1-like signaling cascade-daf-2 (Tatar et al., 2003; Baumeister et al., 2006). This is "the common ancestor of human insulin and IGF-1 receptor genes, and regulates metabolism, development, reproduction and longevity in the worm *C. elegans*" (Katic and Kahn, 2005). Mutations affecting the daf-2 receptor expression/activity of insulin receptor or its downstream signaling targets extend life span. These mutations influence the entry of worms into a dormant, growth-arrested larval state called dauer—much discussed in a comprehensive and fascinating article on the subject by Kenyon (2005)—in which the animal stops reproductive functions and becomes more stress resistant and long-lived, "highlighting the relationship between aging and energy metabolism" (Picard and Guarente, 2005). Typically, extension of the life span and stress tolerance are parallel phenomena, suggesting that a similar genetic program is involved in both aspects, as Baumeister et al. (2006) put it. In a similar way, "CR blunts sexual maturation and fertility, which allows long-term survival through energy sparing" (Picard and Guarente, 2005). Kirkwood (2005b) put forth the "disposable soma theory" whereby the mouse "will benefit" by investing any spare energy into thermogenesis or reproduction rather than somatic maintenance and repair, even though this means that damage will eventually accumulate and cause aging." Perhaps CR is contrary to the earlier ancestral need to provide abundant fat stores to survive in times of famine, and thus introduces an adaptation that would run counter to today's orientation on excessive caloric intake, increased fat mass, and their attendant ills and shortened life span (Chapter 4, section 3).

In summary, the conditions shared by CR and genetic manipulation that seem to promote longevity across species' lines involve (1) reduced function mutations with down-regulation of insulin-IGF signaling pathways, and with low-plasma IGF and insulin; (2) mechanisms dependent on sirtuins or Sir, which function to silence chromatin by deacetylating the histones and reducing gene expression; (3) enhancing stress resistance (antioxidant) factors, and down-regulating NF-kB, thereby diminishing the induction of many inflammatory cytokines (IL-1, IL-6, TNF-α) (Chung et al., 2005); and (4) increasing mitochondrial superoxide dismutase and catalase activities to reduce reactive oxygen species (ROS) (Finkel and Holbrook, 2000; Katic and Kahn, 2005). Sir2 proteins may play a conserved role as sensors of CR from yeast to mammals (Koubova and Guarante, 2003; Guarente, 2005). Already, the longevity factor encoded by SIR2 and its potential role as an endogenous inhibitor of stress-mediated cell death is being translated into studies of disease, for example, with isolated neonatal cardiomyocytes from rats (Crow, 2004), or in chronic obstructive pulmonary disease in which there appeared to be reduced histone deacetylase, "a key molecule in the repressing proinflammatory cytokines in alveolar macrophages" or inducing glucocorticoid insensitivity to inflammation-mediated gene expression (Ito et al., 2006). (See also therapies, Chapter 5, section 5.) Thus, enhancement of longevity occurs with improved glucose homeostasis, reduced aging-acquired body fat, decreased growth, and protection against oxidative damage (Longo and Finch, 2003).

The remarkable advances made across the species' lines described here may still call for further exploration in the world of "social insects to study the molecular basis of ageing." Keller and Jemielity (2006) note that fruit flies, nematodes, yeast, and mice are "short-lived model organisms," and suggest "that ants and social bee species would provide an excellent complementary system to study ageing." Ants and queen bees are extraordinarily long-lived, and a great range of life span variations exists among genetically identical queens, workers, and males. It will be interesting to see if this suggestion gains more widespread use in the future of aging research.

In terms of human application, most persons are not likely to embark on short-term or life-long caloric restriction, and it is not certain if begun in older persons what the outcome might be. In fact, long-term CR in persons over 65 "will most probably have significantly deleterious effects on their nutritional status and quality of life" (Nass and Thorner, 2004). Yet a six-month study of caloric restriction (12–15% of baseline energy requirements) with or without exercise in overweight but nonobese persons (body mass index 25 to <30) showed that two "biomarkers of longevity" (fasting insulin level and body temperature) were decreased (Heilbronn et al., 2006). In an accompanying editorial, Fontana (2006) mentions earlier studies on human volunteers in Biosphere 2 and points out, in the Heilbronn et al. study, the significant body weight reductions and fat loss—as might be expected. Again, decreased core body temperature, serum triiodothyronine (T_3) levels, and less oxidative damage to DNA, as reflected by a reduction in DNA fragmentation, were also observed. However, long-term issues remain unanswered, especially supervision of subjects, compliance, end point, and maintenance. Approaching "ideal nutritional status" in older persons will almost certainly remain elusive and a challenge to define. Yet, as will be discussed in Chapter 6, some lifestyle choices tend to promote conditions that somewhat mimic CR. Moreover, therapies by way of pharmacologic interventions may already be under consideration to produce the same prolongevity effects that CR provides but without substantially reducing caloric intake (Argmann et al., 2005). These "await the confirmation that the response to CR in humans parallels that observed in rodents." This caveat is one of the major limitations to hasty conceptual efforts at translation of results obtained in mice to man—a caution I will express again (concerning adipokines in Chapter 4, section 3, and PPARγ in Chapter 5, section 4).

Perhaps readers of *Science Times* in the October 31, 2006, issue of the *New York Times* will contemplate a lifestyle dietary change aiming at caloric restriction. Two rhesus monkeys—"senior citizens" at age 25—are depicted, one "aging gracefully" on CR, the other "wrinkled with hair falling out, frail, and moving slowly with arthritis," who gets all the food he wants. There are also pictures of mice and worms, and a fine exposition of the subject by Michael Mason, surely comprehensive and accurate, with reference to most of those investigators cited in Chapter 3, section 4. There is also a young man of 36, "on a low-calorie diet for six years, who is 6 feet tall and weighs 135 pounds, with BP of 112/63."

Laboratory-based studies indicate that reduced insulin/IGF activity is associated with extended longevity and that in rodents GH and IGF-1 promote aging and age-related diseases (Richardson et al., 2004). Indeed, long-lived humans also had lower IGF-1 plasma levels, and obesity was associated with higher levels (Corton and Brown-Borg, 2005). So it is necessary to reconcile the clinical studies first set forth by Rudman et al. (1990), showing that injections of GH improved parameters of body composition in men 61 to 81 years of age, findings that were seized on as "antiaging" (Vance, 2003). Of course, studies on longevity were not a factor here. This is also an area in which experimental and clinical studies will need to be reconciled in terms of possible long-term translational therapy to humans, as will be discussed in Chapter 5, section 6 concerning hormonal therapies. Some "optimal setpoint for the IGF-1 axis" may mediate between reducing cardiovascular and neoplastic diseases and enhancing longevity (Yang et al., 2005).

Interrelations of Certain Aging-related Conditions

1. CYTOKINES, INFLAMMATION, AND RESPONSES TO STRESS

Each of the conditions discussed in this chapter has a unique underlying pathogenesis. Some conditions are more diffuse or generalized in the extent of their systemic involvement (metabolic syndrome, frailty), whereas others are organ specific (vascular, bone, joints). Nevertheless, it is possible to consider their interrelated features. Aging, of course, the underlying substrate in all of them, sets the stage for ongoing local pathology that over time leads to organ and tissue damage and the expression of disease. A further unifying feature, and increasingly emerging as crucial in pathogenesis, is an *inflammatory underpinning*. Inflammation has certain features in common: an orchestrated immune cell response to provocative antigenic stimuli, the influx of cells of the monocyte/macrophage lineage responsive to the conditions at the local site, and the cellular release of several inflammatory mediators—especially cytokines, such as interferon, tumor necrosis factor (TNF-α), interleukins (IL-1, IL-6), and other chemoattractants that in turn bring in additional cells, thus maintaining an ongoing process. A great deal of interest has been expressed in IL-6, "a cytokine for gerontologists" (Ershler, 1993), and a "magnificent pathway in aging and chronic disease" (Maggio et al., 2006), because elevations of IL-6 have been documented in premorbid conditions in community dwellers, as I will discuss, its roles in inflammatory and neoplastic diseases (Kishimoto, 2005), and its immunologic functions that induce antigen-specific cytotoxic T lymphocytes and memory cytolytic T cells (Barton, 1997). TNF-α has been implicated in cardiomyopathy and heart failure (Bristow, 1998), in insulin resistance, and in bone loss associated with postmenopausal osteoporosis with diminished estrogens, where an array of systems limit bone formation by osteoblasts and enhance bone resorption by osteoclasts.

Two concepts I have pursued in my publications in linking the biology of aging with clinical practice are the *inflammatory* underpinning of many of the so-called age-related diseases, and the concept of aging over time in the evolution of disease expression (i.e., *chronicity*). To explore the first issue, some years ago I presented a geriatrics clinical case conference on an elderly woman whom the geriatric fellows had brought to my attention. The patient was a previously vigorous woman, age 80, living independently, who began to lose weight and had become anorectic, fatigued, and disinterested in activities and events, with no discernable cause after a physician's review. The evidence of *frailty* that this patient presented spurred my interest and prompted me to review the literature. I wrote a paper entitled "Toward an understanding of frailty," which was published in the *Annals of Internal Medicine,* and which I will discuss in detail in Chapter 4, section 6 (Hamerman, 1999). It seemed to me that frailty represented a state of decline, with no specific measure, and perhaps with an altered metabolic balance over time whereby catabolic processes began to be expressed and exceeded anabolic responses. The further exploration of this view on the "underpinnings of frailty" generated several reviews by others about "cytokine-related aging process" (Morley and Baumgartner, 2004) and "catabolism of aging" as an inflammatory condition (Roubenoff, 2003a).

In subsequent related papers on cancer cachexia (Hamerman, 2002, 2004), perhaps an extreme and irreversible form of frailty, and a paper on two widely prevalent, age-related diseases—atherosclerosis and osteoporosis (Hamerman, 2005b)—I presented evidence for inflammation as part of the basic, underlying process with variability dependent on the disease or condition. The years of preparation and writing about these subjects formed the basis for my present efforts to develop the detailed and more comprehensive discussions in this text in which overall linkages between disease and inflammation could be explored. Several authors eloquently echoed the idea of inflammation as the underpinning of the aging conditions discussed in this chapter. Tracy (2003), in a supplement that appeared in the *International Journal of Obesity and Metabolic Disorders,* writes, "In addition to atherothrombotic disease, markers of inflammation are associated with, or predictive of, virtually all chronic diseases of older age, including heart failure, type 2 diabetes, cancer, frailty, cognitive decline and dementia, and osteoporosis." Weber (2005) proposes the "proinflammatory heart failure phenotype as a case of integrative physiol-ogy: elevations of circulating chemokines, cytokines, and activated lympho-cytes and monocytes, with wasting of tissues that include muscle and bone." Westendorp (2006) notes that "aging is associated with chronic, low-grade inflammatory activity" expressed in Alzheimer disease, atherosclerosis, dia-betes mellitus, sarcopenia, and osteoporosis. These considerations certainly reinforce the value of presenting these mainstream aging-related conditions, highlighting their interactive nature and their fundamental relation to inflam-mation, to the academic geriatrician concerned with education and patient care.

The second issue bears on the *chronicity* of age-related diseases: how long it takes for symptoms ultimately to reach clinical expression, and the nature of physician intervention at the time the patient seeks health care. In an article on osteoarthritis in the *Journal of the American Geriatric Society* (Hamerman, 1993), I adopted a diagram from Fries's (1988) well-known "schematic of hy-pothetical development of chronic illness, focusing on the emergence of joint symptoms in osteoarthritis" (Fig. 4.1). Osteoarthritis, with the manifestation of knee pain, represents part of the broad concept that applies to the "age-re-lated diseases:" the long "preclinical" period that precedes the appearance of symptoms, until these emerge into a threshold of awareness. At this point "aging changes" become expressed as "disease," with symptoms, limitations of activities, and visits to a physician (see Chapter 4, section 5). This would be part of the natural history of many chronic illnesses, those most prevalent in older persons (Verbrugge and Patrick, 1995; Olshansky and Cassel, 1997).

We are entering a new era of geriatric medical practice in which we are beginning to understand that organ pathology overtly expressed as disease may be asymptomatic for long periods, manifest if sought for by inflammatory mediators in the circulation, thus linking chronicity and inflammation. We do not routinely seek to identify these inflammatory mediators or markers in clinical practice, with some exceptions I'll discuss a little later, because most are "nonspecific." Yet these markers are assayed in clinical research stud-ies, which is how we have become aware of them in relation to functional disability in older persons. For example, active community dwellers expe-riencing "stressful" events showed elevated levels of circulating cytokines, especially IL-6, and displayed a variety of symptoms, including fatigue, de-pression, frailty, anorexia, weight loss, failure to thrive, and muscle weakness (sarcopenia) expressed as functional limitations rather than as disease (Cohen et al., 1997). In a study of "healthy" community dwellers age 70–79 years, lev-

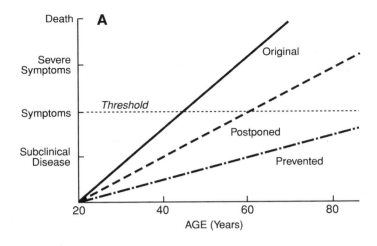

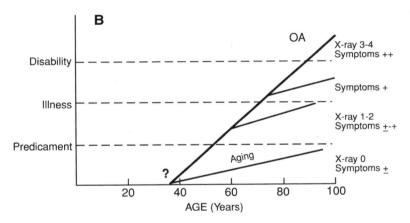

Figure 4.1. (A) Schematic of the hypothetical development of chronic illness; the vertical axis could represent amount of atherosclerotic change, the probability of malignant change, degeneration of articular cartilage, or other chronic processes. Note the onset early in life, the long presymptomatic period, and the different slopes representing either different individuals or the same individual operating under different sets of risk factors. Morbidity and mortality are determined by the rate of accumulation of pathology. Legend and figure from Fries (1988), with permission. (B) Schematic illustration of the "natural history" of cartilage aging and osteoarthritis (OA). Adapted from Fries. Designations on vertical axis adapted from Hadler (1985) to describe joint (hips, knees, feet) complaints: predicament, awareness of joint discomfort; illness, seeks medical attention; disability, ambulation limitation; ?, refers to evolution of cartilage changes to the presumed point of divergence between aging and OA. Lines branching from OA indicate potential for limited progression at each stage. Designation of symptoms (± to ++) and X-rays (0–4) are arbitrary and indicate progressive severity. Reprinted from Hamerman (1993), with permission from Blackwell Publishing.

els of IL-6 and TNF-α were in general associated with lower muscle mass and muscle strength (Visser et al., 2002). Cappola et al. (2003) note that age-related functional disability and decline are poorly understood but may involve both low insulin-like growth factor 1 (IGF-1) and high IL-6 levels. Indeed, women with both of these findings were at greatest risk for death; their functional decline suggested "dysregulation of endocrine and immune systems." Cytokine elevation may account for "tiredness" in daily activities and be an indicator for later disability, mortality, and increased use of social and health services (Avlund, 2004). The so-called proinflammatory state in older persons has been attributed to the high prevalence of cardiovascular morbidity (Ferrucci et al., 2005), but is also likely to be associated with widespread obesity and insulin resistance, as I will discuss shortly.

Chronic stress in caregivers of spouses with dementia was studied using a stress scale; among the symptoms recorded were depression, loneliness, and sleep disorders. In these caregivers, IL-6 levels were elevated and rose with age, a finding that was not observed in controls (Kiecolt-Glaser et al., 2003). Stress, as a response to environmental perturbation or conflict in the encounters of our lives, "translates" (in a "reverse" direction, if you will) into cellular disturbances. The emerging concepts of psychoneuroimmunology, as discussed by Glaser and Kiecolt-Glaser (2005), represent "a broad interdisciplinary research field that addresses the interaction of the central nervous system, the endocrine system, and the immune system." In the context of one of the major themes of this book, the stress response has been considered an integral part of the inflammatory response, with "implications for atherosclerosis, insulin resistance, type II diabetes, and metabolic syndrome X" (Black, 2003). Jacobson (2005) notes that "stress is any actual or perceived threat to well-being" which then induces hypothalamic-pituitary-adrenal (HPA) activity. "This is a highly specific process mediated by stressor-specific afferent inputs to the hypothalamus and transduced into an endocrine signal by specific combination of hypothalamic ACTH-releasing factors," that is, "hypercortisolism, and suppression of the thyroid and reproductive axes as well" (Khosla, 2002). Thus, stress in this way enhances adrenal glucocorticoid secretion (Syed and Weaver, 2005). The HPA axis (Fig. 4.2) is part of a broad system of hypothalamic input of signals (stress, hormonal, cytokines) that feature prominently in this book, and also relate to later discussions of growth hormone activation (Chapter 5, section 6). Indeed, "stress from poor personal health habits, including obesity and cigarette smoking, has well known adverse health risks" (Ezzati et al., 2005; Zamboni et al., 2005), and may activate

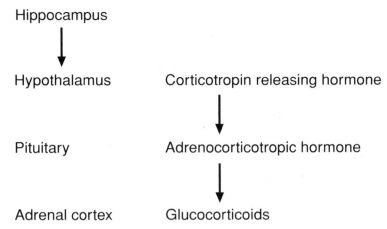

Figure 4.2. The hypothalamic-pituitary-adrenal axis (HPA). Impulses from the hippocampus activate the hypothalamus and the release of corticotropin-releasing hormone (CRH). The CRH releases adrenocorticotropic hormone (ACTH) from the pituitary, which then enhances glucocorticoid secretion from the adrenal cortex. Glucocorticoids exert a negative feedback on the higher centers—hippocampus, hypothalamus, and pituitary—and on the release of CRH and ACTH. The HPA axis is responsive to a range of systemic influences—especially cytokines released from inflammatory and immune cells (monocytes, macrophages, T cells) in vessels, synovium, bone, and adipose tissue. Multiple systemic responses include decline in gonadotropic secretion (Ferrin, 1999), and effects on metabolic, cardiac, behavioral, endocrine, musculoskeletal, and neuroendocrine systems, as discussed in Chapter 3, sections 3, Chapter 4, sections 1–3, and Chapter 5, section 6. Data from Mobbs (1996), Sternberg (1997), Glaser and Kiecolt-Glaser (2005), and Veldhuis et al. (2005a).

the HPA axis (obesity) and, remarkably, appears to accelerate shortening of telomere length (Epel et al., 2004), especially in women (Valdes et al., 2005), as discussed earlier in relation to the inception of cell senescence.

Yet further *molecular* manifestations of stress exist. Wellen and Hotamis-ligil (2005) note that "inflammation, stress, and diabetes" impose metabolic overload with endoplasmic reticulum "stress," leading to the activation of signaling pathways; and there is mitochondrial "stress," with the production of reactive oxygen species (ROS), organelle damage, and inflammation. These aspects of cellular-based stress figure prominently in the discussion of diabetes and the metabolic syndrome (Chapter 4, section 3). Obesity itself, as measured by body mass index (to be discussed), seemed to be associated with fatigue, unrelated to IL-6 levels (Lim et al., 2005). It is not yet certain, however, that claims of causality of cytokine elevation and disease are warranted; instead, they need to be considered in the light of the expected (possibly inevitable) expression of diseases with aging over an extended time. Alzheimer disease expression is one example, considered in Chapter 4, section 8. Other associated factors may also prevail. For example, "stress" may predispose to overeating, weight gain, and obesity, contributing to deleterious health status not so much related to the initial stressors and cytokine responses. Nevertheless, the potential for cytokine elevation as "conditioning" for morbidity gains credibility from its widespread documentation in community dwellers. More to the point, once disease is manifest, inflammatory mediators are *expressed in the circulation and tissues in every condition discussed in this chapter.*

So what are the long-term geriatric implications? "Stress" from many sources—psychological, adverse personal health practices, financial, social, physical, medical—is indeed recognized as "pro-aging" at the molecular/cellular level. Translated to the clinical level, stress is able to promote harmful health events that share cytokine production and release at the cellular level (e.g., by macrophages, adipocytes, etc.), activating the sympathetic and central nervous systems, resulting in a variety of immune, physiologic, neuroendocrine, and behavioral responses (Mobbs, 1996; Sternberg, 1997).

During routine health examinations would cytokine assays be indicated to assess stress before overt disease expression? The interleukins and TNF-α are not assayed in routine clinical practice and have little organ-localizing power. Nevertheless, as discussed in Chapter 5, section 5, therapies to suppress TNF-α have come into use to reduce inflammation in rheumatoid arthritis, the metabolic syndrome, and to treat heart disease as well (Khanna et al., 2004). Perhaps, as Duff (2006) wrote, medical practices in the future will focus on "sequence variations in the regulatory DNA of the genes coding for important members of the IL-1 family" involved in the inflammatory response in many chronic diseases and in diagnosis and therapy (see Chapter 5, section 5).

Other mediators, deemed "risk factors," are indeed readily assayed, however, for they may predict risk that precedes organ damage—in particular, cardiovascular. A whole literature affecting clinical practice is emerging in the realm of biomarkers that may be predictive of cardiovascular disease risk. Especially for those engaged in laboratory medicine (Plebani and Laposata, 2006), new opportunities exist for "translation" of biomarker identification and association in a range of clinical conditions, with the potential for earlier therapeutic intervention and improved patient outcome. Vasan (2006) wrote an extraordinarily comprehensive review of the *molecular* basis for the dis-

covery of these biomarkers "and the practical considerations that are a prereq-
uisite to their clinical use." As he notes, "the expectation of a cardiovascular
disease biomarker is to enhance the ability of the clinician to optimally manage the patient." The list of potential biomarkers for the patient vulnerable for
atherosclerotic cardiovascular disease is extensive, for it includes not only an
array of serologic assays but also functional and structural markers. More accessible blood biomarkers for cardiovascular risk are low-density lipoprotein
(LDL)-cholesterol, high-sensitivity C-reactive protein, and other less familiar
markers that include lipoprotein-associated phospholipase A_2, myeloperoxi-
dase, serum amyloid A, and fibrinogen, among others (Ferri et al., 2006; Tsimikas et al., 2006). Some of the well-known serologic factors have received the
striking designation of "markers of malign across the cardiovascular continuum" (Dzau, 2004). In general, a range of cardiovascular risk factors arise from
adverse lifestyle habits, certain physical findings (visceral obesity), strained
socioeconomic circumstances, and certain genetic markers, but for this discussion the most important *circulatory biomarker* is high-sensitivity C-reactive protein (CRP). At present, CRP assay is done somewhat routinely because
it has—despite its designation as an acute phase reactant—high relevance to
atherosclerosis and cardiovascular disease as an inflammatory marker (Yeh,
2004). Indeed, elevated IL-6 enhances CRP formation by the liver; hence, a
potential provocative cycle for atherogenesis exists (Hansson, 2005). After appropriate clinical assessment, finding a high CRP may prompt added therapies, such as the use of statins, although Sepulveda and Mehta, as of September 2005, believe that "despite a relatively strong epidemiological association
with future adverse cardiovascular events, current knowledge is insufficient
to implicate CRP as a causative factor in atherothrombosis or to test for CRP
as a guide to preventive or therapeutic interventions in cardiovascular diseases." Timpson et al. (2005) came to a similar conclusion with respect to the
phenotypic features of the metabolic syndrome (see Chapter 4, section 3) and
the "lack of causal association with CRP levels." More recently, the debate
about CRP and causality in atherogenesis has been joined by Paffen and De-
Maat (2006) reviewing the evidence implicating CRP in the development of
atherosclerosis; by Scirica and Morrow (2006) expressing reservations about
CRP as an innocent bystander; and by Verma et al. (2006) affirming "active
participation" by virtue of CRP stimulation of plasminogen activator inhibi-
tor-1 (PAI-1), IL-8, matrix metalloproteinases, and inhibition of endothelial
nitric oxide synthase with reduced nitric oxide (NO)—needed for vasodilation
and endothelial integrity. All authors agree more research is necessary. Thus,

the controversy about "the place of CRP in the cardiovascular disease preven-tion pantheon," as Smith et al. (2006) put it in a fine summarizing editorial, continues to engender ongoing debate, a call for further studies, and as of this writing, "remains uncertain." Rather than a "prima ballerina," CRP seems to be a "chorus girl" after all, a marker more than a maker (de Buyzere and Ri-etzschel, 2006).

A final word here on the long-standing classic phase reactant, the eryth-rocyte sedimentation rate (ESR). Elevations may alert the clinician to suspect polymyalgia rheumatica or may be useful to follow the therapy and course of rheumatoid arthritis. The ESR has been used in long-term follow-up of middle-aged men as a significant predictor of heart failure, independent of established risk factors for heart failure, and interim myocardial infarction (Ingelsson et al., 2005).

2. ATHEROSCLEROSIS: RISK FACTORS FOR CARDIOVASCULAR DISEASE

As Libby (2002) put it, "cardiovascular disease, currently the leading cause of death and illness in developed countries, will soon become the pre-eminent health problem worldwide. Atherosclerosis . . . constitutes the single most im-portant contributor to this growing burden of cardiovascular disease." Indeed, this must be so, because the accumulation of lipid in the intima, manifested as the fatty streak, is universal and begins *decades* before atheroma and plaque protrude into the lumen of the vessel, with an acute event, if this occurs at all. Hence, this is another prototype of aging-related changes that may eventuate in disease, as I have discussed, and whose tempo is influenced by so many fac-tors: aging over time, genetics, lifestyle practices, exercise, health care, meta-bolic status, hypertension, etc.

Views about atherosclerosis have altered over the past two decades. Pre-viously, like so many conditions in the realm of aging, atherosclerosis was considered primarily "degenerative" (i.e., inevitable lipid accumulation), per-haps not very interesting. Framed in this way, fundamental inquiry into this aging-related condition did not seem worthy of inquiry. In an interesting par-allel, a discussion by the neurologist Robert Katzman (1988) at a Ciba Foun-dation Symposium on "Research and the Ageing Population" noted how the words "normal ageing and disease are misguided." When people regarded "se-nility" or "senile dementia" as normal (inevitable) aging, he wrote, research on Alzheimer disease was barely evident. But casting Alzheimer disease as a

disease, introduced not only molecular studies of pathogenesis and sophisticated screening modalities, but also the potential for intervention, therapies,
and even reversibility.

These aspects do not solve uncertainty about the boundaries between aging and disease, which I have discussed, but I want to use atherosclerosis as a
model in which inquiry into a disease has also shifted from "degenerative aging changes" to fundamental studies using methods of molecular biology and
clinical application of risk factors. These factors are part of a new dynamic
orientation on aging and the consequences of *senescence* of vascular smooth
muscle cells (Gorenne et al., 2006), and more particularly, of *inflammation*
linked to *metabolic* and *immune* events (Wick et al., 2004). The late Russell
Ross (1999) must be acknowledged as instrumental in establishing this new
dynamic. Libby (2006) more recently noted that the "traditional view of atherosclerosis as a lipid storage disease crumbles in the face of extensive and
growing evidence that *inflammation* (emphasis added) participates centrally
in all stages of the disease, from the initial lesion to the end-stage thrombotic
complications." Koh et al. (2005) have also stated that chronic inflammation is
a pathogenic feature of atherosclerosis; their review is valuable for its illustra-
tive diagrams. But as Steinberg (2002) emphasizes, there are many "intersec-
tions" between inflammation in the artery wall and the effects of oxidized LDL
and its products. In the long controversy about atherosclerosis pathogenesis,
the lipid hypothesis and the response-to-injury hypothesis are compatible,
representing a shared pathogenetic pathway as "partners in crime." It is essential that the geriatrician and other health practitioners understand some
of the basic aspects of atherosclerosis to make clinical practice dynamic and
current and to maintain ties with research that may then be "translatable" to
the bedside. Figure 4.3 diagrams the three "phases" of atherosclerosis over
decades—from early (the fatty streak) to end stage (with protrusion of the atheroma as a thrombus into the lumen), recognizing, of course, the great clinical
variability.

The starting concept for the formation of atheroma is modification in the
integrity of the single layer of vascular endothelial cells that make contact
with the circulation. Endothelial dysfunction comes about by virtue of many
factors interactive with inflammatory mediators: aging, diminished nitric oxide formation that impairs vascular tone and promotes platelet aggregation;
increased synthesis of endothelin-1, which enhances vasoconstriction and release of proinflammatory cytokines; possible circulating antigenic microbes
that activate an immune response with major histocompatibility complex

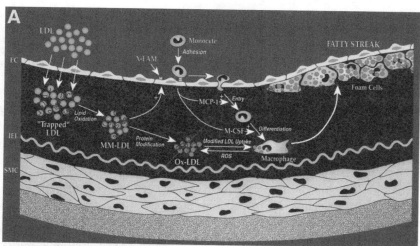

(MHC) molecules; hypertension; the state of insulin resistance (Nigro et al., 2006; Semenkovich, 2006); shear stresses in blood flow; smoking; and lipids—especially LDL and free fatty acids—in the circulation. Also contributing to endothelial cell dysfunction is the up-regulation of nicotinamide adenine dinucleotide phosphate (reduced) (NADPH) oxidases that are a potent source of superoxide radicals and other ROS (as noted in Chapter 3, section 3). The complex biology of the NADPH-oxidases in cardiovascular diseases is reviewed comprehensively by Cave et al. (2006). The endothelial cells are separated by a basement membrane from the underlying intima consisting of connective tissue and smooth muscle cells; as the endothelium is impaired, subsequent events in atherogenesis will take place in the intima. These processes evolve over a lifetime and are described now in abbreviated form.

A cascade of cellular and molecular events arises from the interactions between circulating cells in the vascular lumen and the activated endothelium to which these cells will increasingly adhere. These events are well described by Charo and Taubman (2004), Choi et al. (2004), Quehenberger (2005), and Szmitko et al. (2005), among many others. The circulating monocytes and T cells—mainly CD4+ memory cells—*roll* along the normally smooth-surfaced endothelium until the endothelium becomes sticky by activation of the selectin family of adhesion proteins. The interaction of the selectins and their

Figure 4.3. Molecular and cellular mechanisms involved in the development and progression of atherosclerosis. (A) The fatty streak. Low-density lipoprotein cholesterol (LDL) in the lumen of the vessel penetrates an impaired endothelial cell (EC) layer and is trapped in the intima. Monocytes in the lumen attach to adhesion molecules (X-LAM) that arise from the effects of mildly oxidized LDL (MM-LDL), and under the influence of the chemokine monocyte chemoattractant protein (MCP-1) adhere to the ECs and enter the intima. Macrophage colony-stimulating factor (M-CSF) enhances macrophage development and these cells take up highly oxidized LDL (Ox-LDL) and release reactive oxygen species (ROS). These lipid-laden cells accumulate as foam cells, and constitute the fatty streak. (B) In the intermediate stage, smooth muscle cells (SMC) migrate through the internal elastic lamina (iel); these cells secrete polysaccharides and collagen, which form a fibrous cap on the necrotic core of the plaque. (The pericyte-like cells are mesenchymal stem cells that are also referred to as calcifying vascular cells.) (C) In the advanced lesion, the necrotic core of the fatty streak may calcify. Matrix metalloproteinases released by cells within the plaque contribute to the impending rupture of the fibrous cap. The plaque contents break through the endothelium into the lumen, and being highly thrombogenic, trap platelets; a clot forms that partially to completely occludes the lumen. Reprinted from Berliner et al. (1995), © Lippincott Williams & Wilkins, with permission. (See color gallery.)

ligands is not sufficient for the firm adhesion, or *tethering,* of the monocytes; the initial attachment is followed by a second event—the engagement of the integrins. These integrins on the monocytes interact with a class of ligands called adhesion molecules—intracellular (ICAM-1) and vascular (VCAM-1)— on the endothelium. Chemokines are also a major player in activating the adhesion process and promoting the transendothelial migration of the adherent monocytes. Chemokines are a family of small-molecular-weight secreted chemotactic proteins that activate G-protein-coupled receptors on responsive leukocytes. Among the chemokines playing a key role in the atherosclerotic process are monocyte chemoattractant protein (MCP-1) and its receptor CCR2; these components will also reappear in a discussion of adipose tissue in the next section.

The entry of cells into the intima is of great consequence in terms of the involvement of mediators in various stages of atherosclerosis (Hansson et al., 2002). *Innate immunity* functions by way of phagocytes, leukocytes, complement, and proinflammatory cytokines. *Acquired or adaptive* immunity is represented by the response of CD8+ T cells to antigenic stimuli consisting of oxidized lipids; these cells differentiate into Th1 effector cells, which begin producing the macrophage-activating cytokine interferon-γ. Macrophage colony-stimulating factor (M-CSF) is also involved in this process of monocyte-macrophage conversion; macrophages ingest the oxidized LDL and atherogenic apolipoprotein B (apoB)-containing lipoproteins that enter the intima. These lipid-laden macrophages, called *foam cells,* form the earliest lesion— the fatty streak (Sheikine and Hansson, 2004). Countering this, high-density lipoproteins (HDLs) interact with macrophages to promote the efflux of excess cholesterol from the foam cells, helping to protect against formation, or even inducing regression, of macrophage-foam cells (Rader and Puré, 2005). Chapter 5, section 4 discusses the transit of lipid components in and out of the macrophage in relation to use of thiazolidinediones (TZDs) and activation of PPARγ, and other aspects of reverse cholesterol transport. From recent reviews, especially one in *Current Vascular Pharmacology,* by Boyle (2005), and the other in *Cell Metabolism,* by Rader and Puré (2005), the macrophage emerges as perhaps the key cellular player in the totality of its inflammatory and innate immune responses in the vessel wall in atherosclerosis. As Rader and Puré (2005) note "the lipoprotein-macrophage interactions that go beyond foam cell formation have important implications for the understanding of the pathogenesis of atherosclerosis and for the development of new therapeutic approaches."

Again, Libby (2002) and others point out that the older, traditional concept may be observed less frequently, although many of the pathologic features still prevail: that atheromas inevitably progress by growing in bulk by way of smooth muscle cell proliferation from the media into the intima (Orlandi et al., 2006), lipid accumulation, the death of engorged macrophages in the necrotic core (Steinberg, 2002), with slow narrowing of the lumen reducing blood flow, resulting in angina. Now, reevaluation based on clinical and angiographic studies suggests that the progression of an atheroma is not orderly; rather, the plaque ruptures from erosion in which matrix metalloproteases digest the collagenous and proteoglycan fibrous cap laid down by the vascular smooth muscle cells (Gorenne et al., 2006) or by microvascular hemorrhage. Acute protrusion into the lumen results in thrombosis. These emerging concepts make the identification of risk factors and potential therapeutic intervention earlier even more important so as to forestall acute catastrophic events. Indeed, Tracy (2003) notes that white men dying of sudden cardiac death in their fourth and fifth decades mainly exhibit acute thrombosis. This number drops sharply by the seventh decade, with most deaths then being stable plaque-based coronary deaths. This change has implications for the use of CRP as an assay, which was discussed in Chapter 4, section 1, as a marker of risk. In the younger age range, CRP has long-term predictive potential, but in elderly persons, CRP and fibrinogen "exhibited a marked interaction with time to event from baseline: these markers were much stronger predictors of fatal coronary events when the event occurred close to the baseline." This was termed "proximate pathophysiology to indicate the sensitivity of inflammation markers to changes in the underlying disease process that portend soon-to-occur events."

This chapter has emphasized oxidation of LDL as part of the pathogenesis of atherosclerosis, but as Heistad (2006) points out in a fine review, "oxidative stress also plays a key role in endothelial dysfunction, as superoxide radical (O_2^-) inactivates nitric oxide to form peroxynitrite" (ONOO⁻), and "vascular protection slowly vanishes" (Förstermann and Münzel, 2006). Yet in line with an earlier discussion (see Chapter 3, section 3), antioxidants and vitamins do not consistently reduce the risk of cardiovascular disease; this might be due in part to their nonspecificity (Cave et al., 2006) and to superoxide dismutase (SOD), which dismutes O_2^- to H_2O_2. "Structural changes may not be attenuated if H_2O_2 plays a key role" (Heistad, 2006). Indeed, H_2O_2 has a complex role to play, both in providing an "oxidizing environment" needed for the binding of activated transcription factors to their consensus sequence on DNA; but also in activating mitogen-activated protein kinase (MAPK) signaling cascade and transcription factors, such as NF-kB (Packer, 2005). Although SOD is an important antioxidant, excessive amounts generating H_2O_2 can be harmful, leading to formation of hydroxyl (OH·) radicals that are highly reactive with cellular macromolecules. Therefore, as Surh (2005) notes, SOD induction should be accompanied by up-regulation of a gene coding for an H_2O_2-inactivating enzyme, such as catalases or glutathione peroxidase.

3. THE METABOLIC SYNDROME AND TYPE 2 DIABETES

The metabolic syndrome is both a concept and a multifaceted condition, really a constellation of findings that has undergone an evolution over 40 years, according to Zimmet and Thomas (2003). The recent literature (2003–2006) is extensive, and the reader may wish to pursue several reviews, among others: an issue of *Endocrinology and Metabolism Clinics,* edited by Grundy (2004); Moller and Kaufman (2005) in the *Annual Review of Medicine;* a perspective on diet and genetics by Roche et al. (2005) in the *Proceedings of the Nutrition Society;* an entire issue of the *American Journal of the Medical Sciences* in 2005, edited by Reisin and Alpert; an entire issue of the *Annals of Medicine* in 2006, edited by Laakso and Kovanen; and a supplement to the *American Journal of Medicine* in 2006, edited by Ritz et al.

No single entity among those discussed in this chapter has generated such an evolving and intense debate about its validity as the metabolic syndrome. As Bruce Leslie (2005) put it in a historical perspective, "acceptance of the metabolic syndrome as a disease entity has been hampered by nonuniform criteria for its diagnosis, continuing debate about its etiology and pathogenesis, and uncertainty about its treatment." Edwin A. M. Gale (2005), at the University of Bristol in the United Kingdom, quotes Margaret Thatcher to the effect "when people can't agree about something they reach a consensus," and he illustrates the blind men approaching the elephant—all as part of what he calls the "myth of the metabolic syndrome." The greatest concern he has about the constellation of criteria presented below is the elusiveness and imprecision of *measurements* used to formulate these criteria, especially the centerpiece—insulin resistance. Indeed, Gerald Reaven (2005a), one of the key originators in his Banting lecture in 1988 of "syndrome X"—the triad of diabetes, hyperlipidemia, and hypertension, to which he linked insulin resistance—apparently had sufficient second thoughts to write "The metabolic syndrome: requiescat in pace" (2005b). He states, "it appears that making the diagnosis of the metabolic syndrome does not bring with it much in the way of pathophysiological understanding or clinical utility." In a follow-up published a year later, Reaven (2006) was even more critical of the effort to create a diagnostic category for the metabolic syndrome; "I believe it has little clinical or pedagogical utility, and can potentially do more harm than good." Yet in his paper he does provide an in-depth review of the traditionally accepted components of the metabolic syndrome (see below).

The discussion to follow takes no stand with respect to the *validity* of the metabolic syndrome, and I have not observed practicing geriatricians labeling patients with the term "metabolic syndrome." Rather, in the context of what I have chosen to discuss, the reader will find many aspects that will be enlightening in terms of physiologic linkages, and helpful in terms of clinical appraisal. Ultimately, I hope there will be a greater understanding of the bioscientific basis for therapies, which will be introduced here and discussed further in Chapter 5, section 4. I agree with the joint statement of the American Diabetes Association and the European Association for the Study of Diabetes, as of September 5, 2005, that the "metabolic syndrome has been a useful paradigm" (Kahn et al., 2005). The physician "will always observe and treat the individual components of the metabolic syndrome" and not require "a pick and mix approach" (Roberts, 2006) (i.e., some form of compilation or score). The geriatrician in clinical practice may not see many patients who conform to the definition of "obesity," discussed below, as a result of the real reduction in life expectancy (Haslam and James, 2005). But obesity may have been a forerunner and major contributor to the many other features manifest in the metabolic syndrome most frequently observed by geriatricians in their patients.

The metabolic syndrome is, of course, interconnected with atherosclerosis and cardiovascular disease risk (Haffner et al., 2006), especially in women (Bonora, 2006), with an inflammatory underpinning that predisposes to insulin resistance with impaired glucose tolerance and overt type 2 diabetes, dyslipidemia, hypercoagulability, and hypertension, with progressive target organ damage promoting cardiovascular disease (CVD) risk markers and microalbuminuria with renal impairment. Haffner (2006a) cites several studies demonstrating that the risk of cardiovascular disease starts to increase long before the onset of clinical diabetes. This is important in the light of cardiovascular risk factors that emerge based on the "significant defects in glucose homeostases and fuel metabolism that are detectable long before overt diabetes occurs" (Herman and Kahn, 2006). Prediabetic persons have lower HDL, higher plasma triglycerides, fasting glucose, and insulin levels, and higher systolic blood pressure than persons who remain healthy. DeFronzo (2006) suggests "a whole host of potential mechanisms exist by which the components of the metabolic syndrome, and insulin resistance itself, can exert independent effects to promote atherogenesis"—beyond the current "glucocentric" (i.e., hyperglycemia) view of diabetes. As one example, he points out that small, dense LDL is an important predictor of coronary artery disease.

The accumulation of ROS, especially superoxide ion (discussed in Chapter 3, section 3) induced in cardiac and vascular smooth muscle cells by angiotensin II, may contribute to myocyte apoptosis and heart failure (Griendling et al., 1994; Kajstura et al., 2006). The renin-angiotensin system is controlled through the production of the enzyme renin in the juxtaglomerular cells of the kidney. A fine account of the renin-angiotensin system is presented by Ferrario and Strawn (2006); what follows is an abbreviated version. Renin cleaves angiotensinogen to form angiotensin I, which is split by angiotensin-converting enzyme (ACE) to angiotensin II. The known pressor, proliferative, and profibrotic actions of angiotensin II are mediated through its binding to the angiotensin type I receptor. Because of the pivotal role of the renin-angiotensin system in vascular inflammation and atherosclerosis, neutralization of the actions of angiotensin II by ACE inhibitors and angiotensin receptor blockers exert antihypertensive and anti-inflammatory effects. Ferrario and Strawn (2006) and Tikellis et al. (2006) discuss the widespread proinflammatory and pro-oxidant properties of angiotensin II, reviewing evidence for its possible role in the pathogenesis of diabetes—both the pancreatic effects and vascular complications. Overall, treatment considerations based on risk factor assessment for diabetes development or for components of the metabolic syndrome (Haffner, 2006a; Kurtz, 2006) need to be carefully reviewed, including—perhaps foremost—lifestyle interventions (see Chapter 6, section 2). Després (2005) spoke of "our passive lifestyle" and asks if we can "afford to be sedentary and unfit?" The same early attention needs to be paid to microalbuminuria and reduced glomerular filtration rate even in the earliest and otherwise asymptomatic stages because they also constitute a risk factor for cardiovascular disease (Ritz, 2006).

Goodpaster et al. (2005) reviewed an NIH Consensus Conference on characteristics of the metabolic syndrome and listed the following: (1) waist circumference 102 cm in men, 88 cm in women; (2) serum triglycerides of 150 mg/dl or more; (3) HDL cholesterol less than 40 mg/dl in men and 50 mg/dl in women; (4) blood pressure 130/85 or greater; (5) fasting serum glucose 110 mg/dl or more. One of the unifying pathophysiologic principles I will discuss is that cytokine-like factors ("adipokines") released from central (visceral) fat stores, and C-reactive protein from the liver, are markers for and may contribute to many of the metabolic and pathologic alterations that constitute the syndrome.

Type 2 diabetes and non-insulin-dependent diabetes are used here interchangeably, signifying not so much β-cell pancreatic islet failure as insulin resistance. O'Rahilly and Savil (1997) note that "an increase in the global prevalence of non-insulin dependent diabetes is likely to occur as Westernization of dietary habits and patterns with physical activity become widespread. . . . By 2020 there will be an estimated 250 million people with non-insulin dependent diabetes in the world." In the United States diabetes is the leading cause of blindness among working-age adults, end-stage renal disease, and nontraumatic loss of a limb, and the fifth-leading cause of death, with direct medical costs amounting to $90–130 billion annually (Perlin and Pogach, 2006; Petersen and Shulman, 2006). Furthermore, "in parallel with the increasing incidence of type 2 diabetes there has been a global increase in both obesity and the metabolic syndrome ('diabesity') resulting in increased morbidity and mortality from CVD" (Zimmet and Thomas, 2003), a so-called diabetes and coronary risk equivalency (Grundy, 2006a). To explain this rising and pronounced tendency worldwide, some researchers have put forth the concept of obesity and type 2 diabetes evolving from what Neel more than 40 years ago called the "thrifty" genotype, and more recently reviewed (Neel, 1999; Zimmet and Thomas, 2003). This illustrates the evolutionary aspects—somewhat like antagonistic pleiotropy discussed previously—essentially, that much earlier humans evolved with genes that increased fat storage—useful for those times of frequent famine—and enhanced fertility, hence survival and procreative advantages (Schwartz and Niswender, 2004; Lazar, 2005; Unger, 2005). Now, however, fat accumulation is a maladaptation in the present environment of food aplenty (an "obesigenic environment"), especially the ingredients in so-called fast foods. Fast foods and obesity in the United States have become something of a *cause célèbre*. The problem has now been addressed in children, which seems appropriate for long-term health impact. Even the prestigious *Proceedings of the National Academy of Sciences* has entered the field with an article by Bernstein et al. (2005), whereby "planting false beliefs, that fattening foods (such as chocolate chip cookies) may cause illness, resulted in their avoidance." Yet the effects of telling children what they should eat (fruits, vegetables) and not eat ("bad snacks"), or banning vending machines, "doesn't work" (Klurfeld, 2005). A place to modify traditionally unhealthy food for children is the school lunch program, where choices are usually limited. Valiant efforts to do so in what sounds like a truly uphill battle are vividly described by Burkhard Bilger in "The Lunchroom Rebellion," in the *New Yorker* of September 4, 2006.

The debate about food choices has been joined by several writers in the lay press. Paul Krugman discussed in his columns in the *New York Times* of July 4, 2005, "Girth of a nation," and again on July 9, 2005, "Obesity is America's fastest growing health problem: let's do something about it." Steven Shapin wrote a most interesting essay in the *New Yorker* of January 16, 2006, entitled "Eat and Run," reviewing books and articles on obesity—what it is, what it appears to do to us, and even how to calculate body mass index (BMI—see below). In closing, he adds, "I pass on my way to work a cafeteria whose display features an assortment of doughnuts, brownies, croissants, and pastries. It looks pretty good today, so I pick up a prune Danish." How difficult it is to resist temptation, even among the enlightened! Perhaps Shapin's temptation illustrates the concern of Kim and Popkin (2006) that "lifestyle behaviors are largely affected by the environment in which people live. Providing 'freedom' to choose unhealthy lifestyles that will lead to obesity and related health problems is not what a healthy society should do." Yet perhaps the "exponential increase in mass media attention to obesity in the US and abroad seems to have many of the elements of what social scientists call a 'moral panic' that emerges during times of rapid social change and involve an exaggeration or fabrication of risks, and the projection of societal anxieties onto a stigmatized group" (Campos et al., 2006). The profoundly important societal and public health issues concerning obesity (of which the Kim and Popkin [2006] paper is just one), are surveyed in several articles in a recent issue of the *International Journal of Epidemiology* volume 35, number 1 (2006), which are too extensive to discuss individually here, but are well worth reading. Indeed, according to Flier (2004), who has contributed so much to our understanding of this field, efforts to address the worldwide "obesity epidemic" are being made on the *environmental* front as well as in *research* where we should now focus our attention: basic studies "provide new molecular and physiologic insights that offer exciting possibilities for future development of successful therapies."

For the purposes of this text, I wish to take up the many threads that are interconnected with respect to the biology of type 2 diabetes and the metabolic syndrome and their clinical expression in older patients:

- linking early development with the metabolic syndrome
- what constitutes obesity?
- what is the significance of central obesity?
- adipose tissue as an endocrine organ—release of adipokines
- interrelation of adipocytes and macrophages
- adiponectin in clinical practice
- the evolution of insulin resistance and type 2 diabetes
- the role of PPARγ and its agonists the TZDs

Linking Early Development with the Metabolic Syndrome

The evolution of type 2 diabetes raises an interesting aspect that seems to be ushering in a new concept of medicine, and requires attention on the part of geriatricians who often focus exclusively on persons over 65 years of age. The concept I refer to, in fact, bears on the early origins of adult disease—in this case, the developmental origins of diabetes and the metabolic syndrome. I will give examples and direct quotes in some detail because this subject opens a new way to look at conditions through the generations, predisposing to disease inception, as I discussed in Chapter 1. The "preventive aspects" transcend preventive gerontology; they "reach back" much further than that. Pursuing the developmental origins of disease will require interdisciplinary efforts in the decades ahead. McMillen and Robinson (2005) approach the metabolic syndrome from this perspective, with evidence that an association exists between the perturbation of the early nutritional environment and the major risk factors (hypertension, insulin resistance and obesity), which could evolve into cardiovascular diseases, diabetes, and the metabolic syndrome decades later. O'Rahilly et al. (2005) note that low birth weight and later diabetes are a further example. Even when genetic factors are held constant, intrauterine influences affect fetal growth and may have long-term implications for metabolic health, especially in "those born smaller who become relatively obese as adolescents or adults" (Gluckman and Hanson, 2004). Fagerberg et al. (2004) point out the interaction between low birth weight and accelerated catch-up growth in early adulthood that is associated with the metabolic syndrome in late middle age, findings "in accord with the concept that the effects of genes are conditioned by fetal growth . . . and later environmental factors." Barker (2005) and Simmons (2006) also noted that low birth weight appeared to be associated with increased rates of coronary heart disease and type 2 diabetes. The perspective of "nutritional epigenomics of the metabolic syndrome" has been introduced by Gallou-Kabani and Junien (2005). They write "maternal nutritional imbalance during critical time windows of development may have a persistent effect on the health of the offspring, and may even be transmitted to the next generation." By *epigenetic* (see Chapter 2) they mean changes associated with chromatin remodeling and regulation of gene expression that may "program" the metabolic syndrome but without a change in the sequence of the genetic DNA. "Imprinted genes, with their key roles in controlling feto-placental nutrient supply and demand and their epigenetic lability in response to nutrients, may play an important role in adaptation/ evolution." Finch and Kirkwood (2000, p. 185) also discuss the interaction of environment and genes as follows: "Thus, for each of us, the expression of our inherited genes is subject to immediate and transgenerational environmental factors that can greatly modify outcomes of aging."

Gluckman and Hanson (2004), in evaluating the relative role of genetic and environmental factors, note that "birth size has only a small genetic component and primarily reflects the quality of the intrauterine environment . . . and the sensitivity of fetal growth to adverse intrauterine influences. 'Maternal constraint' encapsulates those environmental factors that influence birth size even in healthy pregnancies." Yura et al. (2005) agree that intrauterine undernutrition predisposes to obesity and the metabolic syndrome later in life by way of "adapting to restricted nutritional supply—the thrifty phenotype." The concept of the "thrifty phenotype" underlies the influence of early environmental factors on later-life health status (Lazar, 2005): that maternal undernutrition results in fetal growth restraint in utero and early life, which may somehow cause β-cell dysfunction and program metabolism in later life toward the metabolic syndrome and type 2 diabetes (Zimmet and Thomas, 2003). Lawlor and Chaturvedi (2006), reviewing the early developmental period, cite evidence that intrauerine *overnutrition* predicts life-long obesity. They go on to say that "the consequences of these findings are potentially formidable: the obesity epidemic could accelerate through successive generations independent of further genetic or environmental factors." Jaddoe and Witteman (2006) examine the overall picture of epidemiologic studies concerning low birth weight and cardiovascular diseases, type 2 diabetes, and their risk factors in adult life. Central roles have been proposed for (1) fetal undernutrition, (2) increased cortisol exposure, (3) genetic susceptibility, (4) accelerated postnatal growth, and (5) mitochondrial dysfunction with oxidative stress (Simmons, 2006). "Further knowledge is needed to develop strategies for identifying groups at risk and prevention focused on early life"

(Jaddoe and Witteman, 2006). Indeed, from the perspective of groups at risk, the most alarming trend is that the prevalence of overweight and obesity in the United States seems to be particularly prominent in children and adolescents, which has profound societal and health consequences (Ogden et al., 2006). Major public health education initiatives are needed to address this situation. "Developmental elasticity provides organisms with the ability to change structure and function in response to environmental cues . . . during critical time windows. Postnatal life-style interventions depend on this. The need to promote the health and nutrition of females of reproductive age is one important element for the prevention of chronic disease in future generations across the globe" (Gluckman and Hanson, 2004).

Westendorp and Wimmer (2005) link early development with aspects of later aging and also point out that "developmental adaptations to pre- and perinatal nutrition influence susceptibility to disease in old age. . . . the process of healthy aging cannot be understood without considering the entire human life history."

What Constitutes Obesity?

Leave gourmandizing. Know the grave doth gape
For the thee thrice wider than for other men.

Did Shakespeare, by way of the newly crowned King Henry V's advice to Falstaff, anticipate that obesity would become the defining health issue for persons in developed countries as this new century begins? But obesity is not unique to our times. Almost two hundred years after Shakespeare died, a striking portrait of "the largest man in England" appeared in 1809; he weighed 739 pounds at the time of his death at age 39 (Fig. 4.4). According to Haslam and James (2005) in an extensive review of obesity, the "WHO describes obesity as one of the most blatantly visible, yet most neglected public health problems that threatens to overwhelm both more and less developed countries. The problems of overweight and obesity have achieved global recognition only during the past ten years, in contrast to underweight, malnutrition, and infectious diseases, which have always dominated thinking." Zamboni et al. (2005) review the health consequences of obesity in elderly people, noting that weight gain or fat redistribution may still confer adverse health risks or functional decline in this cohort.

Although central fat has great significance, as I will discuss, the standard

Figure 4.4. Daniel Lambert, reputedly the largest man in England, died in Stamford, Lincolnshire, in 1809, aged 39 years. At the time of his death, he weighed 739 pounds, and it is said that it took 20 men to lower his coffin into the grave in St. Martin's churchyard. His portrait now hangs in the mayor's office of Stamford Town Hall. This picture was displayed on the cover of the *Quarterly Journal of Medicine,* December 2004, vol. 97, and the legend is taken from there. Reproduced with the kind permission of Stamford Town Council, Lincolnshire, England. (See color gallery.)

of measure for obesity is the body mass index (BMI), calculated as weight in kilograms divided by the square of height in meters, to characterize men and women who are of normal weight (BMI < 25.0), overweight (BMI 25.0–29.9), or obese (BMI > 29.9) (Goodpaster et al., 2005; Haslam and James, 2005). To simplify the calculation of the BMI, Steven Shapin (cited above) suggested "dividing your weight in pounds by the square of your height in inches, and then multiplying the result by 703." In some older adults, loss of skeletal muscle and subcutaneous fat, with or without decline in health, may nevertheless result in the metabolic syndrome with normal weight. Goodpaster et al. (2005) conducted an extensive study of body fat distribution and the metabolic syndrome in men and women 70–79 years of age and noted its prevalence was about 60 percent in those who were obese compared with about 40 percent in those who were overweight and about 15 percent in those of normal weight. This study and many others confirmed that indeed *visceral* adipose tissue is most related to the metabolic syndrome; adipose tissue in the thighs—*leg fat mass,* if you will—was associated with a lower prevalence of the metabolic syndrome, consistent with other reports of an inverse relation between subcutaneous fat and cardiovascular disease risk. Hence, whereas BMI is of value as a clinical measure—and there are now machinelike scales that compute this automatically. (One such is the Tanita® Body Composition Analyzer, Arlington Heights, Illinois.)—BMI may not entirely reflect adiposity because it does not distinguish between fat mass and lean (nonfat) mass, and there may be differences in the elderly, and non-Caucasian and younger Caucasian populations with respect to BMI and body fatness (Snijder et al., 2006). Perhaps measures of waist circumference, or waist-to-hip circumference ratio (WHR) over time may be helpful, even superior to BMI in view of an association with myocardial infarction (Malnick and Knobler, 2006), despite limitations in distinguishing visceral from subcutaneous fat (Janssen et al., 2002; Snijder et al., 2006) and the formidable difficulty in measuring waist circumference, as Reaven (2006) points out. Clinicians who wish to adopt this measure of waist circumference for their use should consult his article for techniques.

What Is the Significance of Central (Visceral) Obesity?

The predominant type of adipose tissue (white adipose tissue) is composed of adipocytes filled mainly with triacylglycerol and embedded in a loose connective tissue meshwork located mostly in the subcutaneous region and around the viscera (Ahima et al., 2006). Central or visceral fat tissue is of course a depot for fat storage and assumes special significance by virtue of "high enrichment" (Fukuhara et al., 2005) in adipokines, released by adipocytes, including leptin, visfatin, adiponectin, resistin, and several potent cytokines (TNF-α, interleukins) (Table 4.1). Overloading of fat in the diet as one ages has serious consequences by virtue of accumulation in the visceral-abdominal area and, more seriously, when lipid surplus forms harmful fatty acid derivatives, such as ceramide and reactive oxygen species (ROS). The accumulation of fatty acids, or steatosis, in liver, heart, skeletal muscle, and pancreatic β cells, contributes to "lipotoxicity" and cell death or "lipoapoptosis" (Unger, 2005). When the release of fatty acids becomes dissociated from energy requirements and they are stored in nonadipose tissues, including muscle and liver, the conditions are set up for insulin resistance (Yi-Hao and Ginsberg, 2005) (see below). Bergman et al. (2006), using an obese-dog model, suggested that "free fatty acids (FFAs) per se were among the most important products of the visceral adipocyte to cause insulin resistance." Portal drainage into the liver also played a part.

Picard and Guarente (2005) summarize the accumulation of adipose tissue in aging in this way. "The mass of adipose tissue results from the balance between adipogenesis and adipocyte apoptosis on the one hand, and on the other hand between the rate of *de novo* fatty acid synthesis and of dietary free fatty acids stored in adipocytes, and fatty acid release (lipolysis) and transport

TABLE 4.1
Proteins secreted by adipose tissue

Adipose-derived protein	General function	Other tissue source
Leptin	Energy homeostasis	None
	Appetite control	
Adiponectin	Insulin sensitivity	None
Adipsin/ASP*	Insulin resistance	None
Resistin	Insulin resistance	None (rodent)
		Macrophage (human)
Visfatin	Insulin sensitivity	Liver, lymphocytes
TNFα	Inflammatory cytokine	Macrophage
IL-6	Inflammatory cytokine	Macrophage
MCP-1	Inflammatory cytokine	Macrophage
Angiotensinogen	Vascular tone	Liver
PAI-1	Phase reactant	Liver
Serum amyloid	Phase reactant	Liver
α-1 acid glycoprotein	Phase reactant	Liver

Source: Data from Lazar (2005), Kusminski et al. (2005), Kanaya et al. (2006), and Sharma (2006).
*Acylation-stimulating protein. For other abbreviations, see Glossary.

to oxidizing tissues. Aging is associated with decreased lipolytic capacity and increased synthesis and uptake in adipocytes, whereas fatty acid oxidation in skeletal muscle appears to be reduced."

There are three forms of lipid overload (Unger, 2005). The first two are rare. Supersizing is a rapid, short-term intense engorgement of fast foods to supra-physiologic levels exceeding satiety constraints and the existing capacity of the adipocytes—fat storage cells—to handle the increase. The liver bears the brunt of the nutrient flood, hyperinsulinemia is induced by the high-carbo-hydrate content, and lipogenic transcription factors are up-regulated with he-patic production of very-low-density lipoprotein (VLDL), which delivers fatty acids to lipoprotein-lipase-bearing tissues throughout the body. The second example of the failure of the lipotoxic protection system is the rare syndrome of congenital generalized lipodystrophy in which adipocytes are absent and there is a paucity of fat. Patients with congenital generalized lipodystrophy have severe fasting hyperinsulinemia, hyperglycemia, hepatic steatosis, and very low leptin levels in accord with virtually no fat stores. Indeed, leptin therapy resulted in near normalization of fasting plasma glucose and im-proved muscle and liver insulin sensitivity (Petersen and Schulman, 2006). The third form of lipid overload, lipotoxic failure, concerns us, and is based on diet-induced obesity. Visceral fat mass is selectively increased relative to subcutaneous fat, and the metabolic syndrome may ensue. Also, ectopic lipids are deposited in skeletal muscle and pancreatic islets with insulin resistance and type 2 diabetes, lipotoxic cardiomyopathy, and hepatic steatohepatitis.

Enter leptin, a key player in regulating the extent of caloric intake, secreted by the adipocytes during overnutrition to act both on appetite centers in the hypothalamus and on various peripheral tissues. This latter action reduces surplus lipids in nonadipose tissue by up-regulating fatty acid oxidation and inhibiting lipogenesis (Unger, 2005). During the development of diet-induced obesity, the caloric intake exceeds the calories used, the adipocyte expands to store the surplus calories, and leptin secretion rises to stimulate oxidation of fatty acids. However, continued diet-induced obesity results in a decompensated state in which high-leptin levels cannot maintain sufficient compensatory oxidation of surplus fatty acids. These accumulate in peripheral tissue, as noted.

Aging may be the most common cause of impaired leptin sensitivity, whereby fat accumulation increases serum leptin levels and yet less leptin passes through the blood/brain barrier to the hypothalamus (central resistance); or there is peripheral resistance. Leptin resistance may be enhanced by a suppressor of cytokine signaling (SOCS-3), increased in the leptin-unresponsive fat cells of aged rats (Unger, 2005). SOCS proteins inhibit components of the cytokine-signaling cascade via direct binding or preventing access to the signaling complexes (Alexander and Hilton, 2004). Also note that visceral adipocytes express leptin at lower levels than subcutaneous adipocytes (Unger, 2005), contributing further to antilipotoxic failure.

Some therapeutic options to address the metabolic syndrome in obese humans evolve from experimental models: to surgically excise visceral fat (Gabriely and Barzilai, 2003) or to seek means for caloric restriction (see Chapter 3, section 4), up-regulating a gene: silent information regulator protein 2 (SIR2) encoding nicotinamide adenine dinucleotide (NAD)-dependent deacetylase, which promotes fat mobilization (discussed in Chapter 3, section 3). Further, in terms of a biological basis for therapy, I will discuss the use of cholesterol-lowering and antidiabetic drugs in Chapter 5, sections 2 and 4, respectively, and in Chapter 6, section 2, healthy lifestyle interventions (Foreyt, 2005; Stone and Saxon, 2005), especially "a moderate weight loss in initially abdominally obese patients associated with a preferential mobilization of visceral adipose tissue, which in turn leads to a substantial improvement in the metabolic risk profile predictive of reduced risk of coronary artery disease and type 2 diabetes" (Després, 2006). New developments relating to the reduction in "high-risk abdominal obesity" (Després et al., 2006), weight loss, improvement in the metabolism of carbohydrates and lipids (Després et al., 2005)—thus lessening the risks of the metabolic syndrome—may be at hand with the endocannabinoid-CB_1 receptor system as a potential therapeutic target for inhibition (Bramlage et al., 2006; Mackie, 2006a). Endocannabinoids are metabolites of long-chain polyunsaturated fatty acids, especially arachidonic acid. Receptors for cannibinoids, called CB_1 and CB_2 are coupled to G proteins, and modulate the activity of adenylate cyclase (inhibition) and mitogen-activated protein kinase (MAPK, stimulation). The CB_1 receptor is the most widespread G-protein-coupled receptor in the central nervous system (Mackie, 2006b), expressed in the arcuate nucleus and elsewhere in the hypothalamus; hence the CB_1 receptor interacts with neuropeptides such as cocaine—and amphetamine-related transcripts (CARTs) and other signals concerned with satiety (see below). Stimulation of CB_1 leads to food intake; however, CB_1 blockade with *rimonabant* decreases the "desire for foods with an intense taste, and subsequently the intake of food" (Bramlage et al., 2006). Rimonabant also induces expression of the adiponectin gene. Potential adverse effects need to be carefully monitored (Jobst et al., 2006). Readers may recall the publicity attendant on use of marihuana to stimulate the CB_1 receptor, enhancing food intake in those with AIDS and cancer. Finally, Chapter 4, section 4 also discusses the endocannabinoid system in relation to experimental studies showing increased bone mineral density (BMD) with blockade of the CB_1 receptor.

Adipose Tissue as an Endocrine Organ: Release of Adipokines

Adipose tissue has undergone a conceptual evolution. Formerly considered to play a "passive" role (Yi-Hao and Ginsberg, 2005), or serve as a "passive fuel depot" (Kahn and Flier, 2000; Lazar, 2005), or "passive repository" (Lau et al., 2005) storing triglycerides, "from which energy is stored and called forth at times of need in the form of free fatty acids and glycerol" (Kahn and Flier, 2000), adipose tissue and the adipocytes therein are the subject of increasingly "intense scientific interest as a result of the emergence of obesity as a serious public health problem" (Rosen and Spiegelman, 2006). The adipocytes integrate a wide array of homeostatic processes—regulating fat mass and nutrient (energy) homeostasis, and for our consideration here, serving as an *endocrine organ* communicating with the brain and peripheral tissues by secreting a host of factors termed adipocytokines, or more simply adipokines, that modulate inflammation (Prins, 2002; Diez and Iglesias, 2003; Frayn et al., 2003; Ouchi et al., 2003; Rajala and Scherer, 2003; Haluzík et al., 2004; Kershaw and Flier, 2004; Yi-Hao and Ginsberg, 2005). Within the adipose tissue, adipokines may influence the same cell from which they arise (autocrine function), influence adjacent cells (paracrine action), or act remotely to fulfill endocrine functions, as discussed in Chapter 2 and Figure 2.1. Further, the inflammatory role of adipokines may also be considered in relation to "special types of *regional* adipocyte, such as retro-orbital, synovial, visceral, subdermal, peritoneal, and bone marrow adipocytes in internal medicine diseases" (Schaffler et al., 2006).

The number of identified adipose-derived secretory proteins is rapidly expanding. Rondinone (2006) describes more than thirty. Table 4.1 lists some of the most important adipokines. Their widespread metabolic effects are ushering in a new era focusing on adipocyte physiology, what Rajala and Scherer (2003) call the "crossroads of energy homeostasis, inflammation, and atherosclerosis." There is also the important sense of "*overlap*" of adipocyte functions. For example, components that affect energy homeostasis also encompass insulin sensitivity (leptin, adiponectin, resistin); the acute phase reactants and vascular regulators affect lipoprotein metabolism and underlie the pathogenesis of atherosclerosis (α-1 acid glycoprotein, serum amyloid A, plasminogen activator inhibitor); and the inflammatory cytokines impact the innate immune system (IL-6, TNF-α, monocyte chemoattractant protein). Recall innate immunity as the cellular capacity to recognize substances that are foreign to self—such as bacterial components—with release of immunoregulatory cytokines (TNF-α, IL-6) into the circulation that cross the blood/brain barrier. Activation of the hypothalamic-pituitary-adrenal axis follows, with enhanced glucocorticoid secretion (Reichlin, 1999; Chapter 4, section 1). The overlap of functions from components that are adipose derived but not adipose specific has important implications, because, for example, those proteins that play a role in innate immunity also arise from macrophages (Lazar, 2005). The links between the adipocyte and the macrophage, discussed below, bear on their immune and inflammatory responses that underlie, in particular, the pathogenesis of atherosclerosis, as discussed in Chapter 4, section 2, and the condition of insulin resistance I will discuss shortly (Lau et al., 2005; Rader and Puré, 2005; Wellen and Hotamisligil, 2005; Shoelson et al., 2006; Tilg and Moschen, 2006).

Adiponectin in particular seems to exhibit remarkable overlap in its functional relations to suppressing atherosclerosis and improving vascular vasomotor function, moderating insulin resistance, inhibiting TNF-α and CRP expression, and improving lipid status (Ouchi et al., 2003; Gable et al., 2006; Kadowaki et al., 2006; Okamoto et al., 2006). For example, just to point out its role in atherosclerosis, its structural resemblance to the fibrillar collagen family suggested that adiponectin might accumulate in the arterial wall after endothelial alteration; indeed, it appears to attenuate monocyte attachment to endothelial cells by reducing TNF-α-induced expression of adhesion molecules; suppress TNF-α-inhibitory IkB-α phosphorylation and subsequent NF-kB activation through the protein kinase A pathway; and diminish vascular smooth muscle cell proliferation.

Kanaya et al. (2006) and Sharma (2006) write specifically about visceral obesity, adipocytokines, and the incidence of diabetes mellitus in older adults. Increased levels of IL-6, leptin, plasminogen activator inhibitor (PAI-1), and TNF-α, activation of the renin-angiotensin system, and decreased levels of adiponectin have all been associated with the development of type 2 diabetes, and other components of the metabolic syndrome. Below I discuss clinical studies of adiponectin in relation to coronary artery disease. Here I focus on PAI-1 because of the three adipokines (leptin, TNF-α, IL-6);, PAI-1 remained independently associated with type 2 diabetes, although its "connection with cardiovascular disease is more firmly established." The point of interest here is fibrin formation as an essential defense mechanism, protecting the body from bleeding (Mutch et al., 2001). Yet fibrinolysis is also essential: it brings together plasminogen—an inactive plasma protein; its activators—tissue plasminogen activator (and urokinase)—and the degradation of the generated fibrin. Thus, PAI-1, present in excess over the activators of fibrinolysis, favors stabilization of fibrin; presumably pathologically *high* levels of PAI-1 impair fibrinolysis, and fibrin accumulating locally in the vessel is prothrombotic. PAI-1 is not easily measured for clinical purposes, which is unfortunate, for there are pathogenetic aspects here: Smith (2006) notes the "ideal in risk assessment is the development of measures that provide prognostic information directly related to atherothrombotic properties of vessels. Improved understanding of underlying disease processes has changed the conception of coronary disease from that of atherosclerotic to that of atherothrombotic disease."

Thus, a "central" issue linking adipokines, inflammatory mediators, and the constellation of conditions that constitute diabetes, atherogenesis, and other components of the metabolic syndrome seems to be *visceral obesity* (Skurk and Hauner, 2004; Després, 2006): a kind of internal circuit exists, in which more fat produces inflammatory components, which in turn stimulate adipocytes to produce more adipokines—a "mechanistic link between obesity and the associated vascular complications" (Juge-Aubry et al., 2005).

In a commentary Arner (2005a) raised a note of caution about transferring studies of adipokine functions from mice to humans. Resistin is an example; released from adipocytes in mice, it appeared to induce insulin resistance and has proinflammatory properties, stimulating the release of TNF-α and interleukins from macrophages via a NF-kβ-dependent pathway (McTernan et al., 2006). But in humans there has been a lingering and increasing uncertainty that resistin plays a comparable role. This uncertainty is well reviewed by Kusminski et al. (2005). Arner (2005a) believes a similar case can be made for TNF-α in humans, in whom, unlike in mice, TNF-α may not enter the circulation but rather act locally on adipocyte lipolysis. The adipokines that seem to be equivalent in terms of mice–man action are leptin, IL-6, and adiponectin.

The mouse, of course, is indispensable to exploring new molecular-based interventions that derive from observations on disease in humans. Two examples have been pioneered by researchers at the Albert Einstein College of Medicine. Both Richard Kitsis in medicine-cardiology and Philipp Scherer in cell biology have shared with me new approaches that depend on *induced apoptosis*—programmed cell death: one for cardiac myocytes, and the other for adipocytes to create an animal model of "fatlessness."

The observation in humans in the first case derives from low—but still abnormal—rates of cardiomyocyte apoptosis in patients with end-stage dilated cardiomyopathy. The pathologic observations of apoptosis include DNA fragmentation, chromatin condensation, membrane blebbing, cell shrinkage, and disassembly into membrane-enclosed vesicles (apoptotic bodies) (Thornberry and Lazebnik, 1998). The molecular mechanisms are extremely complex, and presented here in outline, based on a paper from the Kitsis laboratory (Foo et al., 2005). Apoptosis is mediated by two pathways conserved through evolution. In an extrinsic pathway ligand binding induces cell surface death receptors to recruit a family of proteases that are synthesized as precursors—procaspases—which are converted to active caspases by binding of cofactors or removal of inhibitors. In apoptosis, caspases function in both cell disassembly (effectors) and in initiating this disassembly in response to proapoptotic signals (initiators) (Thornberry and Lazebnik, 1998; Dirks and Leeuwenburgh, 2006).

In contrast to the extrinsic pathway that mediates a specialized subset of death signals, the intrinsic pathway transduces a wide variety of extracellular and intracellular stimuli. These generate signals that activate proapoptotic proteins (the family of Bcl-2) that trigger release of mitochondrial-derived apoptogens into the cytoplasm. Again, procaspases are activated within the framework of neutralizing inhibitory pathways as well as activating effector mechanisms. Recently, the endoplasmic reticulum has been recognized as an important organelle in the intrinsic pathway. Endoplasmic reticulum oxidative stress can induce mitochondrial permeability and apoptogen release, as well as procaspase activators.

All this is relevant to the mouse model whereby the Kitsis group created mice with cardiac myocyte-specific expression of a ligand-activated caspase (extrinsic pathway). Their aim was "to specifically manipulate the central death machinery and quantitatively modulate the induction of apoptosis in cardiac myocytes in vivo" (Wencker et al., 2003). When this system was "turned on," low levels of myocyte apoptosis—even four- to tenfold lower than those seen in human heart failure—were sufficient to cause lethal, dilated cardiomyopathy in these mice. Equally intriguing, the system could be "turned off," inhibiting cell death and curtailing the development of heart failure. The pros and cons of a causal relationship of cardiac myocyte apoptosis to heart failure are discussed in the paper; nor is the possibility overlooked that "inhibition of this cell death may provide a novel target for treatments directed at this common and lethal disorder" (Wencker et al., 2003).

Kajstura et al. (2006) considered "apoptosis and its deregulation" part of a broader picture that contributed to neoplasms (see Chapter 4, section 7), to oxidative stress-mediated pathologic conditions (for which they list diabetes), and to the accumulation of senescent cells that can undergo apoptosis (Fig. 3.3). In the aging heart, cell dropout without replenishment results in poorly contracting myocytes and, over time, leads to end-stage heart failure.

The second model of apoptosis developed in the laboratory of Philipp Scherer has no counterpart in humans, yet many issues of adipokine function are investigated in the FAT-ATTAC mouse. Noting that "new approaches are needed to better understand and appreciate the contribution of adipose tissue towards normal physiology and its dysregulation in a relatively acute setting—rather than addressing lipoatrophy in the chronic fatless mouse models"—Trujillo et al. (2005) developed an inducible fatless model (fat apotosis through targeted activation of caspase 8). The techniques to do so were modeled on those developed by the Kitsis group for the cardiac myocyte, and now applied to the adipocyte. Activation of caspase 8 that induces apoptosis in mature adipocytes is not known to trigger any other signaling event. In these mice, fat loss can be chronicled by the >95 percent reduction in circulating levels of adipokines. These mice are glucose intolerant yet fail to exhibit secondary complications of long-term insulin resistance. Food intake increases dramatically in the fatless state, animals eating twice the amount of food compared with leptin-deficient (ob/ob) mice (to be discussed below). One gets the sense that many new avenues will be opened up for research relating to appetite control, glucose homeostasis without adipocytes, angiogenesis, adipogenesis, and the role of adipocytes in mammary gland development and tumorigenesis—areas the authors mention that need to be explored. Of interest to geriatricians will be longevity (or its lack) in these mice, which seem to be the opposite of CR mice. It will also be of interest to consider whether selective fat apoptosis in this mouse model may be applicable to humans.

Interrelations of Macrophages and Adipocytes

Although this part of the discussion on the metabolic syndrome relates primarily to macrophage infiltration of adipose tissue, it may be useful first to look more generally at an overview of the macrophage. This cell in its multiple expressions in tissues figures prominently not only in relation to adiposity and inflammation, but also as a macrophage-laden foam cell in atherosclerosis (see Chapter 4, section 2) and as a unique multinucleated cell in bone in osteoporosis (Chapter 4, section 4) (Bouloumie et al., 2005; Chitu and Stanley, 2006). The reader may also be aware of tissue-based macrophages that are dendritic cells and interact with T cells in antigen-specific presentation; with microglia—the resident macrophage in the brain; and with synovial membrane macrophages (the type A cells). The underlying theme of macrophage functions appears to be participation in inflammatory and immune responses.

A population of progenitor cells in the bone marrow gives rise to monocytes that circulate in the blood and enter the tissues to become tissue-resident macrophages. The entry into the tissues of blood-borne monocytes requires modulation of their adherent properties as well as those of the endothelial cell layer by way of adhesion molecules, movement of the monocyte through the vessel wall, and differentiation within the tissues under the influences of chemokines, cytokines, and growth factors. Some of these mediators include CCL2, or macrophage chemoattractant protein (MCP), CCL3 or macrophage inflammatory protein (MIP), IL-8, M-CSF-1, vascular endothelial growth factor (VEGF), and plasminogen activator inhibitor-1 (PAI-1)—many of which feature prominently in discussions in this chapter on atherosclerosis and on the metabolic syndrome, as well as in osteoporosis and cancer.

Wellen and Hotamisligil (2005), in considering interrelations of macrophages and adipocytes and the links between obesity, metabolism, and immunity, emphasize the "chronic low-level of inflammation" that underlies these conditions. There is a "high level of coordination and integration of inflammatory and metabolic pathways in the overlapping biology and function of macrophages and adipocytes in obesity." Suganami et al. (2005) point out that weight gain is associated with infiltration of fat by macrophages, suggesting that these cells are an important source of inflammation in adipose tissue. These workers postulate that free fatty acids (FFAs) released from adipocytes may activate TNF-α in macrophages, creating a "paracrine (cell to cell) loop that establishes a vicious cycle and enhances inflammation." Linton and Fazio (2003) discuss "the striking overlap between the biology of adipocytes and macrophages. Several genes that are critical in adipocytes, including those that regulate transcription factors, cytokines, inflammatory molecules, nuclear receptors, fatty acid transporters, and scavenger receptors, are also expressed in macrophages, and play a significant role in their biology." The adipocyte and macrophage are both involved in abnormal cytokine production and the "seemingly surprising ability of fat cells to mount an immune response is a function of its common features with macrophages" (Hotamisligil, 2003). Thus, similarities of *gene expression* include fatty acid-binding protein (FABP aP2), peroxisome proliferator-activated receptor (PPARγ), TNF-α, IL-6, and matrix metalloproteinases (MMP). Similarities of *function* include uptake and storage of lipids in adipocytes, and lipid-laden macrophages as foam cells in atherosclerosis.

Recently, Neels and Olefsky (2006) reviewed the association of macrophage infiltration in fat in obesity: "Inflamed fat; what starts the fire?" They point out that this process may involve increased secretion of chemotactic molecules by adipose tissue, especially "the C-C motif—chemokine ligand 2—CCL2— also known as monocyte chemoattractant protein-1 (MCP-1)." Chemokines, as noted in Chapter 4, section 2 on atherosclerosis, are a superfamily of small proteins that play a crucial role in immune and inflammatory reactions, and especially in chemotaxis of leukocytes (van Damme and Mantovani, 2005; Bachmann et al., 2006). CCL2 and its receptor CCR2 likely are important chemoattractants for macrophages in adipose tissue, as demonstrated in experimental studies by inactivating them and ameliorating vascular disease in mice (Semenkovich, 2006). But the crucial question remains, as Neels and Olefsky (2006) point out, "Is the activation of the inflammatory pathway within adipocytes, and secretion of factors by fat cells due to macrophages acting on them causally related to insulin resistance or simply a secondary, reactive event?" In addition to the proinflammatory role of macrophages and adipocytes themselves, Shoelson et al. (2006) note that the microvasculature "undoubtedly plays a role in adipose tissue inflammation."

Overlaps and links also occur between the macrophage-derived osteoclast in bone and the macrophage in the arterial wall in atherosclerosis, as will be discussed in Chapter 4, section 4 on osteoporosis. Just to mention here (and do consult the glossary for abbreviations), osteoblast-released RANK ligand activates RANK receptor on the osteoclast and up-regulates NF-kB, which also occurs by way of inflammatory mediators acting on the macrophage in the vessel wall in atherosclerosis. (The molecular basis of NF-kB activation is discussed in Chapter 4, section 7.) The cytokines TNF-α and IL-6 figure prominently in the process of osteoclastogenesis and atherosclerosis. Colony-stimulating factor (CSF), which acts on macrophages in the arterial wall in atherosclerosis, also promotes osteoclast development when CSF binds to its receptor c-fms, and associates with and activates phosphatidylinositol-3-kinase (PI3K) (Golden and Insogna, 2005; Lowell and Schulman, 2005). PI3K figures prominently in states of insulin resistance and sensitivity, as I will discuss below.

Overall, "in the course of evolution, in addition to their individual roles, a common network of functional and molecular pathways were retained in cells that assume inflammatory and metabolic roles: the adipocyte (insulin resistance), the macrophage (atherosclerosis and perhaps osteoporosis), and the hepatocyte (hyperlipidemia and phase reactants)" (Hotamisligil, 2003).

Adiponectin in Clinical Practice

Considering that adiponectin was identified only about a decade ago (Rajala and Scherer, 2003; Trujillo and Scherer, 2005; Kadowaki et al., 2006), it has assumed growing importance not only in the fundamental biology of the adipocyte and the metabolic syndrome, but also in its potential for clinical application: adiponectin is measurable and has "profound protective actions in the pathogenesis of diabetes and cardiovascular diseases" (Koerner et al., 2005).

Adiponectin in human blood is readily assayed and its levels are two to three times higher in females than males and are *higher* with low fat mass and *lower* with high fat mass (Haluzík et al., 2004). This is unlike leptin, where blood leptin levels correlate with fat mass. Mean plasma adiponectin levels were 3.7 µg/ml in a group of obese persons and 8.9 µg/ml in nonobese persons. It may be that adiponectin could come to be a "marker" of the metabolic syndrome and cardiovascular risk (Ouchi et al., 2003; Trujillo and Scherer, 2005), but unlike CRP, where *high* levels are correlated with cardiovascular risk factors, *low* levels of adiponectin are found in states of high cardiovascular risk: obesity, hypertension, insulin resistance and diabetes, elevated total triglycerides, and LDL-cholesterol. Levels are particularly low in persons with cardiovascular disease (Chandran et al., 2003; Diez and Iglesias, 2003; Kadowaki et al., 2006; Okamoto et al., 2006). Conversely, high plasma levels of adiponectin are associated with low risk of myocardial infarction over a six-year follow-up in men without previous cardiovascular disease (Kumada et al., 2003; Pischon et al., 2004). Thus, as Pischon et al. note, "low levels of adiponectin may be not only a marker of cardiovascular risk, but also a causal risk." In animal studies, adiponectin may have antiatherogenic and anti-inflammatory properties (Diez and Iglesias, 2003), and in humans it diminishes insulin resistance, improves glucose homeostasis, and may lower inflammatory cytokines (Pischon et al., 2004; Kadowaki et al., 2006). Treatment with thiazolidinediones (TZDs) enhances endogenous adiponectin production (Diez and Iglesias, 2003; Haluzík et al., 2004) by way of the action of PPARγ (Steiner et al., 2006). In cell culture studies, adiponectin up-regulated the anti-inflammatory cytokine IL-10 and tissue inhibitor of matrix metalloproteinase, which may be protective against atherosclerotic plaque disruption (Kumada et al., 2004). Chandran et al. (2003) suggest that adiponectin therapy may be a promising means to manage so many of the lesions of the metabolic syndrome, including hyperlipidemia, insulin resistance, and vascular inflammation.

Philipp Scherer in his Lilly Lecture (2006) further categorized circulating adiponectin with respect to variations in molecular size: a basic unit referred to as a homotrimer that can assemble into higher-order structures, such as a hexamer, and the aggregation of several of these hexamers into a high-molecular-weight complex. The higher levels of adiponectin in females noted above are due to an increase in the high-molecular-weight complex, and this complex also rises with therapies that stimulate PPARγ. Scherer thus believes that the correlations observed by measuring the total levels of adiponectin could be further strengthened by taking this complex size distribution into account. The adiponectin sensitivity index (S_A) reflects the fraction of adiponectin found in the high-molecular-weight form, and Kadowaki et al. (2006) also believe that this is the more active form—more relevant to insulin sensitivity and to protection against diabetes. Thus, S_A is a far superior indicator of improvement in insulin sensitivity with PPARγ agonists than measures of total adiponectin; measures of adiponectin S_A may come to clinical use after further validation.

Factors Contributing to the Evolution of Insulin Resistance and Type 2 Diabetes

Reaven (2005a) traces the history of the convergence of what he called syndrome X in 1988 and the considerations by Himsworth fifty years earlier that introduced the concept of insulin *resistance* rather than insulin *deficiency* in adult-acquired diabetes mellitus. Substantiation awaited the radioimmunoassay developed in 1960 by Yalow and Berson, which showed that insulin levels were on average higher in patients with adult onset diabetes. Barish et al. (2006) note, among other aspects, that hyperinsulinemia enhanced hepatic gluconeogenesis and impaired insulin-stimulated glucose uptake into skeletal muscle and fat. Further, as I will discuss, elevated levels of free fatty acids (FFAs) associated with obesity increased fat accumulation in insulin target tissues (especially muscle) and contributed to defective insulin action. Reaven (2006) presents an enormous list of metabolic abnormalities associated with the "physiologic framework of the insulin resistance syndrome," including, of course, glucose intolerance, dyslipidemia, endothelial dysfunction, enhancement of inflammatory markers, type 2 diabetes, and hypertension; adiposity and lack of physical fitness are mentioned at the end of the list to affirm that the insulin resistance syndrome is the "plague of the twenty-first century."

Picard and Guarente (2005) place the "development of general insulin resistance as a feature of aging, with resulting limitations in glucose uptake in peripheral tissues, impaired hepatic control of glucose production and triglyceride secretion." Reduced insulin sensitivity with aging is closely related to the accumulation of visceral fat, as noted, and Lau et al. (2005) propose that "adipokines are the molecular link between obesity and the metabolic syndrome of insulin resistance and endothelial dysfunction leading to atherosclerosis." The molecular mechanisms for age-induced insulin resistance and fat accretion are, however, far from being completely understood (Picard and Guarente, 2005).

To describe the mechanisms of insulin resistance poses a challenge for my comprehension of the subject and for my capacity to make a presentation that will not overwhelm the general reader, as mentioned in the preface. We must deal with what Taniguchi et al. (2006) describe as the "complexity of the insulin-signaling network" and its many outcomes. Perhaps no subject considered in this book illustrates the great breadth of interactive factors as do those that influence insulin resistance. The reader might begin with an overall orientation by way of Figure 4.5. I have found several reviews helpful, including those by Kahn and Flier (2000), White (2002), Brady (2004), Hawkins and Rossetti (2005), and Petersen and Shulman (2006), and others will follow below. I also quote extensively from a symposium entitled "Complications of Obesity: the Inflammatory Link" in the *International Journal of Obesity and Related Metabolic Disorders* (vol. 27, Suppl. 3, December 2003); a series of papers in *Science* (vol. 307, January 21, 2005) and in the *Journal of Clinical Investigation* (vol. 116, July 2006)—the latter introduced by Masato Kasugo in a review describing aspects of insulin resistance and the contributions of obesity and pancreatic β-cell failure.

Activation of the insulin receptor by insulin stimulates glucose uptake in skeletal muscle and adipose tissue by way of the glucose transporter, GLUT4, which in this process must translocate from the intracellular storage compartment to the plasma membrane at the cell surface (Kanzaki, 2006). The insulin receptor (IR) promotes tyrosine phosphorylation of the insulin receptor substrate proteins (IRS) that are linked to the activation of two main signaling pathways: the phosphatidylinositol-3-kinase (PI3K) family that generates phosphorylated intermediates leading to the activation of protein kinase B/Akt—the pathway responsible for most of the metabolic actions of insulin; and the Ras-mitogen-activated protein kinase (MAPK) pathway, which regulates expression of some genes and cooperates with the PI3K pathway to control cell growth and differentiation (Long and Zierath, 2006; Taniguchi et al., 2006). In Chapter 3, section 4, we encountered the PI3K (class I) signaling pathway, for it is highly conserved in evolution from worms to mammals, and influences the downstream expression of protein kinase B/Akt that is so important, among other biological functions, in FOXO regulation (Engelman et al., 2006). Taniguchi et al. (2006) regard the identification of the IR and IRS, PI3K, and protein kinase B/Akt, as three critical cell-signaling components, or "nodes," part of the "incredible diversification and fine-tuning of insulin's signal in normal physiology and disease states." The evidence shows that IRS-PI3K signaling is "severely impaired" in diabetes, obesity, and other insulin resistance states, whereas the mitogenic pathway retains its sensitivity to insulin, and indeed may be atherogenic (DeFronzo, 2006).

Herman and Kahn (2006) discuss the "metabolism of glucose and the accompanying signals that are generated which alter the availability of other fuels, such as fatty acids, that are also thought to play a major role in the pathogenesis of diabetes." To use their analogy, it seems that "the individual instru-

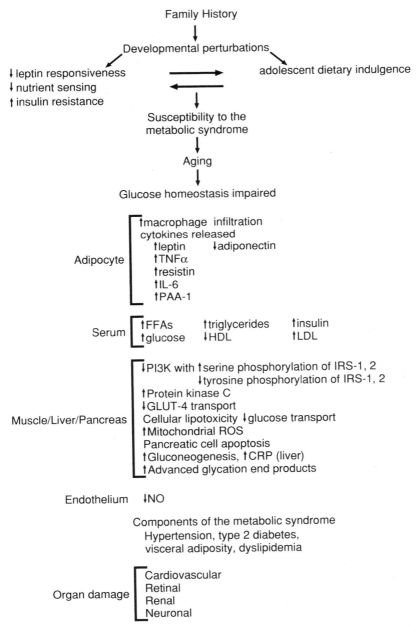

Figure 4.5. Events leading to insulin resistance and the development of the metabolic syndrome. See text for discussion, explanation of abbreviations, and literature citations. ↓ = decrease; ↑ = increase.

ments (as in an orchestra) fail to maintain glucose homeostasis and metabolic harmony," and many facets contributing to this disarray will be presented. To understand the development of insulin resistance it is important that we look in some detail at the conditions that appear to *alter* what would normally be phosphorylation of *tyrosine* residues of the insulin receptor substrate (IRS-1), promoting instead phosphorylation of *serine* residues, causing dissociation of the p85 subunit of PI3K (Morisco et al., 2006), blunting the insulin signals, inhibiting glucose transport activity (Kovacs and Stumvoll, 2005), increasing gluconeogenesis, and reducing hepatic glucose uptake via lowered Akt and increased FOXO transcriptional effects on several key gluconeogenic rate-controlling enzymes (Steiner et al., 2006). As Zick (2003) notes, "while the underlying molecular pathophysiology of insulin resistance is still not well understood, serine phosphorylation of IRS protein represents an important new mechanistic theme." In recent publications, several authors attribute blunting of these dual processes (phosphorylation and glucose transport) to elevated serum free fatty acids (FFAs), part of the link between obesity and insulin resistance (Golay and Ybarra, 2005; Petersen and Shulman, 2006). Elevated plasma FFAs, in fact, *impair* the insulin-induced phosphorylation of IRS-1, attenuate expression of PI3K, and activate protein kinase C, which also blunts insulin-stimulated IRS-1 phosphorylation. Impairment in insulin-signaling pathways contributes to "synergistic coupling of insulin resistance and endothelial dysfunction" by way of reduced protein kinase B/Akt, which then leads to decreased endothelial nitric oxide (NO) synthase and reduced NO expression, normally necessary for vascular vasodilatation (Kim et al., 2006), hence further associations with atherogenesis.

Petersen and Schulman (2006) believe that obesity per se is not the driver of insulin resistance, but rather the accumulation of intracellular lipid metabolites (e.g., diacylglycerol) is. FFAs and glycerol are released from lipoprotein lipase-induced hydrolysis of triglycerides stored in fat cells, more so from adipocytes that are larger in obese women, and especially from visceral fat stores (Arner, 2005b). According to Greenberg and Obin (2006), one cause for the increased release of FFAs is alterations in perilipin expression. Perilipins are phosphoproteins found in adipocytes on the surface of triacylglycerol droplets that act as gatekeepers, preventing lipases from hydrolyzing triacylglycerol to facilitate the release of FFAs. Obese individuals have a deficiency of perilipins even if their fat cells are larger, hence their increased basal rate of lipolysis. Yet even lean young offspring of persons with type 2 diabetes, whose lifetime risk for the development of this disease is 40 percent, have elevated plasma fatty acids—significantly correlated with insulin resistance (Petersen and Shulman, 2006).

Another cause for enhanced release of FFAs from adipocytes is higher expression of inflammatory cytokines from the abundant macrophages in adipose tissue. Inflammatory mediators, especially TNF-α, appear to play a prominent role in the inception of insulin resistance, as described by Hotamisligil in 1993, and recently reviewed (Hotamisligil, 2003). Anti-inflammatory agents, such as salicylates, reverse insulin resistance via an effect on inhibition of IKK-β- and NF-kB-signaling pathways (discussed in detail in Chapter 4, section 7) that lead to activation of a range of inflammatory mediators. TNF-α promotes adipocyte lipolysis and reduces adiponectin expression, itself an inhibitor of TNF-α; low adiponectin levels have cardiovascular consequences, as discussed above. TNF-α stimulates lipolysis "through so-called MAP kinases," and pharmacologic agents such as the glitazones (see below and Chapter 5, section 5) decrease FFA output from adipose tissue and normalize blood glucose levels (Arner, 2005b). But a range of other cytokines are released from the macrophages in "older" adipose tissue as well, including IL-1, IL-6, MCP-1, PAI-1—which promote systemic inflammation, prothrombotic responses, enhance CRP production in the liver, and act on endothelial cells and osteoclasts.

Muscle cells respond to high FFA levels by preferentially using lipids rather than glucose as an energy source, with elevations of plasma glucose and insulin. Glycogen synthase activity is inhibited. As Golay and Ybarra (2005) note, glycemia and insulinemia reflect impaired glucose tolerance in an obese but nondiabetic person that may gradually lead to type 2 diabetes.

And aging itself, of course, is a key contributor to the development of insulin resistance, as I noted. Ye et al. (2006) looked at aged insulin-resistant rats. Compared with their young counterparts, expression of PPARγ at both mRNA and protein levels in adipose tissue of those rats was "dramatically decreased," as was the expression of its target gene lipoprotein lipase mRNA. In an extension to humans, PPARγ mRNA was reduced in omental adipose tissue in older men with "a tendency to insulin resistance" compared with younger men. Clearly, the TZDs, as discussed in Chapter 5, section 4, play a crucial role in improving the diabetic condition by enhancing PPARγ.

The pancreatic islet β cell, the source of insulin, must be considered. Following an initial compensatory increase in β-cell mass that may meet the increased demand for insulin in the face of obesity and insulin resistance, those individuals destined to develop type 2 diabetes incur a progressive decrease in β-cell mass and function (Rhodes, 2005; Prentki and Nolan, 2006). One cause of β-cell loss appears to be programmed cell death, or apoptosis. This is exacerbated by the formation of amyloid plaque deposits in islets, and ultimately a "point of no return" with a permanent type 2 diabetic state requiring insulin. IRS-1, discussed above, is not involved in control of β-cell mass, but IRS-2 is. This insulin receptor substrate-2 is also found in the hepatocyte. Thus, factors that increase phosphorylation of IRS-2 serine/threonine expression in β cells may be linked to β-cell apoptosis and reduced IRS-2 action. In addition, chronic hyperglycemia (glucotoxicity) may diminish β-cell mass by virtue of "stressful" (the cellular or metabolic variety) local conditions, including ROS, endoplasmic reticulum overload, and cytokines. "Lipotoxocity" is also a factor.

Lowell and Shulman (2005) describe defects in mitochondrial oxidative phosphorylation activity that impair fatty acid oxidation, leading in turn to increases in intracellular fatty acid metabolites that disrupt insulin signaling. Glucotoxicity and lipotoxicity may also impair a mitochondrial membrane protein called uncoupling protein-2. This in turn affects the crucial oxidative metabolism of glucose—the actual transfer of energy stored in the carbon bonds—to adenosine triphosphate (ATP) in a process known as respiration. ROS, generated as electrons, are transferred from nutrients to molecular oxygen. Mitochondria are likely to be particularly susceptible to damage and this translates into insulin resistance in skeletal muscle (Semenkovich, 2006). The role of mitochondria as a source of ROS has "also opened new research vistas in vascular biology" and potentially in a "host of other pathologies, including Alzheimer's disease, degenerative changes in aging, Parkinson's disease, and type 2 diabetes" (Gutterman, 2005). Indeed, Brownlee (2005) in his Banting lecture described what he believes to be a unifying mechanism based on *hyperglycemia*-induced metabolic events on selective cell types (endothelial, retinal, renal, mesangeal, neuronal) that sustain diabetic complications: the overproduction of the reactive oxygen species (free radical) *superoxide* (O_2^-). As noted (in Chapter 3, section 3), the mitochondria, as the principal energy-generating organelles in the cell, initiate this reaction, and superoxide production can in turn activate many other reactive oxygen species pathways. Increased free fatty acid oxidation in mitochondria also produces superoxide. Chapter 3, section 4 noted that mutations that *increased* activity of the antioxidant enzyme superoxide dismutase extended longevity by simulating caloric restriction (Longo and Finch, 2003). Indeed, "caloric restriction exhibits an especially broad protective action against oxidative threats, while maintain-

ing counter-acting antioxidant defense systems to sustain a well-balanced redox state during aging" (Chung et al., 2005). Ritz and Berrut (2005) also link
diabetes and "mitochondrial disorders." They note that a defect in oxidative
phosphorylation reduces the capacity of energy-using cells to oxidize fuel and
generate ATP. In the face of a high caloric diet, tissue use of fuel is reduced,
resulting in hyperglycemia and the initiation of the insulin resistance cascade.
Kaneto et al. (2006) agree that oxidative stress plays a role in the progression
of diabetes, leading to β-cell deterioration. Several signal transduction pathways, noted in Chapter 2, Figure 2.3, and discussed above, are activated, including c-Jun N-terminal kinase (JNK), p38 mitogen-activated protein kinase
(MAPK), and protein kinase C (PKC), again with reduced insulin gene expression as a result.

Schwartz and Porte (2005) focus on the brain and central control of energy homeostasis; Morton et al. (2006) and Schwartz (2006) present further
valuable updates. A background for central nervous system regulation of food
intake and adiposity could begin with a brief discussion of mouse models:
the genetically obese mouse (*ob/ob*) homozygous for a point mutation that
results in the elaboration of a *biologically inactive* leptin molecule; and the
other obese mouse phenotype (*db/db*) that arises from a mutation in the leptin
receptor gene. Thus, genetic deficiency of either leptin itself or its receptor is
sufficient to induce a severe obesity phenotype in mice, for there is no check
on food consumption (see below). Leptin deficiency in humans comparable
to the *ob/ob* mouse also displays a severe obesity phenotype with dramatic
normalization of this phenotype in response to leptin replacement. However,
common forms of obesity in humans are rarely responsive to leptin replacement, are associated with increased rather than decreased leptin levels, and
are more likely to be related to *resistance,* which I will discuss in a moment.

With this background we need to consider leptin and insulin circulating
in proportion to body fat mass as "obesity-related signals." Insulin is secreted
in proportion to visceral fat, whereas leptin reflects total fat mass; elevated
visceral fat carries a greater risk for insulin resistance (Woods et al., 2006).
Porte (2006) notes that "insulin becomes the key explanation for the coupling
of type 2 diabetes and obesity." Figure 4.6 shows elements of these adiposity signals to which hypothalamic arcuate nucleus neurons are responsive.
High circulating leptin and insulin levels in obesity stimulate arcuate nucleus
neurons to release proopiomelanocortins (POMCs), the polypeptide precursors from which melanocortins—such as α-melanocyte-stimulating hormone
(αMSH)—are derived and bind to and activate neuronal melanocortin-4 re-

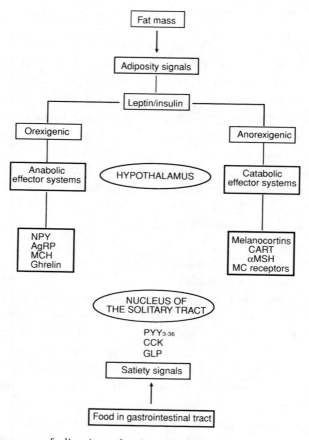

Figure 4.6. Summary of adiposity and satiety signals that regulate appetite through CNS effectors (see text for discussion). In a state of increasing adiposity, higher levels of insulin (secreted from the endocrine pancreas in proportion to visceral fat mass) and leptin (secreted from adipocytes in proportion to total fat mass) constitute *adiposity signals* that turn on the catabolic effector systems, reduce appetite, increase energy expenditure, and turn off the anabolic systems. The converse occurs in a state of diminishing adiposity; the levels of the signals fall, the anabolic pathways are turned on, appetite increases, energy expenditure diminishes, and the catabolic systems are turned off. The hypothalamus is the major site for these systems. *Satiety signals* arise from the presence of food in the gastrointestinal tract, with release of PYY_{3-36}, CCK, and GLP, which convey messages to reduce food intake by way of vagal afferent fibers to the nucleus of the solitary tract. The gastric-derived hormone ghrelin opposes the hypothalamic actions of leptin, insulin, and PYY_{3-36} and powerfully stimulates food intake (Morton et al., 2006). AgRP, agouti-related protein; CART, cocaine- and amphetamine-regulated transcript; CCK, cholecystokinin; CNS, central nervous system; GLP, glucagon-like peptide; MC, melanocortin; MCH, melanin-concentrating hormone; αMSH, α-melanocyte stimulating hormone; NPY, neuropeptide Y; ORX, orexin; PYY_{3-36}, peptide YY_{3-36}. Adapted from Hamerman (2002), © The Gerontological Society of America. Adapted by permission of the publisher.

ceptors in adjacent parts of the hypothalamus. These constitute a system of *anorexigens* that limit food intake and raise metabolic rate—a combination of events that promote weight loss.

Adjacent to the melanocortins in the arcuate nucleus are cells that express *orexigenic* molecules—neuropeptide Y (NPY) and agouti-related protein (AgRP). In times of food scarcity, when fat mass shrinks, leptin and insulin levels fall, and weight loss activates these neurons through a combination of reduced inhibitory (leptin and insulin) and increased stimulatory (ghrelin) input (Morton et al., 2006). The orexigens inhibit melanocortins—stimulate appetite, reduce energy expenditure, increase hepatic glucose production, and promote weight gain. Keep in mind that high serum leptin levels in obesity, with incomplete central nervous system penetration, could be interpreted as a drop in leptin levels; but this is really a form of leptin resistance, resulting in increased NPY and AgRP signaling, leading to hyperphagia, further weight gain, glucose intolerance, and exacerbation of insulin resistance.

Schwartz and Niswender (2004) and Morton et al. (2006) explore a unifying basis for obesity and insulin resistance, implicating impaired signaling of hypothalamic neurons by insulin and leptin. These adiposity signals have overlapping *activating* signaling pathways: for leptin, the Janus kinase (JAK)–signal transducer and activator of transcription (STAT3) pathway, common to many cytokine receptor systems (discussed in Chapter 2 and Chapter 5, section 5); and for insulin, the insulin receptor substrate (IRS)–phosphatidylinositol-3 kinase (PI3K) pathway, discussed above, critical in the activation of the insulin receptor. In some cells, insulin can activate JAK-STAT signaling, whereas leptin can activate the PI3K pathway (Kitamura et al., 2006)—"points of convergence and synergism between intracellular signaling pathways used by leptin and insulin" (Morton et al., 2006)—and the ability of either hormone to reduce food intake may require both pathways to be intact to target hypothalamic neurons. Thus, the two known adiposity-related signals (insulin and leptin) seem to reduce food intake through a mechanism that requires intact IRS-PI3K and JAK-STAT signaling; impairment of these signals may clearly contribute to the pathogenesis of obesity and common forms of insulin resistance in peripheral tissues (Schwartz, 2006). Furthermore, nutrient excess activating cytokine signal transduction pathways, and macrophage infiltration of adipose tissue releasing TNF-α and other cytokines, may lead to phosphorylation of the insulin receptor on serine residues reducing the capacity to activate PI3K and GLUT-4, as discussed earlier.

Adiposity signals can also be unified by mechanisms responsible for *termi-*

nating signals for receptors via leptin and insulin. One such way is by protein tyrosine phosphatase-1B (PTP-1B), an enzyme implicated in the enzymatic cleavage of phosphate from key tyrosine residues of molecules in the signaling cascade-IR, IRS, JAK-2. Another way of terminating cell signaling by both insulin and leptin is suppressor of cytokine signaling-3 (SOCS3), which may dampen signaling via the insulin receptor–IRS-PI3K pathway as well as the JAK-STAT pathway (Alexander and Hilton, 2004).

Schwartz (2006) also points out that, in addition to input from insulin and leptin, the arcuate nucleus senses changes in energy balance conveyed by the gastric hormone ghrelin. Chapter 5, section 6 discusses ghrelin in its role as a growth hormone secretagogue. Ghrelin arises primarily from endocrine cells in the stomach, circulates as a hormone, rising before meals, activating its receptors as well as NPY/AgRP neurons (as I noted above), and thus stimulating food intake; that is, ghrelin is orexigenic in its hypothalamic-induced (central) effects, and its peripheral actions on use of glucose and on storage of fat in adipose tissue may subsequently be mediated by the sympathetic nervous system (Theander-Carrillo et al., 2006). The extent to which normal meal initiation depends on an increase of ghrelin signaling, however, still remains uncertain and requires further study.

Work originating in the laboratory of Luciano Rossetti at the Albert Einstein College of Medicine, which he has kindly discussed with me, has demonstrated a "brain-liver circuit that regulates glucose homeostasis" (Pocai et al., 2005). In experimental studies in the rat, inhibition of a membrane-bound, carnitine-dependent long-chain acyltransferase, also known as carnitine-dependent palmitoyltransferase (CPT1), in the arcuate nucleus of the hypothalamus, leads to accumulation of long-chain fatty acyl-CoAs. Rossetti and his group postulate that "this represents a central signal of nutrient abundance" and results in inhibition of food intake. Furthermore, ATP-dependent potassium channels (K_{ATP}) are activated, stimulating vagus nerve efferent outflow to the liver, which depresses hepatic glucose production. It may be possible to "selectively modulate vagus nerve activity and one day provide therapeutic advantages in diabetes and the metabolic syndrome." The most recent addition to this work is that the biochemical mechanisms by which the hypothalamus senses lipids are rapidly impaired in response to short-term overfeeding (Pocai et al., 2006).

The Role of Peroxisome Proliferator-activated Receptor (PPARγ) and Its Agonists, the Thiazolidinediones

The therapeutic use of the thiazolidinediones (TZDs) will be considered in Chapter 5, section 4. Here it is necessary to mention the biochemical basis for PPARγ action in terms of the continuity of the discussion on insulin resistance and the metabolic syndrome. This section, then, reflects the linkages emphasized in this book between the background of *biological events* that underlie the metabolic syndrome, and the geriatrician's *clinical application* of therapies designed to counteract components of this aging-related condition.

The peroxisome proliferator-activated receptors (PPARs) are members of the nuclear receptor family, and three isoforms, or variants, have been identified: PPARγ, PPARα, and PPARδ (Li and Palinski, 2006). (See Chapter 5, section 4.) These are transcription factors stimulated by components called ligands; natural (endogenous) ligands include fatty acids and prostaglandins; synthetic (therapeutic) ligands and agonists for PPARγ useful in treatment of diabetes are the thiazolidinediones (TZDs) (or glitazones) and fibrates. The PPARs then bind to specific response elements in enhancer sites of genes they regulate.

Systemic activation of PPARγ by glitazones occurs in adipocytes, and secondarily in liver and muscle. PPARγ has a wide range of actions on adipocytes: increasing the uptake and storage of free fatty acids (FFAs), lowering circulating FFAs, and gradually reducing liver and muscle steatosis, increasing insulin sensitivity, suppressing hepatic glucose production, and enhancing insulin-mediated glucose uptake and storage in muscle. In addition, adiponectin is increased, TNF-α is diminished, and NF-kB activation is reduced. The overall effects are to diminish insulin resistance as well as a variety of atherosclerotic mediators, including cytokines and phase reactants. However, fat is redistributed and the potential for weight gain exists (Chinetti et al., 2003; Moller and Kaufman, 2005).

A summary of views expressed recently that seek to reconcile the multifaceted aspects of the metabolic syndrome may be in order here. Certainly, a host of international organizations have struggled with this task. Here is a partial list (Haffner et al., 2006; Laakso and Kovanen, 2006; Roberts, 2006; Sarafidis and Nilsson, 2006):

- World Health Organization (WHO)
- European Group for the Study of Insulin Resistance (EGIR)

- American Heart Association (AHA)/National Heart, Lung and Blood Institute (NHLBI)
- National Cholesterol Education Program—Third Adult Treatment Panel (NCEP-HTP III)
- American Association of Clinical Endocrinology (American College of Endocrinology)
- International Diabetes Federation (IDF)
- American College of Endocrinology

Whereas all definitions of the metabolic syndrome include central obesity, insulin resistance and raised blood glucose, dyslipidemia, and elevated blood pressure, disagreement about the "driving force" or "common pathophysiologic mechanism" for its development still exists. As Sarafidis and Nilsson (2006) state in a "glance at the history of the metabolic syndrome," the acquisition of the components dating back more than 80 years is an "indication that there is probably a long way to go towards understanding its pathogenesis."

Haffner (2006b) considers evidence that elevation of CRP levels may be predictive of the metabolic syndrome, at least in women, and reviews the overall conceptual framework of the metabolic syndrome oriented toward a management approach. Thus, if the basic conditions are attributable to environmental causes (e.g., the approach of the NCEP-ATP III), lifestyle modifications are needed to reduce obesity and increase activity. If the syndrome is viewed primarily as a result of insulin resistance (WHO), then treatment needs to include insulin sensitizers in addition to lifestyle modification. And if inflammation is considered the underlying cause, all therapies noted above, together with other agents (especially statins), are likely to be used. As Grundy (2003) wrote, "atherosclerotic research has traditionally been divided between arterial researchers who focus on injurious aspects that provoke inflammation in the arterial wall and metabolic investigators. Increasingly, however, research of both types is moving toward the interface between inflammation and metabolic disturbances (a theme of this book), which is where arterial disease occurs." Indeed, as I said at the outset of this section, geriatricians are likely to take the "holistic approach" when they survey the overall aspects of the metabolic syndrome, and intervene appropriately to address the totality of its facets with the aim to "reduce all the risk factors simultaneously" (Grundy, 2006b). That is why I have emphasized interactive therapies: exercise in Chapter 4, section 6; statins, the TZDs in Chapter 5, sections 2 and 4; and healthy lifestyle practices throughout the text, especially in Chapter 6.

The widespread public health issues that have been aired about hypertension, dyslipidemia, obesity, and diabetes—hence the metabolic syndrome—seem to have raised *awareness* of this among health care providers. Perlin and Pogach (2006) asked whether "the unprecedented momentum created by publicizing evidence and measuring outcomes of metabolic conditions have resulted in translation of evidence into practice?" The papers they cite suggest that, despite some improvement in outcome, "millions of Americans remain at high risk" for the complications of the metabolic syndrome despite all that has been presented. They state that "clinical inertia"—the failure of health care providers to alter therapy in the face of clear indications for changes—needs to be overcome. Among their suggestions to do so, surely appraisal of self-knowledge and the quest to gain new and broader understanding for actionable interventions remain high priorities "to improve the prospect of healthier lives for tens of millions of Americans."

4. OSTEOPOROSIS

It is understandable that women in the postmenopausal period have dominated considerations relating to the inception of osteoporosis. Awareness is increasing, however, that osteoporosis in men is important, perhaps not as much based on clinical grounds (e.g., lower fracture incidence), but based on hormonal considerations (both androgen and estrogen); these aspects introduce a uniqueness and complexity beyond the traditional involutional or age-related state of bone loss—as occurs in postmenopausal women. For the purposes of my discussion, this section will relate almost entirely to osteoporosis in women, and Chapter 5, section 6—dealing with hormonal therapies—seems to be a more appropriate section to discuss the biology of osteoporosis in men.

As part of a brief introduction on the rationale for treatment, it is interesting to trace an attempt to define osteoporosis that is widely accepted. Osteoporosis, according to a consensus conference in 1993 is "characterized by low bone mass and the microarchitectural deterioration of bony tissue, with a consequent increase in bone fragility and susceptibility to fracture" (Nishizawa et al., 2005). In 2001 the definition was "significantly changed" at a National Institutes of Health (NIH) consensus conference, where osteoporosis was defined as "a skeletal disorder characterized by compromised bone strength predisposing a person to an increased risk of fracture" (Nishizawa et al., 2005). Despite these apparent differences, the *concept* of the two definitions relating to osteoporosis emerges clearly. Figure 4.7 vividly portrays images of the comparative conditions of normal and osteoporotic bone.

Perhaps for the purpose of *clinical* diagnosis and the approaches to management, osteoporosis is best defined by the WHO classification (T score) of reduced BMD that on a dual energy X-ray absorptiometry (DEXA) study is equal to or exceeds 2.5 standard deviations below the mean value observed in women at about 30 years of age. This is a single measurement, a quantitative figure, as a blood pressure reading may be an indication of hypertension, or a fasting blood glucose of diabetes, to mention conditions that I have already discussed. I would like to make a few comparisons between osteoporosis, which is almost always first manifest in postmenopausal women, and those other age-related conditions—diabetes and hypertension:

Osteoporosis is usually asymptomatic (indeed, often referred to as a silent epidemic) even when identified in an older patient after a DEXA study. However, factors contributing to bone loss may be traced back as far as early development, an aspect I have also discussed in relation to the metabolic syndrome with respect to diabetes and obesity. Osteoporosis becomes clinically manifest after a fragility fracture of the spine, hip, or lower radius. Likewise, hypertension and diabetes are silent, and may be asymptomatic even when picked up on physical examination or laboratory tests; or, a clinically related event may be their first manifestation.

The inception of osteoporosis in postmenopausal women may be delayed by healthy lifestyle practices begun years earlier, including diet, calcium intake, smoking cessation, and exercise, all collectively aimed at maintaining "bone health." These health practices, including weight control, may also reduce the risk of inception of hypertension or diabetes, or if already present, delay symptomatic expression.

Treatment may be begun to *prevent* fragility fractures in *osteoporotic* post-

Figure 4.7. Scanning electron micrographs of normal (A) and osteoporotic (B) human trabecular bone in the iliac crest. Note in the normal sample the volume of bone and the thick interconnecting trabecular plates. In osteoporosis, the bone volume is markedly reduced, the trabeculae are thin and rodlike, and one is disconnected because of excessive osteoclastic resorption. These figures illustrate the key aspects of osteoporosis: bone loss and microarchitectural decay contributing to the poor quality and fracture potential of the bone. Reproduced from Dempster DW, Shane E, Horbert W, Lindsay R. 1986. A simple method for correlative light and scanning electron microscopy of human iliac crest bone biopsies: Qualitative observations in normal and osteoporotic subjects. *Journal of Bone and Mineral Research* 1986, 1: 15–21, courtesy of Dr. David Dempster, with permission of the American Society of Bone and Mineral Research.

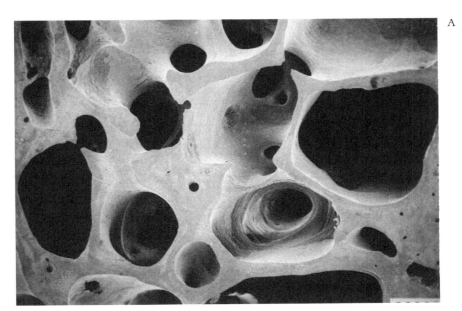

A

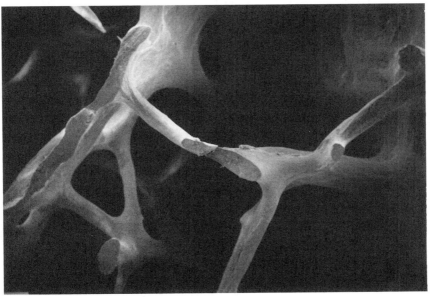

B

menopausal women at risk (Wilkins and Birge, 2005) and must be given to *treat* women presenting with the first evidence of a fragility fracture. Bisphosphonates, the standard of therapy, in general would not be given "prophylactically" for a DEXA reading that is in the *osteopenic* range (i.e., T score of −1 to −2.4). However, decisions concerning treatment in this situation perhaps rest as much (or more) on *fracture risk* than on the absolute T score. In fact, McClung (2005) suggested that "the diagnostic category of osteopenia in individual patients does not serve the clinical community well and should be abandoned," and Rosen and Brown (2005) do not consider osteopenia "a clinically useful term for it does not allow selection of patients at greatest risk; previous fractures far outweigh low BMD in this age group as a risk factor for subsequent fractures." Likewise, guidelines have now been sharply drawn as to what constitutes indications to treat elevated blood pressure to reduce the risk of a stroke or heart attack; the level of the fasting blood sugar is among many parameters that will guide treatment in terms of potential or overt type 2 diabetes.

Yet, there are important differences in the clinical appraisal of osteoporosis, hypertension, and diabetes. Measurements of blood pressure and blood glucose are far more "accessible" to physicians, done almost invariably during the patient encounter. Hence, initiation of appropriate therapies or recommendations for lifestyle-improving habits may be instituted at an earlier stage of hypertension or diabetes ("prevention"). Short of a positive DEXA study or an actual fracture, considerations of osteoporosis are based on several historical risk factors that have limited value—including age, history of fracture in patient's parents, the time of menopause inception—aspects that many physicians may not inquire about (except age) and patients may not consider worth mentioning. Thus, mutual recognition of the potential for osteoporosis prevention may be limited. Moreover, physician awareness of diagnostic and treatment modalities relating to osteoporosis is often deficient. This is in part because osteoporosis management often falls between specialties—including orthopaedic surgery, gynecology, endocrinology, or rheumatology, rather than where it belongs, within the province of the primary care physician. Other reasons for limited early intervention include failure of radiologists reviewing "routine" lateral chest X-rays to report evident thoracic vertebral fractures (Vogt et al., 2000; Mui et al., 2003), and physicians not starting bisphosphonate therapy even in patients who present with an actual vertebral fracture (Ross, 1997; Gehlbach et al., 2000), or especially at a younger age with a wrist fracture (Freedman et al., 2000). Patient and physician awareness of the pa-

Figure 4.8. © The New Yorker Collection 2006 David Sipress from cartoonbank.com.
All rights reserved.

tient's diminishing height is an important manifestation—often overlooked—
that is likely to be related to thoracic vertebral compression fractures, as the
cartoon from the *New Yorker* shows so well, correlated with age (Fig. 4.8).
Compliance with osteoporosis medications on the part of the patient is also a
problem—a trend for many asymptomatic or chronic conditions. In one study
by Solomon et al. (2005a) almost half of persons 65 years or older who initiated therapy for osteoporosis were not continuing to fill prescriptions. The
problem of compliance is especially tenuous with bisphosphonates because
of complex dosing schedules and the potential for adverse effects, although
recent introduction of weekly (risedronate, alendronate) or monthly (ibandronate) doses may improve compliance, and yet "missed doses or improper dosing may have greater consequences with extended dosing intervals" (Emkey
and Ettinger, 2006). Physician–patient educational interactions and communication become especially important in beginning treatment of osteoporosis

before fragility fractures occur and in maintaining patient compliance with bisphosphonate therapy (Cramer and Silverman, 2006; Gold and McClung, 2006).

In this age-related condition, almost as widespread as hypertension or diabetes, appropriate treatment for osteoporosis may be more limited and inconsistent, or it may not be initiated in a timely way (Melton, 1995). There is even a new situation relating to use of estrogens for *short-term* relief of menopausal symptoms rather than deriving "bone benefits" from its more *long-term* use based on "fear of cardiovascular disease, but more especially of breast cancer" (Stopeck, 2005). All these aspects relating to the timing and duration of osteoporosis treatment have important implications for public health: while deaths from cardiovascular disease are declining, projections are for a markedly increased incidence of osteoporosis and related fragility fractures, especially hip fractures, with attendant high mortality in the first year (at least 20 percent), serious impact on a formerly independent lifestyle, and the potential for institutionalization in a long-term care facility, all diminishing quality of life and markedly augmenting health care costs measured in the multiple billions (Braithwaite et al., 2003; Reginster and Burlet, 2006; Gass and Dawson-Hughes, 2006). Although hypertension and diabetes are risk factors for acute and fatal outcomes, osteoporosis is important for the geriatrician because it is one of the chronic, nonfatal conditions that are increasingly prevalent in our aging society and thus in clinical practice.

Much as I did in developing discussion points relevant to the metabolic syndrome, osteoporosis can be considered here based on certain defining issues:

- the biology of endochondral bone formation
- distinctions between cortical and trabecular bone
- "bone health" across the generations
- cellular mechanisms for bone formation and resorption
- inflammatory basis for osteoporosis: linkages with atherosclerosis
- assessment of osteoporosis in geriatric practice

It is a tribute to the basic scientists working on the biology of bone that so much new information has emerged in little more than a decade, for "the skeleton is an organ of unappreciated complexities" (Karsenty, 2003). It is within the scope of laboratory studies of cell culture behavior, the effects of hormonal mediators on these cells, and especially techniques of molecular genetics using mouse models of enhanced or deleted genes, that new insights have been gained on the complex processes of bone development and the interactions of osteoblasts and osteoclasts in bone formation and resorption. This is what Karsenty (1999) calls the "genetic transformation of bone biology." I will concentrate first on cellular events that regulate skeletal development to introduce nomenclature, or "the cast of characters" (Karsenty and Wagner, 2002). Continuing the theme that underlies this text, it seems desirable for the geriatrician to gain some familiarity with the basic mechanisms of bone accrual and loss, the latter dominant in osteoporosis, so that the science can be linked to clinical management and therapies. As Lawrence Raisz (2005) put it, reviewing "concepts, conflicts, and prospects, it is now possible to diagnose osteoporosis, assess fracture risk, and reduce that risk with antiresorptive or other available therapies. However, new and more effective approaches are likely to emerge from a better understanding of the regulators of bone cell function."

The Biology of Endochondral Bone Formation

Most bone is formed from a cartilage model, a process termed endochondral ossification. (Bones of the skull develop directly from mesenchymal cells, a process called intramembranous bone formation.) Mesenchymal cells are the progenitor, or stem cells, from which the cells involved in endochondral ossification are derived. Because the mesenchymal cells are multipotential, they may differentiate into chondrocytes, osteoblasts, myocytes, or adipocytes. This is important because molecular modulation of these mesenchymal cells by local or systemic factors will enhance the development of one cell rather than another, and will ultimately determine the final identity of the cell line and hence the tissue characteristics. The processes that influence *chondrogenesis*—or the earliest phases of skeletal development—are extremely complex, and some of these have been set forth recently by Goldring et al. (2006). My intent in this brief presentation is to shift the emphasis from development—although this needs to be introduced—to the unfolding events that lead to *osteogenesis* as ultimately more germane to geriatric bioscience. Cohen (2006) has brought together in the "new bone biology" the pathologic, molecular, and clinical associations of disorders of cartilage *and* bone.

In the developing limb mesenchymal cells destined to become chondrocytes condense, and under the control of transcription factors of the Sox family, differentiate into type II collagen-producing cells; they proliferate and hypertrophy in the extracellular calcified cartilage matrix they produce (Smits et al., 2004) and as hypertrophic chondrocytes express type X collagen as well

(Provot and Schipani, 2005). The balance between chondrocyte proliferation and hypertrophy is controlled by a negative feedback loop involving Indian hedgehog (Ihh), which induces parathyroid hormone-related peptide (PTHrP), formed in the proliferating chondrocyte, to signal to its receptor (PTHR1) to suppress differentiation to hypertrophic cells (Lai and Mitchell, 2005). The negative feedback means that PTHrP stimulation of its receptors in prehypertrophic chondrocytes can inhibit Ihh expression. PTHrP resembles parathyroid hormone (PTH) but circulates as a hormone that causes hypercalcemia in patients with certain cancers. Indeed, PTHrP "should not be regarded as a hormone at all in its normal role"; rather it functions locally on cells (paracrine effects). PTHrP is produced very early in cells of mesenchymal lineage, acts on chondrocytes, as noted, and also promotes osteoblast development and reduces cell death (apoptosis) (Martin, 2005). Fibroblast growth factor receptor (FGFR3) may suppress PTHrP from promoting further Ihh expression (Amizuka et al., 2004). Bone morphogenetic protein (BMP-2) stimulates the differentiation of mesenchymal cells into osteoblasts by binding to its receptor, a serine/threonine kinase, which phosphorylates and activates the intracellular signaling molecules Smad 1 and Smad 5 (Mundy, 2006; Phimphilai et al., 2006). This in turn leads to the expression of transcription factor Cbfa/Runx2 (that belongs to the runt transcription factor family), and its cofactor core-binding factor β (CBF-β) required for hypertrophic chondrocytes to release vascular endothelial growth factor (VEGF) (Provot and Schipani, 2005). This growth factor induces vascular invasion and entry of the actual bone-forming cells—the osteoblasts—also under the control of Cbfa/Runx2. Runx2 functions to promote chondrocyte maturation by inducing Ihh expression, and is essential for the commitment of multipotent mesenchymal cells into the osteoblastic lineage while inhibiting adipocyte differentiation (Komori, 2005). Other factors secreted from the hypertrophic chondrocytes, such as Wnt/LRP 5 (see below), appear to act in synergy with and downstream from Ihh to induce osteogenesis. Cbfa1 continues to be necessary for osteoblast differentiation and function, and controls the expression of other downstream genes that encode osteoblast-specific transcription factors, in particular, a zinc finger–containing transcription factor called osterix (Nakashima et al., 2002). Thus, chondrocyte evolution in cartilage is clearly a key part of an extensive sequence and progression of events that ultimately lead to the replacement of cartilage by bone.

An additional genetic process in osteogenesis involves a gene-encoding low-density lipoprotein-receptor–related protein (LRP). Apolipoprotein E and

factors called Wnt are ligands (signals) for LRP expression in skeletal development. Wnt molecules exert their functions by activating several distinct
intracellular pathways, including one mediated by β-catenin, known as the
"canonical pathway" (reviewed by Hu et al., 2005; Krishnan et al., 2006). In
this pathway, Wnt proteins signal through the Frizzled family of receptors and
LRPs. Patel and Karsenty (2002) commented that although one member of the
LRP family "appeared to have none of the hallmarks of an interesting gene,
it has emerged as an important molecule." Indeed, "LRP5 and Wnt signaling
constitute a union made for bone" (Johnson et al., 2004). In humans, LRP5
polymorphisms contribute to normal variation in BMD (Koay et al., 2004), but
there are mutations in the LRP5 gene that result in diseases associated with
low bone mass, fractures, and severe ocular pathology (Gong et al., 2001); or
conversely, high bone mass (Boyden et al., 2002; Little et al., 2002; Babij et al.,
2003). The Wnt-LRP5 signaling pathway is likely to be explored more extensively because of its actions on osteoblast functions; perhaps this will result
in the development of novel therapeutic approaches for osteoporosis. Indeed,
Wnt signaling may direct mesenchmal cells to osteoblast differentiation and
enhanced ossification and suppression of chondrocyte formation (Day et al.,
2005; Glass et al., 2005). Yet, in their review, Ferrari et al. (2005) add a word of
caution in that disrupting "the delicate balance between the multiple components of the Wnt-LRP5 system could result in osteosclerosis and extra-skeletal
manifestations as well, including changes in lipid and glucose metabolism."
Harada and Rodan (2003) present an extensivoe write-up on new genes involved in bone formation and high bone mass. In line with the aims of this
text, these authors point out that the mechanisms of the LRP5 mutation that
enhance bone formation "need to be understood in terms of the tissue specificity and how scientists might try to mimic it pharmacologically."

Distinctions between Cortical and Trabecular Bone

This explanation follows from Parfitt's (1988) contribution to the text *Osteoporosis*, edited by Riggs and Melton. Two types of bone structure are present
in the adult skeleton: cortical (or compacta) bone and trabecular (cancellous,
spongiosa) bone. Cortical bone is more dense and provides about three-quar-
ters of the total skeletal mass, but only one-third of the total surface. It forms
the outer wall of all bones—mostly the shafts of the long bones, called the appendicular skeleton. The surfaces of cortical bone are those that face the marrow space (endosteal) and the outer (periosteal) surface. Postmenopausally,

cortical bone is lost more rapidly from the endosteal surface than it is gained periosteally, with a marked reduction in thickness of the shafts, but the extent of this change varies by skeletal site (e.g., greater in the lower extremities, greater in the bone nearer the joint—metaphysis—rather than the shaft—diaphysis). Critical thinning as a result of endosteal resorption may be viewed in radiographs of the hands.

Trabecular bone has a high degree of porosity, proximity to bone marrow and blood supply, and a high degree of metabolic turnover, perhaps accounting for its initially accelerated loss (resorption) in those sites where it is present: the vertebral (axial) skeleton and the extreme distal end of the radius. A distal radius or Colles' fracture is often the first presentation in early postmenopausal women with hitherto unsuspected or undiagnosed osteoporosis. Figure 4.7 shows thin, disconnected trabeculae and the potential for fragility fractures in osteoporotic bone.

"Bone Health" across the Generations

In a paper on "Bone health across the generations" for *Maturitas* (Hamerman, 2005a), I mentioned that bone health is not a measurable entity but rather an overall state of bone quality that applies to bone, much as "general health" applies to overall well-being. Seeman and Delmas (2006) discuss bone quality as the "process of bone modeling and remodeling throughout life to adapt the material composition and structure of bone to prevailing loads." Sambrook and Cooper (2006) also list bone macroarchitecture (shape and geometry), bone microarchitechture (both cortical and trabecular), matrix and mineral composition, degree of mineralization, microdamage accumulation, and the rate of bone turnover as among the additional aspects that influence bone quality. This section is not strictly biological, the overarching theme of this text, but I thought it appropriate to present a concept of bone health because of my prior comments relating to the early origins of disease.

In the *Maturitas* article I wrote that "in the realm of osteoporosis prevention it is necessary for physicians to review those *early* aspects that might influence the *present* state of bone health in their female patients and to follow these women over time." In fact, I had experience in organizing a unique means to do so by virtue of a Bone Center representing an interdisciplinary model that enlisted health care providers from three major departments—pediatrics, obstetrics/gynecology, and medicine—and their respective divisions of adolescent medicine, reproductive endocrinology, and geriatrics and en-

TABLE 4.2
Guide for the health care provider assessing risk factors for
osteoporosis in postmenpausal women based on age transitions:
bone health across the generations

Adolescence
 Conditions limiting attainment of peak bone mass or predisposing to accelerated bone loss
  Late age at onset of menarche
  Personal health practices
   Limited dietary calcium and protein intake
   Cola drinks
   Smoking
   Amenorrhea secondary to intense exercise (ballet, marathon), eating disorders (anorexia
    nervosa)
   DepoProvera birth control injections
  Heritable disorders
   Polycystic ovary syndrome
   Turner syndrome
   Growth hormone deficiency
   Idiopathic juvenile osteoporosis
  Acquired disorders
   Disease requiring corticosteroid use
   Cancer
The reproductive years
 Parity and lactation seem not to be accompanied by persistent bone loss, although the
  trend is for vertebral (trabecular) bone loss.
The perimenopause
 A time of accelerated bone loss. Earlier age at menopause and surgically induced
  menopause may enhance bone loss up to age 55 years. A history of maternal fractures is
  important, as is use and duration, or nonuse, of hormone replacement therapy.
The postmenopausal years
 Age
 History of fracture
 Years since menopause
 Maternal history of osteoporosis and fractures
 Body mass index
 Medications
 Cigarette smoking
 Exercise

Source: Reprinted from Hamerman (2005a), © 2005, with permission from Elsevier.

docrinology. Because of our educational and clinical interchange, the status of bone health could be surveyed in women across the generations as the patients presented for evaluation in adolescence, the reproductive years, the menopause inception, or at more advanced age decades after the menopause.

A useful knowledge base, although not exhaustive, is summarized in Table 4.2. Recall that the previous section dealt at length with obesity and its negative features. It is worthwhile to point out that in postmenopausal women "body weight is one of the most significant factors that determines bone density and fracture risk"; in this case, higher body weight may mean higher BMD, a "biological tradeoff" between a greater risk for the cardiovascular system (obesity) and a lower risk for future fracture occurrence (thinness) (Ott, 2004).

Cellular Mechanisms for Bone Formation and Resorption

Although the molecular aspects of bone formation by osteoblasts and bone resorption by osteoclasts—a lifelong process of tissue renewal called remodeling—are complex and perhaps unfamiliar to geriatricians, some awareness of remodeling is needed as advances in the science of bone contribute to therapeutic breakthroughs applied to the treatment of osteoporosis in the next decade (Ross and Christiano, 2006).

One of the two functions of the osteoblast is bone formation. Bone matrix is composed of collagen type 1 (note the transition from type 2 collagen in cartilage), several growth factors (see Chapter 4, section 7), and a variety of noncollagenous proteins that act as the site for mineralization by hydroxyapatite, a calcium-containing salt. One noncollagenous protein is osteocalcin, an osteoblast-specific protein. It contains γ-carboxylated glutamate residues called gla that bind and incorporate calcium into hydroxyapatite crystals (Bügel, 2003). Osteocalcin is of particular interest because vitamin K is a coenzyme for glutamate carboxylase, which converts glutamate to γ-carboxyglutamate. Spontaneous or warfarin-induced vitamin K deficiency results in undercarboxylated osteocalcin, with low mineral binding; the presence of undercarboxylated osteocalcin in the circulation has been associated with low BMD and a risk for fragility fractures (Siebel et al., 1997; Olson, 2000). Conversely, a review and meta-analysis of randomized control trials of adults taking oral vitamin K preparations for longer than six months suggested that bone loss was reduced, with "a strong effect on incident fractures among Japanese patients" (Cockayne et al., 2006). Osteocalcin may be measured in the blood as a clinical marker of bone formation. Bone matrix and the arterial wall share many features, and similarities between the biological mechanisms of osteoporosis and atherosclerosis have often been cited (Hamerman, 2005b; Tanko et al., 2005a). The bone matrix Gla proteins figure prominently in the control of calcification in the arterial wall. Another noncollagenous protein is osteopontin "thought to act as a bridge between the cell surface and hydroxyapatite crystals" (Karsenty, 1999). Ovariectomized mice that are also osteopontin deficient seem protected from accelerated bone loss, perhaps because of altered osteoclast function due to impaired cell surface activity of $\alpha V\beta 3$ integrin, and the hyaluronic acid receptor (CD 44) needed to bind osteoclasts to bone.

Osteopontin is also a major component of the intima of human arteries, but whether it promotes or limits calcification in the arterial wall is not clear.

Besides bone formation, the other major function of the osteoblast is to control osteoclastic bone resorption by releasing mediators that influence osteoclast development and functions (i.e., osteoclastogenesis) (Fig. 4.9). The system of interaction I am about to describe focuses on *local regulation* of bone remodeling by factors secreted by bone cells; after that, I will discuss *systemic regulation* of bone remodeling by hormones "that often impinge upon the expression of local factors" (Chien and Karsenty, 2005).

Local Regulation of Bone Remodeling

The osteoclast is derived from the monocyte/macrophage hematopoietic cell lineage and differentiates from a common progenitor under the influence of the transcription factor PU-1 and the early response gene c-Fos (Karsenty, 1999; Zaidi et al., 2003). Maturation of the osteoclast depends on two mediators released by the osteoblast/stromal cell "which are necessary and sufficient for ostoclastogenesis" (Boyle et al., 2003; Xing and Boyce, 2005). One of these mediators is macrophage colony-stimulating factor-1 (M-CSF). An indication of the crucial role of M-CSF in osteoclast development is shown by what happens with a loss-of-function mutation of the M-CSF gene in mice designated *op/op* (Pollard and Stanley, 1996). These mice exhibit a dramatic increase in bone mass, or osteopetrosis, due to the absence of functional osteoclasts. The M-CSF released by osteoblasts binds to the osteoclast receptor c-fms, with activation of phosphatidylinositol-3-kinase (PI3K)—important in osteoclast differentiation and bone attachment (Golden and Insogna, 2004). (We have encountered PI3K before in terms of the insulin receptor substrate activation.)

A second mediator arising from the osteoblast that influences osteoclast maturation is the *receptor activator of nuclear factor k-B*, called RANK-ligand, a member of the TNF superfamily that binds to a specific receptor on the osteoclast called RANK. A signaling cascade then occurs involving TNF receptor-associated factor (TRAF 6), which activates downstream signaling pathways, including c-Jun N-terminal kinase (JNK) and NF-kB (Ross, 2000; Steeve et al., 2004; Teitelbaum, 2004a). Chapter 4, section 3 discussed activation of the JNK system by inflammatory cytokines in states of obesity and insulin resistance, with "endoplasmic reticulum stress" and pancreatic β-cell apoptosis. As Liu and Rondinone (2005) point out, "JNK may be a promising drug target for its

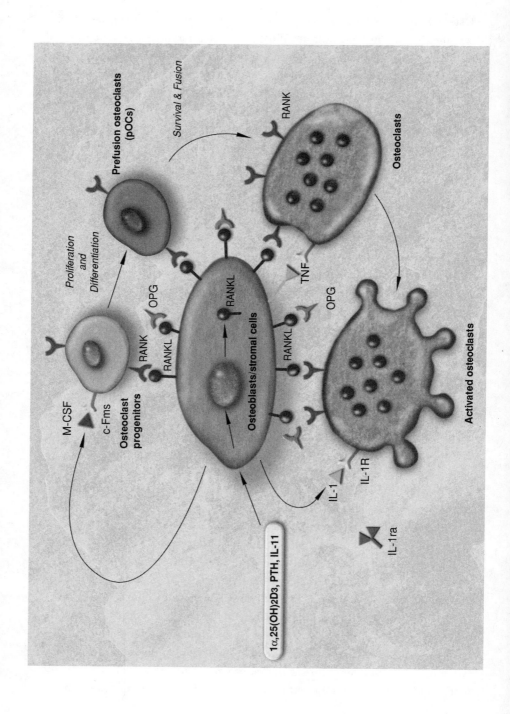

M-CSF

c-Fms

Osteoclast progenitors

RANK

RANKL

OPG

RANK

Proliferation and Differentiation

Prefusion osteoclasts (pOCs)

Survival & Fusion

RANK

Osteoclasts

RANKL

Osteoblasts/stromal cells

TNF

OPG

RANKL

IL-1

IL-1R

IL-1ra

Activated osteoclasts

1α,25(OH)2D3, PTH, IL-11

Figure 4.9. Osteoblast/stromal cell involvement in osteoclastogenesis through cell–cell interactions with osteoclast progenitor cells. Osteoclast progenitors derived from hematopoietic cells of the monocyte-macrophage lineage evolve under their genetic direction and from factors arising from the osteoblast: macrophage colony-stimulating factor (M-CSF) and receptor activator of nuclear factor k-B ligand (RANK-L); these bind to c-Fms and to RANK receptors on the osteoclast, respectively. The osteoblast also releases osteoprotegerin (OPG), which acts as a decoy to block RANK-L-RANK binding, thus potentially limiting osteoclastogenesis. Additional cytokines arising from the osteoblasts that promote multinucleate osteoclasts include tumor necrosis factor α (TNF-α), and interleukin-1α, (IL-1α) that attach to their respective receptors on the osteoclast. Osteoclasts induced by TNF-α form resorption pits on bone surfaces in the presence of IL-1α, demonstrating the key independent role of these cytokines in bone resorption due to inflammation. Interleukin-1 receptor antagonist (IL-1ra) can inhibit IL-1 signals (Suda et al., 2001). Reprinted from Suda T, Takahashi N, Udagawa N, Jimi E, Gillespie M, Martin TJ. 1999. Modulation of osteoclast differentiation and function by the new members of the tumor necrosis factor receptor and ligand families. *Endocrine Reviews* 20: 345–57.

Copyright 1999, The Endocrine Society. (See color gallery.)

roles in other inflammatory conditions, including stroke, rheumatoid arthritis, Alzheimer's disease," and, of course, osteoporosis (see Chapter 5, section 5 on therapies). c-Jun is paired with c-Fos to form an activator protein-1 (AP-1) complex, an essential transcription factor for osteoclast development. c-Jun also interacts with nuclear factor of activated T cells (NFAT), a process crucial for RANK L-induced osteoclast differentiation (Ikeda et al., 2004). RANK L requires costimulatory signals provided by immunoreceptor tyrosine-based activation motifs (ITAMS) to further trigger osteoclast differentiation (Chien and Karsenty, 2005). Also, some signaling cascades are mediated by protein kinases and a role for the receptor, dendritic cell-specific transmembrane protein (DC-STAMP), which promotes the uniquely multinucleated osteoclasts. The osteoclast resorbs bone by rearrangement of its actin cytoskeleton and forms a tight junction, by way of its integrins, between its basal membrane and the bone surface. Within a sealed compartment acidification takes place by the export of hydrogen ions generated by an adenosine triphosphate (ATP) complex and lytic enzymes: tartrate-resistant acid phosphatase (TRAP) and cathespin K are secreted into the bone (Blair et al., 2005). Products of bone matrix dissolution—collagen fragments and calcium and phosphate—are processed within the osteoclast (vacuolar transcytosis) and released into the circulation. One of the bone-resorption products—N-terminal telopeptide of collagen (NTX)—may be measured in the urine as a clinical marker of bone resorption.

An additional osteoblast-secreted factor is called osteoprotegerin (OPG), a TNF receptor-related protein that can act as a *decoy* to block RANK L and prevent its union with the RANK receptor on the osteoclast (see Fig. 4.9). In experimental studies in mice, gene deletion of OPG resulted in overexpression of RANK L, osteoclastic bone resorption, and profound osteoporosis. But the converse is also important; OPG enhancement in mice blocked RANK L from binding to RANK, limited osteoclastic bone resorption, and resulted in osteopetrosis (Simonet et al., 1997). The RANK L-OPG system actually "empowers" the osteoblast to control osteoclastic functions to enhance (RANK L, M-CSF) or diminish (OPG) bone resorption. OPG has been tested in clinical trials for osteoporosis (Bekker et al., 2001), and at the other end of the age spectrum, has been used in the treatment of juvenile Paget disease of bone (Cundy et al., 2005) in which "abnormal arterial calcification" has been observed (Laroche and Delmotte, 2005). Yet OPG therapy "seems to have fallen by the wayside, partly because neutralizing antibodies to it could develop in patients" (Whyte, 2006). So it is encouraging to observe a new entry into the therapy of osteoporosis that is "OPG-like," namely, *denosumab* (formerly known as AMG 162, Amgen), a fully human monoclonal antibody (IgG_2) that binds to RANK L with high affinity and specifically blocks the interaction of RANK L with RANK, "mimicking the endogenous effects of OPG" (McClung et al., 2006). Denosumab has a significantly longer half-life than OPG that allows less frequent administration (Kosteniuk, 2005); it is administered subcutaneously either every three months or every six months, as was done in a one-year trial in postmenopausal women with low BMD. As Whyte (2006) points out, this treatment may go far to improve the traditional poor compliance with osteoporosis medications. Yet in the close interplay between bone and the immune system that led Walsh et al. (2006a) to write of "osteoimmunology," Whyte (2006) expressed "concern that denosumab could globally disrupt the signaling pathway that involves RANK L, OPG, RANK, and NF-kB, since RANK is expressed on cells other than osteoclast precursors, including dendritic cells and T and B cells."

I have mentioned my interest in the relationship between osteoporosis and atherosclerosis even though two different "systems"—bone and blood vessel—are involved. But many factors also make their association (what I called their biological linkages) important in considering aging-related conditions. These linkages include (1) the clinical association of vascular calcification and low bone density; (2) noncollagenous proteins present in bone and the arterial wall: the Gla-proteins and osteocalcin, osteopontin, bone morphogenetic protein, and OPG, among others, which appear to play roles in vascular and bone matrix calcification. In particular, mice with deletion of the OPG gene develop arterial calcification and osteoporosis with multiple fractures (Bucay et al., 1998); (3) calcification and bone formation in the arterial intima by way of transcription factors Cbfa, Msx2, and Sox9 that also act on bone (Tyson et al., 2003); (4) the key roles of macrophage-derived cells in intima and in bone; (e) the array of mediators that affect bone and blood vessel (IL-6, TNF-α, M-CSF-1, adiponectin, estrogen); (5) nitric oxide (NO) produced by endothelial NO synthase, as well as osteoblastic NO, mediate the anabolic estrogen response and mechanical stretch response in bone, and inhibit osteoclastogenesis (Blair et al., 2005). Indeed a final point is a "new and exciting area in bone biology" (Ross and Christiano, 2006): a subset of osteoblasts occupies a hematopoietic stem cell "niche" in bone, and those osteoblasts interact with endothelial cells which reside in a vascular "niche." These side-by-side relationships have important implications for cellular self-renewal—perhaps for both bone and vascular functions that may apply to "human diseases—including hematopoietic and bone-related disorders" (Yin and Li, 2006). Moreover, Veillette and von Schroeder (2004) described "a mutual signaling axis" between endothelial cells and osteoblasts, sharing receptors and cell-derived factors, especially endothelin-1 and vascular endothelial growth factors during bone formation and repair. These cell–cell interactions are relevant for the atherosclerosis–osteoporosis interchanges discussed so often in this section and also may pertain to bone metastases mentioned in Chapter 4, section 7.

OPG involvement is especially prominent in this array of interactive (vascular-bone) factors. OPG appears to be a regulator of calcification in the vessel wall. The arteries exhibiting calcification in OPG-deleted mice are normally sites of endogenous OPG expression (Simonet et al., 1997), suggesting that OPG may protect arteries from pathologic calcification (Min et al., 2000). Indeed, Schoppet et al. (2002) note that "OPG could represent the long sought-after molecular link between arterial calcification and bone resorption, which underlies the clinical coincidence of vascular disease and osteoporosis." The mechanism by which OPG regulates calcification in arteries is not known.

Systemic Influences on Bone Remodeling

Besides the local factors affecting osteoblast–osteoclast interactions that I have discussed, another level of regulation of bone remodeling is "exerted systemically by hormones, such as sex steroids, PTH, and leptin, and by neural inputs" (Fu et al., 2005). When parathyroid hormone (PTH) secretion is sustained it binds to its specific receptor (PTHR1) on the osteoblasts and upregulates RANK L expression and decreases OPG, enhancing osteoclastic bone resorption. Yet as is now observed clinically, intermittent PTH injection of the therapeutic antiresorptive agent teriparatide enhances bone formation and is used in treatment of severe osteoporosis, although the mechanism is poorly understood (Tashjian and Gagel, 2006).

The roles of estrogen in bone enhancement, and estrogen deficiency in bone loss, merit great attention, but it seems that Karsenty's prediction in 1999, that estrogen action will be an extremely active area of research in the future, perhaps "the most important issue in the field of endocrine control of bone biology," may now be more restricted because of the risk of cardiovascular disease with conjugated estrogens, and thus their curtailed use. I will discuss estrogen action on bones in the section on inflammation that follows.

Leptin is perhaps the most recently described systemic regulator of bone remodeling, for it was only in 2000 that the work of Karsenty, Ducy, and colleagues brought leptin—previously widely studied in appetite control and reproduction—to the field of bone (Ducy et al., 2000). Their initial observation of a high bone mass in obese, leptin-deficient mice (*ob/ob*) was striking in view of coexisting hypogonadism and hypercortisolism, conditions that should predispose these mice to low bone mass. Yet the story that emerged of leptin's action on bone was independent of leptin's other effects on appetite and reproduction, and moving to the present, the research, like many discoveries, has opened up new and fascinating complexities.

Leptin affects both arms of bone remodeling—bone formation and resorption by way of homeostatic and neural mechanisms that act on the osteoblast, which in turn controls osteoclast function, as we have seen. The antiosteogenic arm of leptin evolves from its binding to receptors on hypothalamic neurons which relay signals via the β-adrenergic impulses of the sympathetic nervous system (SNS) to their receptors on the osteoblast. As a result, osteoblast proliferation is inhibited, and RANK L is expressed, with enhancement of osteoclastogenesis (Elefteriou et al., 2005). This arm is counterbalanced by leptin-induced expression of a hypothalamic neuropeptide anorexigin (Fig. 4.6) called cocaine- and amphetamine-regulated transcript, or CART, which appears to inhibit the expression of RANK L in osteoblasts. Thus, the original observation of high bone density in leptin-deficient *ob/ob* mice derives from absence of SNS activation of osteoblastic RANK L despite concomitant downregulation of CART (Jobst et al., 2006). I will go into more detail below about leptin's complex role on bone.

The dual arms that leptin seems to display in remodeling suggested to the Karsenty group a *homeostatic* function of the type controlled by circadian (diurnal) rhythms (Fu et al., 2005). Genes that control these rhythms are generally called *clock* genes. Clock genes seem to mediate two opposing mechanisms: enhancing the leptin-dependent SNS inhibition of bone formation by suppressing cell cycle regulatory components (such as the *c-Myc* promoter), thereby limiting osteoblast proliferation; and alternatively, activating AP-1 genes that stimulate proliferation of osteoblasts. As de Crombrugghe (2005) states in his accompanying editorial, "work by Fu et al. (2005) provides evidence that the inhibition of osteoblast proliferation by clock proteins is the dominant effect." He goes on, "perhaps the decrease in bone mass produced by leptin signaling is such a critical function that both the inhibition of osteoblast proliferation and the increased degradation of bone by osteoclasts need to be buffered and counterbalanced by opposing mechanisms."

From a clinical point of view, leptin has not gained prominence in terms of correlation with BMD, and assays of leptin blood levels are largely restricted to research studies. Perhaps the "earliest" assays of leptin in terms of the "life cycle" are those in umbilical cord venous blood. In line with what I wrote on the early origins of disease in Chapter 4, section 3, Cooper's group (Javaid et al., 2005) at Southampton in the United Kingdom continues to explore environmental influences on bone development during intrauterine and early postnatal life. They reported earlier, regarding a group of term infants, that maternal birth weight, smoking, fat mass, and exercise during late pregnancy independently predicted neonatal bone mass. Now to this is added a strong positive association between umbilical vein leptin predicting the size of the neonatal skeleton and its estimated volumetric mineral density. Perhaps maternal fat stores mediate their effect on fetal bone accrual through variations in fetal leptin concentration.

High fat mass is associated with high leptin levels, as I've noted, but clinical studies in adults attempting to correlate leptin blood levels with bone mass, corrected for fat mass, have been inconclusive. One of the problems is that total fat mass per se appears to be an important determinant of BMD in postmenopausal women (Douchi et al., 2000). Leptin levels may also depend on ethnicity, and sex: for after adjustment for BMI, leptin correlated with BMD in white postmenopausal women but not in black postmenopausal women (Arabi, 2005; Jen et al., 2005), and in postmenopausal women but not in men (Weiss et al., 2006). It is also possible that leptin resistance or leptin receptor insensitivity would contribute to high BMD in *obese* women, especially diabetics with concurrent insulin resistance. Elevated leptin levels reflected cardiovascular morbidity with ventricular hypertrophy, and evidence of "autonomic overactivity." At least in the studies in mice, leptin's role to promote bone loss by way of β-adrenergic action may thus be countered by the clinical effectiveness of β blockers that diminish "autonomic overactivity," suppress bone loss, reduce the risk of fractures, and improve cardiac function (Pasco et al., 2004; Schlienger et al., 2004).

The protective action of obesity against bone loss and osteoporotic fractures is due to many factors, including mechanical loading, increased aromatization of androgens to estrogens by adipose tissue, and increased insulin levels (Khosla, 2002). The *clinical* influence of leptin on bone mass, which could be related to higher leptin levels in obesity, has not been resolved. The clinical relationship between fat mass, bone mass, and leptin turns out to be a complex one (Cock and Auwerx, 2003); it is not easily resolved because many of the findings derive from animal studies (Khosla, 2002; Reid et al., 2006). It

appears that leptin decreased bone marrow adipocytes and enhanced osteoblasts from precursor cells, increasing bone mass. Yet Hamrick et al. (2005)
reported that this occurred in leptin-deficient *ob/ob* mice that are highly sen-
sitive to leptin and was not observed in leptin replete mice. So leptin sensitivity may be an issue, and of course, human obesity raises concerns about
leptin resistance. There is also the issue of whether leptin's effects on bone
are observed from *systemic* (peripheral) administration—where it seemed to
enhance bone density; or from *central* (intracerebroventricular) administration—where leptin was shown to diminish bone density by the mechanisms
of β-adrenergic activation of osteoblast's release of RANK L, promoting osteoclastogenesis, as I discussed previously. Elefteriou et al. (2005) showed that
this sympathetic activation of RANK L was countered by the hypothalamic
neuropeptide CART, which inhibits bone resorption, as I mentioned.

The paradox, then, was whether leptin had opposing peripheral and central actions. I go into this in some detail because we are on the threshold of an
unfolding story with important implications for leptin's roles in a more general sense. In fact, Khosla (2002) points out the fascinating potential biological significance of leptin based also on a broader earlier perspective by Flier
(1998). In its integration of energy homeostasis and neuroendocrine functions,
leptin is a "starvation signal:" in times of deprivation, peripheral leptin levels
fall, the HPA axis is activated by this stress response (Chapter 4, section 1),
and reproductive and somatotrophin hormones decline, with bone loss.

Further, under conditions of high caloric intake, the "thrifty genotype"
(Chapter 4, section 3) promotes increasing energy intake and fat storage, and
leptin levels rise peripherally and enhance bone formation. Central resistance
to leptin exists because of reduced entry from the blood to the central nervous
system (CNS), and thus diminished SNS activation of bone loss. Admittedly,
all this is an extremely complex way to account for leptin's dual (peripheral,
central) effects on bone, and time will test its validity.

Fat mass is inversely related to blood adiponectin levels, and as I've discussed regarding the metabolic syndrome, low adiponectin levels associated
with obesity are a cardiovascular risk factor but a "tradeoff" for possible higher
BMD (Ott, 2004). In a group of healthy pre- and postmenopausal women, Jurimae et al. (2005) recently found that circulating adiponectin levels had an inverse association with BMD independent of other body compositional factors.
Whether low adiponectin levels in women with a high fat mass adds to our
understanding of apparent higher BMD that may be observed in this situation
is not clear. In the present state of our knowledge, higher plasma adiponectin

seems to diminish cardiovascular risk and may be a "trade-off" if a tendency for lower BMD exists.

An Inflammatory Basis for Osteoporosis: Linkages with Atherosclerosis

Although inflammation is a dominant aspect in atherogenesis, it may be much less invoked in the pathogenesis of osteoporosis. Yet inflammation is likely to be one of the unifying processes that influence atherogenesis and bone loss (Hamerman, 2005b). Many of the inflammatory mediators driving atherogenesis in the arterial wall are known to be in the circulation as markers of cardiovascular risk (Willerson and Ridker, 2004) and could gain access to bone, where, with local cytokines, they enhance osteoblastic release of factors that in turn promote osteoclastogenesis. Horowitz (2003) emphasized the roles cytokines play in regulating osteoblast functions in bone and pointed out that bone-resorbing osteoclasts need to be "kept under control, in part by suppressive factors," of which OPG is a good example. In turn, mediators from these bone cell activities would gain access to cells in the arterial intima by way of the circulation.

Ferrari and Rizzoli (2005) reviewed "genetic variations that may explain as much as 70% of the variance for BMD in the population," including the vitamin D receptor (VDR), collagen 1 alpha chain, estrogen receptor alpha, IL-6, and LDL receptor-related protein 5." The authors note that the VDR and IL-6 gene variants have also been associated with "other aging-related complex disorders, including cancer and coronary heart disease, representing common genetic susceptibility factors exerting pleiotropic effects during the aging process."

Estrogen prevents bone loss by enhancing osteoclast-programmed cell death (apoptosis), and decreasing the capacity of mature osteoclasts to resorb bone (Cenci et al., 2000). Estrogen also appears to increase OPG, which would inhibit RANK L–RANK interaction. Further, Pacifici and colleagues (Cenci, 2000; Teitelbaum, 2004b) have proposed an inflammatory basis for postmenopausal bone loss due to estrogen deficiency. They showed that activated T cells promoted bone loss in ovariectomized estrogen-deficient mice. Ovariectomy induced rapid bone loss in *T-cell-replete* mice, whereas ovariectomized, athymic *T-cell-deficient* mice were completely protected against the increase in bone loss. The mechanism of bone loss in estrogen deficiency seems to be by enhanced T-cell production of TNF-α that binds to the p55 TNF-receptor in osteoblasts and potentiates RANK L-induced osteoclast bone resorption. This interplay among estrogen, T cells, and osteoclasts reinforces what I cited above as "osteoimmunology" (Walsh et al., 2006a). Indeed, Weitzmann and Pacifici (2006) raise the possibility that postmenopausal osteoporosis should be regarded as the product of an inflammatory, immune disease of bone, if studies in mice are relevant in humans. Cenci et al. (2000) also conclude: "Whether TNF-α is central to the pathogenesis of estrogen deficiency-induced bone loss (as observed in mice) is applicable to humans, remains to be determined"— a caveat often repeated in this book (see resistin, Chapter 4, section 3).

This caveat may also apply to two recent demonstrations in mice that pertain to bone and have potential relevance to osteoporosis in humans. One example is a "brain-to-bone" pathway of bone loss involving the proinflammatory cytokine IL-1 in conjunction with the hypothalamic IL-1 receptor type 1 (IL-1R1). Low bone mass occurred in genetically modified mice where the IL-1 receptor was "silenced" (i.e., knockout of IL-1rKO); or in mice with transgenic overexpression of the IL-1 receptor antagonist (IL-1raTG). The authors (Bajayo et al., 2005) conclude "although the pathway connecting the central IL-1R1 signaling to bone remodeling remains unknown, the outburst of osteoclastogenesis in its absence suggests that normally it controls bone growth and mass by tonically restraining bone resorption."

A second example of mouse studies that shed new light on control of bone involves the cannabinoid receptors, first discovered as the molecular target of the psychotropic component of the plant *Cannabis sativa* (Pagotto et al., 2006). Idris et al. (2005) observed that mice with knockout of the CB_1 receptor had significantly increased BMD compared with wild-type littermates and were completely protected against ovariectomy-induced bone loss. Indeed, synthetic cannabinoid receptor antagonists inhibited osteoclast formation in vitro, but agonists enhanced osteoclast function. Perhaps cannabinoid receptor antagonists "represent a promising new class of antiresorptive drugs for treatment of osteoporosis."

That mouse studies on systems and factors controlling bone may be applicable to humans is, of course, highly relevant, as, for example, how the OPG story emerged (Simonet et al., 1997). Equally important are those studies that open new domains of bone biology for further exploration; evidence for hypothalamic controls (leptin and IL-1 receptor) and the effects of CB_1 receptor knockout, are examples. These studies give some indication of the profound extent to which bone biology has been "transformed" by application of molecular genetics in mice, as Karsenty wrote so perceptively in 1999. Certainly, without actually being sure of the application to humans, the potential for "translation" must be tempered. But the *principles* by which the hormone leptin, the cytokine IL-1, and the CB_1 receptor influence bone will surely be considered in relation to osteoporosis, the dominant bone disorder geriatric clinicians encounter in their practice.

Assessment of Osteoporosis in Geriatric Practice

This section is not intended to provide guidelines for clinical care. This has not been part of the conception of this book; standard textbooks of geriatrics provide a much more applicable and authoritative account. Yet certain issues that relate to biological considerations of osteoporosis overlap with geriatric practice.

In the initial encounter with a postmenopausal woman, it is presumed that considerations of the diagnosis of osteoporosis and fracture risk are prominent parts of the evidence gained by the history and physical, as I have discussed in "bone health." Criteria for the assessment of fracture risk include appraisal of prior fractures and the status of BMD, best arrived at by an initial DEXA (the use of peripheral measures of bone density have been reviewed by Siris et al., 2001). A key element that needs to be evaluated is risk of falling, which is unique to the geriatric population (Wilkins and Birge, 2005).

Then there is the issue of measuring bone turnover markers, which are important as adjunctive means to assess bone status, yet may not be entirely familiar to geriatricians or widely available. Serum markers that reflect bone formation include two in general clinical use—bone-specific alkaline phosphatase and osteocalcin—and one often used in research studies—amino-terminal propeptide of type I collagen (PINP). For bone resorption, the carboxy-terminal telopeptide of type I collagen (CTX-1) may be measured in serum, or more widely available, the amino-terminal portion (NTX) assayed in urine (Sambrook and Cooper, 2006). A group that reviewed various attempts to reach a consensus about a definition of osteoporosis also sought to establish guidelines for the use of biochemical markers of bone turnover in osteoporosis (Nishizawa et al., 2005). Although the authors are exclusively part of the Japan Osteoporosis Society, publication of their paper in the *Journal of Bone and Mineral Research* assures a wide audience. The authors propose that measurement of bone turnover markers "is now used to assess bone quality and the risk of future fractures." In addition, "in prescribing newer antiresorptive agents that significantly suppress bone turnover markers, their measurement has been an effective method to evaluate the efficacy of the drugs." Certainly, measures of bone markers are valuable in monitoring the course of therapy because they may be used on a more frequent basis than DEXA measures.

5. OSTEOARTHRITIS

Perhaps most representative of age-related conditions is osteoarthritis (OA). In my role as a preceptor for geriatric fellows in their ambulatory care experience, I could not fail to notice that virtually every patient they presented bore the diagnosis of "OA." It did not seem that this was a *new* diagnosis they had made, but one *carried over* from successive generations of fellows, who acquired the patient during the time the fellow spent in the program. Moreover, a repetitive pattern also occurred in the patients themselves: women in their seventies, decidedly overweight, with "knobby," painful knees. I often pointed out the association of Heberden's nodes, almost always present, and these are indeed correlated with underlying radiographic changes of knee OA, especially the bony spurs called osteophytes (Thaper et al., 2005). The following discussion will focus clinically almost entirely on the knee, for which David Felson (2006) recently presented a fine *clinical vignette.*

Earlier in this chapter (sections 3 and 4) I also wrote on other age-related conditions regarding a time of indeterminate length without clinical symptoms, yet in which a measurement—blood pressure, serum glucose, DEXA study—indicated hypertension, diabetes, or osteoporosis, respectively. However, for OA, there is no "preclinical" measurement (see also Fig. 4.1). Joint X-rays, imaging, or arthroscopy, would not be done on asymptomatic individuals as part of a work-up. Thus, the cogency of the suggestion by Norton Hadler (1992) many years ago that "knee pain is the malady, not OA," has reverberated to an extent in the literature because this created uncertainty about the actual diagnosis of "OA." Furthermore, most people with knee pain have multiple other joint-site pain as well (Croft et al., 2005), so-called pain elsewhere, with diminished physical function, anxiety, depression, and perhaps even a "link" with atheromatous vascular disease (Conaghan et al., 2005). To compound the issue of relating "knee pain" to OA, X-rays of the knee in those patients with knee pain do not invariably confirm evidence of OA: radiographic features that include joint space narrowing due to cartilage loss, subchondral bone sclerosis, and marginal osteophytes, and what has come to be designated as a Kellgren–Lawrence grade relating to these features (Spector and Cooper, 1993; Dieppe, 2004; Bauer et al., 2006) (Fig. 4.10). Thus, discor-

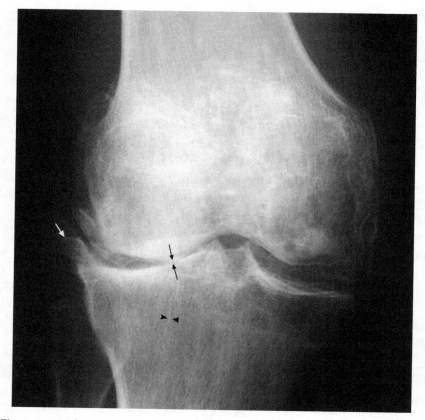

Figure 4.10. Radiographs of the knee in osteoarthritis. This is a frontal X-ray of the right knee of an 80-year-old woman, taken with the subject standing, and represents changes of severe osteoarthritis. There is marked joint space narrowing at the lateral tibial plateau (black arrows) owing to cartilage loss; osteophyte formation (white arrow) is present at the articular margin. The vertically oriented weight-bearing trabeculae (arrowheads) in the subchondral bone are prominent and heavily mineralized. Hamerman (1997).

dance between the presence of radiographic changes and symptoms (Hamerman, 1995; Dieppe, 2004) does not provide reassurance for the diagnosis of OA despite a radiograph as the initial and most convenient diagnostic modality for knee pain. The time-honored standing X-ray to better discern narrowing of the joint space more recently has been deemed unreliable, and either X-rays of the knee in flexion or going directly to the MRI has been proposed (Vignon, 2004). Even in the presence of knee symptoms, arthroscopy, the presumed "gold standard" for demonstrating cartilage changes in early OA, may in fact fail to show lesions when radiographs of the knee appear to show joint space narrowing (Hamerman, 1997). Perhaps all this uncertainty about the actual diagnosis was captured by Paul Dieppe (1984), some time ago, when he commented, "Osteoarthritis remains an enigma; everyone recognizes it when they see it, but no one can define it." Yet with due respect, I'm not sure what there is to recognize clinically, except perhaps in advanced cases with knee deformity and malalignment. Presumably, by that point, the underlying condition has existed for years.

A task force funded by the NIA-OA Biomarkers Network represents a multidisciplinary group that has sought to develop and validate OA biomarkers (Bauer et al., 2006). The designation of "BIPED" has been formulated to represent *b*urden of disease, *i*nvestigative, *p*rognostic, *e*fficacy of intervention, and *d*iagnostic markers. This broad assessment of present and future prospects that may help to better define OA uses only "protein and nucleic acid-based biomarkers" (many dependent on cartilage breakdown products, discussed below). This is an interesting attempt to approach "state of the art" biomarkers in OA that have so powerfully advanced cardiovascular disease detection and risk assessment (see, e.g., Vasan, 2006, and discussion in Chapter 4, section 2).

The medical community, in response to their patients' complaints, has followed the lead of the pharmaceutical industry to treat the knee pain: "ask your doctor if . . . is right for you." Initially, nonselective cyclooxygenase (COX-1) inhibitors were the choice, and after that the selective COX-2 inhibitors were the rage—the era of the coxibs did not last long. OA is fairly widespread in the community, but severe knee pain with limitation and disability that may raise considerations for surgery is a relatively rare observation (Peat et al., 2001).

The extent to which "arthritis" in the population—28 percent in those 65 to 70 years old—is related to OA is not clear, but it was indeed the "first" of the seven chronic conditions listed by Verbrugge and Patrick (1994) in their important paper more than a decade ago. They also wrote of the orientation of current health care on fatal diseases and of the sense that, in general, nonfatal conditions and their impact to diminish mobility in late life required more proportionate attention of health services and better research funding. This was echoed by Jack Guralnick in the foreword of *Osteoarthritis: Public Health Implication for an Aging Population* (Hamerman, 1997), where he noted: "Osteoarthritis affects the lives of a large percentage of older people . . . but the attention it has gotten in the research community is still relatively small in proportion to its tremendous impact on functional abilities and quality of life in the older population."

In seeking the biological links that might help to define the clinical entity of OA more precisely, the place to begin is the histology of the joint. Thirty-four years after Mankin et al. (1971) published their classic work on the histopathology of articular cartilage in OA, a group from the Osteoarthritis Research Society International (OARSI) (Pritzker et al., 2005) published an extensive reevaluation and grade assessment. Their article presents diagrams and corresponding prints of the cartilage and subchondral bone that show a progression from an intact surface (grade 1), cartilage loss with denudation (grade 5), and deformation (grade 6). Figure 4.11 captures the extensive cartilage breakdown (fibrillations) and the chondrocyte clusters that appear to be making an enhanced but futile attempt to replenish the degraded collagen and proteoglycans lost from the matrix (Sandell and Aigner, 2001). But many questions arise. What prompts the low-grade inflammation in the joint (Pelletier et al., 2001; Loeser, 2006), reflected systemically by an elevated serum high-sensitivity CRP (Spector et al., 1997), or by scintography of the joint (Dieppe et al., 1993)? Is the basis for knee pain the marrow edema visualized on MRI (Felson et al., 2001; Sowers et al., 2004)? Is there an abnormality in collagen turnover (Bailey and Knott, 1999)? Why does doxycycline appear able to slow the progression of joint space narrowing in knees with "established OA" (although it did not reduce pain), yet with no change on joint space narrowing in the contralateral knee, "suggesting pathogenic mechanisms in that joint were different from those in the index knee" (Brandt et al., 2005)? What is the basis for the density of subchondral bone? Eric Radin (2004) proposed that the repetitive mechanical impact of the cartilage over the sclerotic bone was a basis for cartilage breakdown and thus an initiating event.

For some time the view has prevailed, championed by Jan Dequeker (1985), that high BMD is protective against osteoporosis and predisposes to OA. Others agree (Hamerman and Stanley, 1996; Hart et al., 2002; Dieppe, 2005). In a recent paper, increased femoral neck BMD was observed in those with radiographic OA but a prevalent osteoporotic vertebral fracture seemed much less associated with OA (Bergink et al., 2005). Some time ago, Richard Stanley and I considered using the osteopetrotic (op/op) mouse lacking CSF-1, without osteoclasts and with dense subchondral bone, as a model in which to explore whether cartilage erosions occurred in the joints of these mature mice compared with wild-type nonosteopetrotic (control) mice. We found no evidence in experimental studies to support cartilage erosions in the op/op mice. We next monitored CSF-1 blood levels in humans to see if these might be lower in OA, but we observed no difference between persons with knee OA and

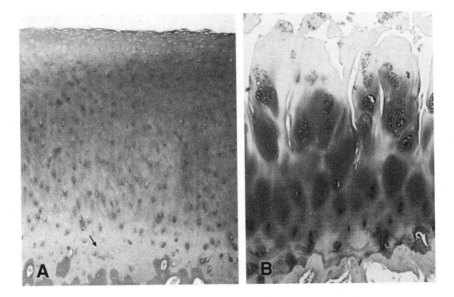

Figure 4.11. Comparison of normal and osteoarthritic cartilage stained with the red dye Safranin O. (A) Normal. The surface is stained green by the counter stain (Fast Green), while the cartilage below stains rather uniformly red as a result of the binding of Safranin O to proteoglycans in the matrix. Note uniform distribution of chondrocytes in the matrix. The arrow indicates the zone of calcified cartilage (tidemark) separating the cartilage from the underlying subchondral bone, which stains green. (B) Osteoarthritis. The surface, depleted of proteoglycans, shows severe disruption with fissures and fibrillations extending deep into the remaining cartilage where the chondrocytes are gathered into clusters, synthesizing proteoglycans in an attempt to replace the depleted matrix (×100). Reprinted from Hamerman (1993), with permission from Blackwell Publishing. (See color gallery.)

those with traumatic knee injury, who served as controls (Hamerman et al., 1998). We speculated that perhaps *local* production of CSF-1 in the subchon-dral bone of patients with OA was somehow deficient and meant diminished numbers of osteoclasts and hence dense bone. We were not able to pursue this further. Yet the subject is important. Indeed, in prevailing views about the in-ception of OA, thinking seems to be divided between those who propose carti-lage loss as the primary event, and those who consider enhanced subchondral bone density as predisposing to cartilage erosions. Yet surely these changes are interactive (Felson and Neogi, 2004). There is agreement that contributory factors to the joint pathology in OA are obesity (Felson, 2004), genetic makeup (Zhang and Doherty, 2004), weaker lower-limb muscles (Jones et al., 2004), and acquired or hereditary femoral-tibial malalignment. Mechanical forces result in microcracks in the subchondral bone and enhanced bone turnover occurs as part of a reparative process (Burr, 2004).

Obesity is of special interest as a known risk factor for OA of the knee. Aspden et al. (2001) explored several interesting speculations about the relationship of obesity to OA. As discussed in the section on osteoporosis, leptin resistance may decrease the sympathetic nervous system activation of RANK L with diminished osteoclastogenesis. However, in obesity a fatty acid surplus provides endogenous ligands for PPARγ that may direct mesenchymal progenitor cells to adipocytes rather than to osteoblasts in the marrow. Adipocytes seem to be more numerous in the marrow in postmenopausal women as part of estrogen deficiency. Obesity may be a risk factor for knee OA not only by virtue of the impact of extra weight on joints but also as a multifaceted and undefined "metabolic precursor" of OA. Each pound of weight lost in a trial of diet and exercise resulted in a fourfold reduction in the load exerted on the knee, with improved knee stability (Messier et al., 2005).

At this point it seems appropriate to describe my own interest in osteoarthritis research, which began when I became director of the geriatrics division. OA and geriatrics seemed, in retrospect, a natural union for me, with my background in rheumatic diseases. Here was a prototypical aging-related condition that evolved over time with changes in cartilage and bone that could be studied at many levels—clinical, radiographic, histologic, and biochemical. I was also inspired by my friendship with Leon Sokoloff, first formed at New York University's Study Group in Rheumatic Diseases in the early 1960s, and whose book *The Biology of Degenerative Joint Disease* (Sokoloff, 1969) remains a classic introduction to the subject from a pathologist's perspective.

My first attempt to bring together the biological basis of OA was to write a review in the *New England Journal of Medicine* (Hamerman, 1989). At about this time, the Albert Einstein College of Medicine was awarded one of the first grants from the NIA, designated "The Teaching Nursing Home," whereby investigators entered this unfamiliar domain for research on its residents. We were funded to do studies on dementia, gait alterations and falls, and my own pursuit—osteoarthritis. I was joined by a nurse practitioner, orthopaedic surgeon, and epidemiologist to assess the diagnosis and impact of OA from a multidisciplinary perspective in women older than 80 years residing in the nursing home (Hamerman et al., 1988). Learning of this, Paul Beeson, one of the founding fathers of geriatrics in the United States, asked me to prepare a *geriatrics bioscience* article for the *Journal of the American Geriatrics Society* on the biology of OA from the perspective of aging (Hamerman, 1993). In this article I discussed how chondrocyte senescence might be linked to aging and OA. Evidence exists for senescence of chondrocytes in aging osteoarthritic cartilage (Price et al., 2002), and Martin et al. (2004) showed that altered mechanical loading appeared to induce oxidative damage in these senescent cells, as I'll discuss further.

To approach newer concepts of the biological underpinning of OA, Sandell and Aigner (2001) proposed five categories of altered chondrocyte and other cellular reaction patterns in osteoarthritic cartilage. (1) Proliferation of chondrocytes to a hypertrophic phase with production of type X collagen, followed by programmed cell death (apoptosis), due in part to factors that gain access to fissured cartilage from the synovial fluid (or to inherent factors in the "senescent" chondrocytes themselves) (Chapter 3, section 3). (2) Enhanced synthesis of extracellular matrix components, principally in the lower zones, yet insufficient to replenish losses (Fig. 4.11). (3) Increased chondrocyte release of cytokines and matrix metalloproteinases, with reduced tissue inhibitors, leading to subsequent degradation of collagen. Two new enzymes have been isolated called ADAM (*a d*isintegrin-like *a*nd *m*etalloproteinase-like domain) capable of cleaving the proteoglycan aggrecan. Figure 4.12 is an illustration of a chondrocyte and the macromolecular components it elaborates that form the cartilage matrix—chiefly aggrecan and type II collagen. Aggrecan consists of a protein core with chondroitin sulfate and keratan sulfate as glycosaminoglycan side chains. Multiple aggrecan units are joined by link protein to hyaluronic acid (hyaluronan) to form an enormous aggregate interwoven with collagen. These aggregates, with their entrapped domains of water, act like coiled springs responding to compression and relaxation, and providing elasticity to cartilage on weight bearing. (4) Modulation of the chondrocyte phenotype with reduced synthesis of typical matrix components. Once again, upper, middle, and lower zones of cartilage show modification of collagen types, although "an overall shift in the chondrocyte phenotype is not observed" (Aigner et al., 2004). (5) Osteophytes or bony protrusions on the margins of the bone, which are a radiologic "signature" of OA, may arise from a population of pluripotential cells, although other studies have found reduced chondrogenic differentiation capacity (Murphy et al., 2002). The progenitor cells may form new cartilage and bone by virtue of mechanical impact and humoral factors, especially transforming growth factor-beta (TGF-β) and BMP-2 (Fukui et al., 2003).

Pluripotential cells and growth factors enter into considerations of tech-

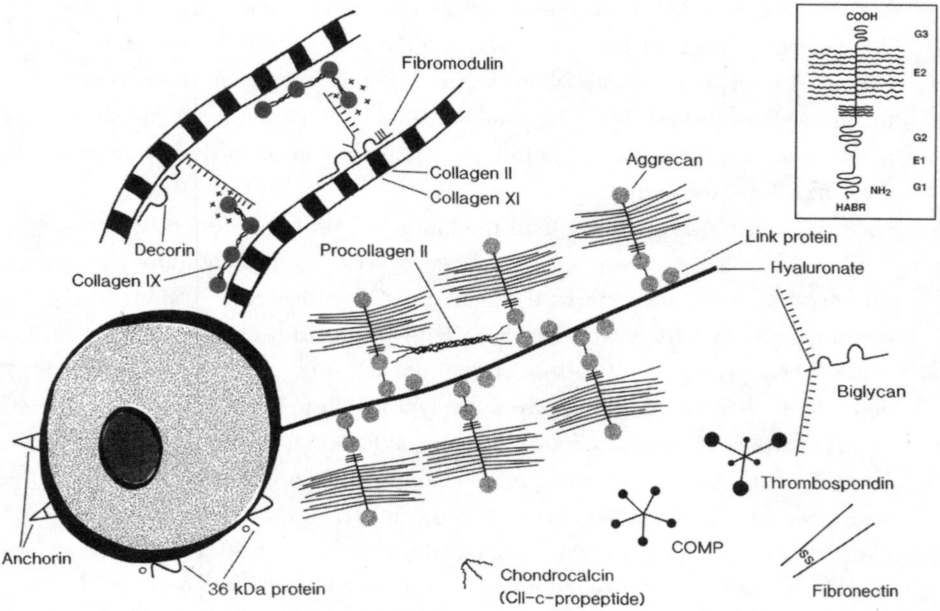

Figure 4.12. Schematic illustration of a chondrocyte and the interactions of macromolecular components in the cartilage matrix. Note multiple aggrecan units linked to hyaluronate (hyaluronan) to form an aggregate. The insert is an enlarged version of aggrecan, indicating its hyaluronan-binding region (HABR) at the amino-terminal end (NH_2); the three globular (G1, G2, G3) and two extended (E1, E2) domains; and the carboxy-terminal end (COOH). The shorter side chains in aggrecan represent keratan sulfate, and the longer chains represent chondroitin sulfate. COMP, cartilage oligomeric matrix protein. Reprinted from Hamerman (1993), with permission from Blackwell Publishing.

niques to repair cartilage lesions. A recent review of autologous chondrocyte implantation to repair cartilage defects was not encouraging (Ruano-Ravina and Jato Diaz, 2006). Stem cells can be isolated, preserved, and added to joints, where TGF-β or BMP may modulate their growth and differentiation into chondrocytes (Bruder et al., 1997). Goldring (2006), with a strong molecular understanding of chondrogenesis (see Chapter 4, section 4), wrote a brief appraisal of cellular methods to repair cartilage defects. The earlier use of autologous chondrocyte transplants for repairing small cartilage defects in the knees of humans did not fare well, with the joint undergoing morbid and osteoarthritic changes. Goldring's current appraisal is that local delivery of BMP-4 by genetically engineered muscle-derived stem cells appears to enhance chondrogenesis and improve repair of articular cartilage in rats. Based on experimental studies, Togo et al. (2006) observed that perichondrocytes from rabbit ear perichondrium are capable of regenerating cartilage on collagen sponge scaffolds in a manner superior to mesenchymal stem cells from bone marrow. These and other approaches to what I would call "molecular bioengineering" will likely enter mainstream use in OA in humans within a few years, and therefore be required knowledge for geriatricians.

The challenge for cartilage repair in vivo will be to promote and maintain "articular chondrocytes" rather than their tendency for cellular dedifferentiation to fibroblasts, and further, the need to limit vascular invasion that might lead to endochondral ossification. Indeed, Drissi et al. (2005) noted that "articular chondrocytes (normally) are constrained from completing the maturational program" toward terminal hypertrophic differentiation and mineralization before endochondral ossification, a constraint necessary if articular cartilage is to be restored. Yet this constraint can be breached in attempts at cartilage repair when chondrocytes respond to signals that activate or repress specific genes (e.g., type X collagen) that may or may not lead to endochondral ossification (discussed in Chapter 4, section 4).

The preceding account may fail to mark progress in our understanding of the biology of OA and its relation to aging. In fact, I suspect there would be a cantankerous but deeply felt commentary about this by Eric Radin, whom I mentioned earlier, such as one he made at the time of an International Workshop on Osteoarthritis Outcomes, and published as a supplement in the *Journal of Rheumatology* (Radin, 2004): "When, in 1966, I began to try to understand the relationship between the various causes of OA, the field was going around in concentric circles. Sokoloff, in 1969, in a brilliant monograph, established that OA was not a process of senescence. Today very competent scientists interested in OA are still studying the aging of articular cartilage and its cells. The reason there has been so little progress is that the OA research is still going around in circles."

So it may be true at present, as Chien and Karsenty (2005) wrote in *Cell,* that "the underlying knowledge of the pathophysiology of OA is very limited." In their article they noted that there is a "lack of specific molecular markers of articular chondrocytes," although three models may be emerging. In one model, mice with deletion of a receptor for a bone morphogenetic protein (BMPpr1a) in chondrocytes developed "an osteoarthritis phenotype after birth, thereby establishing that BMP signaling is required to maintain the integrity of adult joints." In a second model, Smad3, one of the transcription factors involved in TGF-β signaling, was deleted, and mice "developed a post-natal form of degenerative joint disease, similar to osteoarthritis." In a third model, Chien and Karsenty comment on mechanisms that may induce loss of the proteoglycan aggrecan in OA cartilage due in part to chondrocyte release of proteases, such as aggrecanase and matrix metalloproteinase (MMP). Indeed, the aggrecan fragment pattern present in OA *synovial fluid* was similar to cartilage aggrecan cleaved in vitro by aggrecanase-1 (ADAMTS-4) and MMP (Struglics et al., 2006). In a test of aggrecan breakdown by mechanical injury or by the cytokine IL-1, enzymatic activity of ADAM with thrombospondin (ADAMTS 5) mediated aggrecan turnover; ablation of ADAMTS 5 protected mice against erosion of cartilage (Karsenty, 2005). Thus, "the experiments point to a single molecule (ADAMTS 5) as a precise pharmacologic target to prevent OA" (Chien and Karsenty, 2005).

Attur et al. (2002) believe that "in the genomic era of molecular medicine" OA requires a complete rethinking of the traditional concept of inflammation. In this new light, the OA chondrocyte appears to behave like an activated macrophage (indeed, in one figure, the two cells are depicted as shaking hands!), and molecular-based techniques demonstrate "up-regulation" of many inflammatory mediators capable of degrading the cartilage matrix. It is certainly in accord with a theme of this book that inflammation, perhaps newly defined in this molecular era of medicine, is now part of OA (often considered as osteoarthrosis or "noninflammatory") and underlies all the conditions discussed in Chapter 4 as well as in aging itself in Chapter 3.

Loeser (2006) also considers the molecular mechanisms of cartilage breakdown in OA, reviewing evidence for excess production of ROS that could be an important mechanism tying together mechanical forces, genetic predisposition, and aging changes in chondrocytes to induce a proinflammatory state, in which catabolic activity exceeds anabolic activity. We can also make the theoretical leap invoking senescent chondrocytes as contributory to cartilage degeneration. Senescent chondrocytes may indeed synthesize altered matrix molecules and release proteases that degrade the matrix. Mazieres et al. (2006) assayed molecular markers of cartilage breakdown in the blood and urine of patients with hip OA in an attempt to predict structural loss of cartilage as well as synovial inflammation. These markers included various fragments of collagen, cartilage oligomeric matrix protein (COMP) (Fig. 4.12), hyaluronic acid, and matrix metalloproteinases. Some of these markers do seem to be predictive of OA progression "more rapidly than radiographic assessment."

Despite the claim that these markers are "easy to measure with available commercial tests," I doubt if this will be widely practiced.

Finally, Loughlin (2005) points to "considerable success in the identification of genes harboring susceptibility for primary osteoarthritis":

- An association of the FRZB gene with hip OA in females. FRZB codes for secreted frizzled-related protein 3, an antagonist of Wnt signaling. In the discussion on endochondral bone development in the section relating to osteoporosis, I mentioned the Wnt signal transduction pathway as critical for normal bone development. The secreted frizzled-related protein 3 helps to maintain articular cartilage and certain alleles (polymorphism) of FRZB may reduce Wnt activity.
- A possible association of the asporin gene ASPN with knee and hip OA. Asporin is a cartilage extracellular protein that regulates the activity of TGF-β.
- The calmodulin gene CALM1. Calmodulin is an intracellular protein that interacts with several proteins involved in signal transduction. Associated alleles of ASPN and CALM1 reduce the ability of chondrocytes to express the genes encoding aggrecan and type II collagen—essential for cartilage structure. Thus, in a broad sense *polymorphism* of certain genes influences the tempo and occurrence of many of the age-related diseases discussed in this chapter and in the epilogue.

Multiple interactive events in the evolution of the pathophysiology of OA involve the entire joint: the articular cartilage, subchondral bone, synovial fluid, synovial membrane, capsule, ligaments, and muscle (Fig. 4.13). The joint is indeed an integrated unit. Aspects of molecular biology presented here are clearly fragmentary and preliminary, and some of them are not likely to have received mainstream attention in rheumatologic teaching, much less in geriatric education. I hope this discussion provides some biological relevance to the clinical observation of patients with OA that began this section. What ultimate direction research on the biology of OA will take in relation to clinical symptoms and physical findings I cannot decide at this writing. Defining the experiments to do so will be a challenge for scientists and clinicians working together in this area over the next decade. Perhaps a breakthrough will occur in our understanding of the fundamental biology of this key nonfatal chronic condition that will permit early detection of OA, translated to improved management, with widespread public health implications.

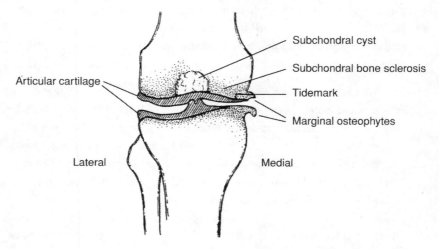

Figure 4.13. A composite diagram of the osteoarthritic knee. The features illustrated include medial joint space narrowing due to severe articular cartilage loss, subchondral bone sclerosis, marginal osteophytes, and a subchondral cyst. The tidemark separates the cartilage from underlying bone.

6. FRAILTY AND RELATED CONDITIONS: ANOREXIA, SARCOPENIA, EXERCISE
Frailty

At the outset of this section on frailty, I wish to pose a contrast with the more well-defined "diseases" discussed in sections 2–5. When I reviewed the subject of frailty in the *Annals of Internal Medicine* (Hamerman, 1999), I wrote that "health providers who care for an aging population inherently associate the word *frailty* with patients whom they *perceive* as frail." That is, at a glance, the patient appears weak and listless, with evidence of substantial weight loss (although I would have to amend that now because Blaum et al. [2005] suggested that obese persons can be frail). But more typically, on questioning, the patient may confirm loss of weight, anorexia, exhaustion on slight effort, and much diminished physical activity and strength. These characteristics were part of the "phenotype of frailty" proposed by Fried et al. (2001) and reaffirmed by that group in a follow-up (Fried et al., 2005), when the likelihood of developing dependence in activities of daily living became evident (although the study cohort at the outset consisted of moderately disabled women). This study also emphasized that hospitalization was a particularly vulnerable time and it was important to target these persons for rehabilitative therapies to limit further decline (Boyd et al., 2005). The Fried group's most recent report (Bandeen-Roche et al., 2006) presents an "operational definition of frailty"—verified across many study groups—which I quote in full: "Frailty is a syndrome of decreased resiliency and reserves, in which a mutually exacerbating cycle of declines across multiple systems results in negative energy balance, sarcopenia, and diminished strength and tolerance for exertion. Accordingly, it proposes exhaustion, weight loss, weak grip strength, slow walking speed, and low energy expenditure as frailty-identifying characteristics."

The reader will note that although the "concept of frailty has emerged as a central and critical aspect of geriatrics" (Cohen, 2000), it resists definition, indeed it is "in desperate need of description and analysis" (Gillick, 2001). Kaufman (1994), a social gerontologist, wrote "frailty is one of those complex terms—like independence, life satisfaction, and continuity—that trouble gerontologists with multiple and slippery meanings." A valuable and comprehensive review of frailty has been prepared by Hogan et al. (2003) representing a Canadian initiative, with models, definitions, criteria, and 277 references. Perhaps frailty remains so elusive because it is a "geriatric condition or syndrome" (Fried et al., 2004) rather than a disease, part of other age-related conditions, such as falls (Kinney, 2004) that merge into disability and comorbidity, which Fried et al. (2004) and Ferrucci et al. (2004) attempted to disentangle and sort out. The prominence and importance of all these overlapping conditions in geriatric practice, despite their nonspecificity, may have prompted Mary Tinetti and Terri Fried (2004) to call for "an end of the disease era." By this they meant to emphasize "the attainment of patient goals and the identification and treatment of modifiable biological and non-biological factors rather than a focus on diagnosis, treatment or prevention of individual diseases." In response, Daniel Foster (2004) wrote, "the demise of disease? I don't think so." He goes on, "it would seem imprudent to pronounce that the disease era is over."

What may be emerging in medical practice is a coalescence of approaches that seeks "integrated interventions": the earlier biopsychosocial model of illness and the broader one now of "acceptance by one individual clinician of responsibility for assessment, planning, linking, monitoring, advocacy, and outreach with respect to all factors that are pertinent to meeting an individual's health care needs" in the context of evidence-based medicine and cost-effective outcomes (Smith and Clark, 2006). This is a tall order to embrace in the light of current concerns about the survival of primary care and the future of geriatric practice. In addition, I have further encouraged clinicians to be aware of the precepts of basic biology in this book. Nevertheless, the interested reader may wish to gain a perspective on *Integrated Care for the Complex Medically Ill,* edited by Huyse and Stiefel (2006). In this volume, the description of the challenges posed by care of the "frail vulnerable elderly" is especially compelling (Slaets, 2006).

Frailty illustrates well what geriatric medicine has introduced into the supposedly more rigorous and identifiable category of "diseases" in internal medicine that internists considered a strength of their discipline and a potential deficiency in the scope of geriatrics (Chapter 1). "Functional decline, frailty, anorexia, sarcopenia (muscle loss and weakness), along with co-morbidities" (Morley et al., 2002) provide a constellation that often is present in older persons whose condition may not provide a pinpoint designation of disease. Frailty is just such a geriatric condition, a vulnerable state, part of a continuum that begins with independence at an earlier time and now one that often portends a *potential* for adverse outcomes, especially a fall, that escalates to a hip fracture (Hogan et al., 2003; Schwartz et al., 2005), thus linking frailty with more specific and devastating morbid events. Many aspects of decline may be verbalized by the frail patient, and in these instances it is indeed appropriate to focus more on the whole person rather than on a disease, as Tinetti and Fried (2004) propose. It is also worth repeating that what is occurring in decline is associated more with *aging* than with a distinct disease, hence a conception that is a strength, rather specifically, of geriatric practice (Ershler, 2004).

It is crucial for the geriatrician to place frailty as part of an *evolving* geriatric functional continuum midway between independence and predeath (Table 4.3). Frailty overlaps with a range of other geriatric clinical correlates that broadly represent a constellation (Hamerman, 1999): functional decline; being underweight or having anorexia or cachexia; moving slowly or taking to bed; responding only partially to nutrients or failing to thrive; coping poorly or not at all (acopia); demonstrating sarcopenia and weakness with limited mobility and a tendency to fall; or requiring limited assistance or being dependent. Slowing of mentation may be evident.

With the perception of frailty, and perhaps added evidence for diminished grip strength, difficulty in rising from a chair, recent increase in falls, visual impairment (Guralnik 2005; Klein et al., 2005), decrease in self-management (Schuurmans et al., 2004), the health care provider must make critical decisions about a patient that relate to "medicalization" of a condition that, as

TABLE 4.3
A geriatric functional continuum

Stage for potential intervention	Clinical correlate
Primary	Independence
	Coping
Intermediate	Independence with difficulty
	The "dwindles"
	Functional decline
Secondary	Frailty
	Failure to thrive
	Disability
	Failure to cope
Tertiary	Dependence
	Taking to bed
	Cachexia
	Predeath

Source: Adapted from Hamerman (1999), with permission from the American
College of Physicians.

I have said, may not necessarily be a manifestation of disease (Hamerman, 1999). For persons who seem to fit the designation of frail, how extensive should the workup be, which is often poorly tolerated, in an effort to uncover underlying, progressive disease (cancer is the most obvious in view of the invariable weight loss)? Should a condition that "tips the balance" (such as stroke or hip fracture after a fall) be considered an accountable or "organic" basis on which to diagnose frailty? When can a frail person, his or her family, or more often, his or her physician, accept frailty as "nonorganic"—a progressive loss of homeostatic and stabilizing mechanisms and a primary failure to thrive—and recommend comfort care? Clearly, these choices will require individual considerations and a review of the place of the frail person within the geriatric continuum from apparently early to very late. Proposed interventions aimed at improving or stabilizing the existing quality of life raise medical decisions, ethical issues, and particularly in this era, aspects of cost and benefit. Defining a rational way to intervene may also be difficult in "many elderly patients where illness does not fit a classic disease model" (Jarrett et al., 1995) and when the presentation is atypical (Resnick and Marcantonio, 1997). Involuntary weight loss alone without apparent disease or vague symptoms may indicate a risk for death in some older persons. In one study, persons in an intermediate care facility who took to bed because of fatigue or nonlocalizing weakness had significantly shorter survival than the group as a whole (Clark et al., 1990). "Tiredness" is not a favorable sign in older persons, as mentioned in Chapter 4, section 1.

Defining frailty may be like attempting to "know the dancer from the dance"; nevertheless, I attempt to present an understanding of frailty by describing many of its clinical correlates and considering evidence for its biological underpinnings. Frailty might emerge more sharply as a clinical entity if it were described in biological terms, perhaps putting it on a "par" with the molecular insights that have been achieved in "classical diseases," such as atherosclerosis, or diabetes (sections 2 and 3). Both Joseph et al. (2005) and Walston (2004) raise the importance of "translational research" to connect basic "mechanisms related to aging, such as oxidative damage and telomere shortening, to clinical aging-related syndromes" (Walston, 2004) with the goal of improving health outcomes (Joseph et al., 2005). Ferrucci and coworkers (2005), who have contributed so much to our understanding of the "biological instability" of aging, functional decline, and frailty, wrote thoughtfully about "oxidative stress, rather than inflammation, residing at the 'core' of frailty." Of course, a profoundly important association exists between oxidative stress and inflammation, whereby the former in its many origins provokes cytokine activation—in particular, by way of the NF-kB pathway.

Translation is indeed a theme of this book, yet advances in the molecular understanding of aging (see Chapter 3, section 3) will be less immediately applicable to addressing frailty than the role of "primary care physicians meeting their obligation to prescribe physical activity to their sedentary patients" (Chakravarthy et al., 2002). So is frailty inevitable, as I wrote about aging in Chapter 3? Jennifer Wilson (2004), a science writer, articulated the view in the *Annals of Internal Medicine* "that frailty—and its dangerous effects—might be preventable." To this, the medical ethicist Joanne Lynn (2004) responded: "delaying the onset of frailty can be important, *but nothing now on the horizon is likely to prevent frailty as part of the last phase of life for those who survive into old age*" (italics added). Perhaps the sense of inevitability of frailty (if one lives long enough) arises from biological changes of the aging process itself— the systemic effects associated with an altered metabolic balance, manifested by cytokine overexpression and hormonal decline.

Chapter 4, section 1 discussed increased blood levels of cytokines and disturbed metabolic balance as part of the response to stress. The idea is now gaining wide acceptance that low-grade inflammation and a "dysregulated" (Krabbe et al., 2004) cytokine output from visceral fat, adverse lifestyle habits (smoking), immune alterations (Joseph et al., 2005), or subclinical infection may accompany *aging* and perhaps contribute to frailty with other factors inherent in the host that we do not yet understand but that are under intensive study (Ferrucci et al., 2004). In participants (mean age, 78 years) in the Framingham Heart Study, production of interleukin-6 and the anti-inflammatory cytokine interleukin-1 receptor antagonist by peripheral blood monocytes markedly increased with age. These data suggested to Roubenoff (1997) that aging is associated with a dysregulation of cytokine formation that could lead to loss of muscle strength (sarcopenia), which may be one of the earliest manifestations of frailty. I have previously mentioned studies by Cohen et al. (1997) of a large community-based population that showed higher levels of serum and mononuclear cell production of interleukin-6 related to participants' self-reports of fatigue and functional decline. Elevated plasma levels of interleukin-6 were found in community-dwelling persons 70 years of age or older who were depressed (Dentino et al., 1999). Healing of punch-biopsy skin wounds was impaired in persons 60 to 90 years of age who did not have apparent disease compared with younger persons; modifications in cytokines released at the inflammatory site in older persons altered endothelial cell-adhesion molecules and enhanced the cellular infiltrate in the wound (Ashcroft et al., 1998). (This is reminiscent of poor wound healing in aged mice with short telomeres, discussed in Chapter 3, section 3.) Elevated levels of cytokines, especially TNF-α, have been observed in the final stages of advanced debilitating diseases that are often associated with frailty in elderly persons: weight loss and malnutrition, chronic inflammation, cardiac cachexia, cardiomyopathy, emphysema, infections, and pressure sores (Hamerman, 1999). More recently, Sambrook et al. (2005) observed elevated levels of markers of bone turnover associated with all causes of mortality—especially cardiovascular causes—in frail persons (mean age, 86 years) living in residential care. It seems that these findings may reflect the effects of increased IL-6 and TNF-α on bone and the blood vessel.

On the other side of the metabolic balance are hormonal mediators. Their link with cytokines, especially TNF-α, IL-1, and IL-6, occurs by way of cytokine interactions with the central nervous system, resulting in various physiologic, neuroendocrine, and behavioral responses (Sternberg, 1997). Elevated levels of IL-6 represent an important stress response that activates the hypothalamic-pituitary-adrenal axis (Mobbs, 1996; Papanicolaou et al., 1998), as presented in Chapter 4, section 1. Aging and disease may subsequently impair the neural-immune, neuroendocrine, and endocrine pathways that control the production and peripheral actions of hormonal mediators, with the decline of endocrine functions described as menopause, andropause, adrenopause, and somatopause (Lamberts et al., 1997), as I've noted. Hence, in a later section

(Chapter 5, section 6), I will discuss the potential for hormonal replacement therapies.

In summary, what further complicates specificity and the attempts to define frailty is its *interconnectedness* with so many conditions associated with the aging process, and ultimately with risk for and then emergence of diseases. As Rockwood (2005) put it, "there is insufficient evidence to accept a single definition of frailty." A spectrum of "common pathophysiologic mechanisms" are related to frailty (Joseph et al., 2005), and the clinical and biological relationships can be viewed once again as a continuum, where the factors are interactive and tend toward patient decline. I have incorporated these interactive aspects, as I see them, into the text below for I wish to emphasize their relation to aging, functional decline, and the potential for disease and their almost-universal representation in this section and in virtually all others in this chapter:

- Lack of exercise
- Hormonal decreases with aging
- Muscle loss: sarcopenia
- Bone loss: osteoporosis
- Visceral fat accumulation: obesity
- Insulin resistance: diabetes
- Knee pain: osteoarthritis
- Heightened inflammatory mediators
- Vascular changes: atherosclerosis
- Falls, fractures

Thus, in this example of interconnectedness of conditions associated with aging and contributing to frailty, led by lack of exercise and diminished circulating hormones, disease outcomes can be projected: osteoporosis with fragility fractures, osteoarthritis, diabetes, and atherosclerosis, all of these in turn leading to organ and systems damage and reduced functional capacity. At present, it seems that approaches to limiting or delaying frailty require a "preemptive" approach (Ferrucci et al., 2004), and are best *begun* by healthy lifestyle choices discussed in Chapter 6, section 2.

Anorexia of Aging

Whether anorexia of aging progresses to a profound state of weight loss, malnutrition, and the visible appearance of cachexia depends, of course, on the duration of loss of appetite and weight, on contributing conditions, and on the outcome of interventions. Even weight loss of 5 percent in older persons reflected poor health, hospitalization, and a higher rate of mortality, so clinical assessment of anorexia of aging must be taken seriously (Newman et al., 2001). Profound weight loss (more than 10 percent of body weight within a period of 6 to 12 months) may be encountered in up to 13 percent of elderly outpatients, and in much higher frequency—up to 65 percent—in nursing home residents (Morley and Kraenzle, 1994; Bouras et al., 2001). In analyzing weight loss as a predictor of death in patients with wasting syndromes (AIDS, cancer cachexia), body weight at death can be extrapolated to be 66 percent of ideal body weight when weight loss is analyzed as a function of time before death (Grunfeld and Feingold, 1992). Weight loss without apparent underlying cause(s) implies that one has not been found either because there has been a choice not to explore possible causes—as may be the case in some frail individuals who have taken to bed—or because comprehensive studies have failed to turn up a contributory cause (Hamerman, 1999).

Wilson (2001) notes that "the diminution of appetite with aging has been recognized for centuries." The basis for a "physiologic" diminution of appetite and subsequent weight loss in older persons is complex and arises from multiple determinants. Perhaps at the outset, the spectrum of palatability— or the hedonic (pleasurable) qualities of food—may be diminished (Morley, 1997). Changes in perceptions of taste and smell lead to dietary deficiencies. Wilson (2001) wrote thoughtfully of "bitter-sweet memories" that relate to appetite and aging. She refers to "a sensory-specific satiety that normally encourages *variation* between food items before postabsorptive palatability wanes." Sensory-specific satiety is diminished in older adults and may explain the tendency, by choice, for the diet to become increasingly monotonous and less varied. Hence, both the quality and quantity of food intake are diminished, and health care providers "may overlook the possibility of intentional dietary restraint in the evaluation of anorexia in the elderly population." Anorexia associated with deficiencies in zinc or magnesium may prevail in many clinical conditions (Caddel, 2000). Satiety signals, principally cholecystokinin (CCK) (Fig. 4.6), released when food enters the stomach and duodenum, may be enhanced with aging and delay gastric emptying (MacIntosh et al., 1999). Aging may impair production of nitric oxide synthase, and diminished nitric oxide formation also delays gastric emptying (Morley, 2001). Adiposity signals— leptin arising from peripheral adipose tissues, and insulin from β cells of the pancreas—induce multiple hypothalamic responses involving systems of appetite suppression (anorexigens) or stimulation (orexigens), as mentioned in Chapter 4, section 3 (Fig. 4.6). Although leptin has multiple physiologic roles (Harris, 2000), it is one of the prime regulators of food intake and other metabolic responses. Morley et al. (1999) linked decline in testosterone in aging males with an increase in leptin levels, thus promoting hypothalamic anorexigens. Indeed, low testosterone levels in hypogonadal men were associated with elevated leptin (Jockenhövel et al., 1997).

Sarcopenia

Rosenberg (1989) coined the term *sarcopenia* to indicate loss of muscle mass and weakness, or more specifically, a sense of effort for a given exercise intensity (Marcell, 2003). "It cannot be considered a disease or a condition that has a clear diagnostic marker" (Nikolić et al., 2005). Sarcopenia, like frailty, also represents a loss of reserve capacity (Marcell, 2003), and while generalized decline may occur, the focus on muscle is more specific than frailty. "Muscle is the source of strength and therefore of independence" (Roubenoff, 2003b). Rosenberg is part of the U.S. Department of Agriculture Human Research Center on Aging at Tufts University, in Boston, a source of so many papers relevant to nutrition and aging by scientists who have underscored the dynamic relationship between undernutrition and sarcopenia. Evans (2004) also notes that reduced protein intake may "accelerate the rate of sarcopenia," and conversely, protein caloric supplements increased strength and muscle mass. Perhaps this is so for community dwellers, but Fiatarone et al. (1994) earlier reported that in nursing home residents multinutrient supplementation without concomitant exercise did not reverse muscle weakness "or physical frailty." Both Wilson (2001) and Morley (2003a), who have written much on nutrition in older persons, note that most physicians will claim awareness of the fact that poor nutrition, weight loss, and decline in strength promote functional deterioration, "although few will intervene." One of the problems, however, that limits intervention for health care providers caring for older patients is how to assess malnutrition. Harris and Haboudi (2005) wrote a brief account of malnutrition screening in elderly people, recognizing the limitations of available "screening efforts," including BMI, anthropometry, and biochemical markers. They mention two "screening tools"—the Malnutrition Universal Screening Tool, and the Mini Nutritional Assessment. Readers may wish to consult this article and other references for nutritional appraisal (e.g., Huls, 2001), but it seems that obvious weight loss must prompt a dietary history, whether vitamin supplements are used, and the nature of medications being taken. The physician should check visual acuity, tongue papillation, blood levels of folic acid, calcium, 25-hydroxy vitamin D, and vitamin B_{12}, and hand grip strength. A measure of BMI is desirable (Iannuzzi-Sucich, 2002). Low serum levels of testosterone in men were associated with impaired mobility and low muscle strength (Schaap et al., 2005).

Early developmental conditions such as infant size and weight at one year influence adult muscle strength (Sayer et al., 2004), another example of "early origins of disease." Most persons past the age of 30 show loss of lean body mass and gain of fat, and nearly everybody loses muscle strength and muscle mass as they age (Roubenoff, 2003b; Nair, 2005). Again, this seems to be an example of the *inevitable* but *variable* effects of aging discussed in Chapter 3, section 2, where, on average, persons past the age of 50 years lose muscle mass at about 1–2 percent per year (Dirks and Leeuwenburgh, 2006). Older men (over 80) appear to be at greater risk for muscle loss than women (Iannuzzi-Sucich et al., 2002). The relationship between muscle mass and muscle strength is linear, but the relationship between muscle strength and physical function is difficult to define because of the influence of other factors relating to extent of exercise, nutrition, comorbidities, motivation, and so forth. Most older people do not exercise and "the majority of people over 75 do absolutely nothing in terms of physical activity" (Roubenoff, 2003b). Sarcopenia may lead to physical decline, falls, and the potential for institutionalization, but studies by Newman et al. (2006) indicate that muscle *strength* is more important than *quantity* in estimating mortality risk. Grip strength provided risk estimates similar to those of quadriceps strength. Muhlberg and Sieber (2004) describe a series of "vicious loops in frail elderly patients, forming a typical pattern in geriatric medicine." Among these "loops" that again show interconnectedness (as in frailty) are: sarcopenia → neuromuscular impairment → falls and fractures → immobilization → sarcopenia. Another "loop" is nutritional, i.e., after immobilization → decline in nutrition → malnutrition → impaired protein synthesis → sarcopenia. In fact, Leveille (2004) noted the widespread impact of "musculoskeletal aging," and Janssen et al. (2004) wrote more specifically on the health care costs in the United States on the basis of increasing risks of physical disability in older persons. The estimate was $18.5 billion, about 1.5 percent of total health care expenditures for the year 2000. A 10 percent reduction in the prevalence of sarcopenia would result in savings of $1.1 billion.

If health care providers are to interact with their older patients to learn of their exercise practices and encourage their patients to be active, then means to evaluate muscle mass and strength are needed. This is an important issue and may not be practiced clinically, like nutritional assessment, and for the same reasons, it is difficult to do. Sarcopenia can be linked to many important functional limitations, as noted, such as deficits in rising from a chair, gait, mobility, and activities of daily living, but to identify sarcopenia quantitatively requires an assessment by DEXA and several calculations. This is not within the realm of clinical evaluation. As Melton et al. (2000) wrote, "It remains to be determined whether a DEXA measurement of muscle mass adds to the prediction of adverse events once strength has been assessed directly" (e.g., using a hand dynanometer). Muscle mass has no well-characterized clinical relationship as bone density has to a single important morbidity—namely, fracture.

I reviewed six epidemiologic studies (Baumgartner et al., 1998; Melton et al., 2000; Iannuzzi-Sucich et al., 2002; Cesari et al., 2005; Szulc et al., 2005; Walsh et al., 2006b) that used DEXA to measure percentage of body fat and appendicular (legs and arms) skeletal muscle mass in their respective surveys. Complex calculations (see description in Baumgartner et al., 1998, and Melton et al., 2000) are used for the relative skeletal muscle mass analogous to the estimate of body mass index (weight [kg]/height^2 [m^2]) for grading adiposity. Sarcopenia was defined as appendicular skeletal muscle mass in kg/height in m^2 greater than two standard deviations below the sex-specific mean of a young healthy reference group. Hence, the analogy with osteopenia based on WHO standards. Sarcopenia has been linked with poor structural parameters of bone (Szulc et al., 2004) and reduced BMD (Walsh et al., 2006b); sarcopenia and the related tendency to fall create a real risk for fractures.

The studies of sarcopenia that used a DEXA analysis on community-dwelling research volunteers older than 80 years revealed prevalence rates of 31 percent and 53 percent in women and men, respectively (Iannuzzi-Sucich et al., 2002). It was also noted that BMI itself is a strong predictor of skeletal muscle mass and might be a useful diagnostic tool.

Table 4.4 summarizes some biological findings reported in sarcopenia. A metabolic overlap (Karakelides and Sreekumaran, 2005) occurs here beyond the effects of limited exercise per se, with relevance to diabetes (fat accumulation, consequences of insulin resistance, cytokine release, mitochondrial DNA mutations, and oxidative stress), osteoporosis (NF-kB activation and osteoclastogenesis), and atherosclerosis (diminished NO production and vascular endothelial injury). Moreover, exercise can markedly diminish the progression

TABLE 4.4

Factors contributing to sarcopenia

Underlying aging, sedentary lifestyle

Loss of skeletal mass and fiber number, especially type 2; decrease in myosin heavy chains; decrease in satellite cells

Diminished rate of muscle protein synthesis due to amino acid deficiencies

Accumulation of noncontractile proteins explaining greater decline in muscle strength than total muscle mass

Insulin resistance with age, with inhibition of NO cascade and decreased amino acid transport into cells

Decreased GH, IGF-1, testosterone, limiting muscle growth and repair

Mitochondrial DNA mutations with oxidative stress

Cytokines—TNF-α and IFNγ suppress myogenesis by activating NF-kB in muscle cells and promoting apoptosis

Decline in central motor system α motor neurons

Source: Data from Bautmans et al. (2005), Cesari et al. (2005), Dirks and Leeuwenburgh (2006), Evans (2004), Hamerman (2004), Kamel (2003), Karakelides and Streekumaran (2005), Marcell (2003), Morley (2003b), Nair (2005), Nikolić et al. (2005), and Roubenoff (2003b).

Note: See Glossary for abbreviations.

of these diseases; apparently this is true for sarcopenia. Inflammatory markers (IL-6, TNF-α, CRP) now figure prominently in association with sarcopenia (Visser et al., 2002; Kamel, 2003; Bautmans et al., 2005; Cesari et al., 2005), perhaps approaching a "molecular signature" of aging skeletal muscle (Giresi et al., 2005). The "signature" in aging muscle revealed increased expression of several genes involved in mediating cellular responses to inflammation and apoptosis. These authors propose that their "molecular model" (exclusively a research modality at present) might at some future time be used "to judge the effectiveness of exercise and other therapeutic treatments aimed at ameliorating the effects of muscle loss associated with aging" (Giresi et al., 2005). Referring to the considerations of aging discussed earlier (see Chapter 3, section 4), Alfred Fisher (2004) in a geriatric bioscience article in the *Journal of the American Geriatrics Society* discussed work of Herndon et al. concerning "sarcopenia" in the worm *Caenorhabditis elegans.* This is a most interesting "translation" of basic biology from worm to women, as Fisher puts it. Mutations affecting the age-1 gene slow aging of the worm and delay sarcopenia. This suggested that in "humans and worms, sarcopenia may represent a biomarker for *biological* age as opposed to *chronological* age, of the individual." Hence, *C. elegans* may "develop into an important model system in which to study the biochemical and genetic events responsible for sarcopenia and to test therapeutics designed to prevent or reverse sarcopenia." In the epilogue, I note that Cynthia Kenyon (2006) also refers to sarcopenia in aging worms, while long-lived mutants do not develop this condition until they are much older.

Exercise, a Necessity

If uncertainty exists about the manner of physician intervention with respect to a patient's evidence for nutritional deficiency or low muscle mass, what might be said about a physician's incentive to provide exercise recommendations (Anderson et al., 1997; Chakravarthy et al., 2002)? Again, to cite Tinetti and Fried (2004), an orientation on "the whole person" as contrasted with a focus on disease can be useful in recommending exercise to counter the constellation of "systems decline" that include frailty, anorexia and malnutrition, and sarcopenia, discussed here, and even depression, and perhaps delay the onset of Alzheimer disease (Larson et al., 2006). James Judge (1997), in a fine review of the principles and prescription of exercise for frail older persons in a book on osteoarthritis that I edited, noted "some controversy about whether loss of mobility is disease specific (Fried et al., 1994) or part of a syndrome of failing function" (Tinetti et al., 1995), as I have discussed. Thus, the disease-based model suggests that intervention targeted at the disease process will be most effective in preventing disability; the syndrome approach suggests that a multifaceted intervention to prevent disability will be most effective. Of course, these are not mutually exclusive, and indeed this calls for an awareness of both approaches, rather than the neglect of one.

All physicians and nurses providing health care for older persons should be aware that exercise may even have more salutary effects on those more sedentary—such as nursing home residents—than on healthy persons of a comparable age. Exercise has been proposed as a model for successful aging and for the treatment of sarcopenia to recover muscle mass and muscle function in older adults (Hawkins et al., 2003). Hunter et al. (2004) are specific about increases in strength and muscle mass following *resistance training* in older adults, perhaps by recruitment "of satellite cells to support hypertrophy of mature myofibers." Aerobic training aims more to improve many metabolic functions (Nair, 2005). In an editorial accompanying articles on types of physical activity to improve cardiorespiratory fitness (aerobic power) and longevity, Blair and La Monte (2005) wrote, "while recommendations can be made, there is still an incomplete understanding as to the specific dose of physical activity in terms of frequency, intensity, and duration, and the related volume of energy expenditure that is effective in achieving specific biological or clinical outcomes."

Most applicable to sedentary older persons is resistance or power exercise and training. In nursing home residents (mean age, 87 years), Fiatarone et al. (1994), in a paper that attracted much attention, applied intense resistance training (for example, leg press and extension), which increased muscle strength, gait velocity, stair-climbing power, and the level of spontaneous physical activity. In a life-care community-residential facility, combined resistance-balance training increased gait velocity (Judge, 1997). Resistance training in relatively healthy community-dwelling persons did not increase gait velocity. Power training improves balance, in particular, a low-load, high-velocity regimen in older adults with initial lower muscle power and slower contraction (Orr et al., 2006). Singh et al. (2005) noted that high-intensity progressive resistance training (80 percent maximum load) tended to reduce depression, improve sleep, and enhance the perception of quality of life in older community dwellers. The sobering additional note is that advances in physical performance were often lost in weeks without repetitive and continuing exercise. Health care providers, in deriving lessons from all this, might set up a plan for patients at home to practice straight leg raises against resistance (e.g., with the extended leg to lift a chair). Ideally, consultative services of physical therapy may be part of the geriatric team plan.

In conclusion, my view of exercise as a necessity is also expressed, although more moderately, by W. J. Evans (1995), a leading researcher in this field, as a call to all health care providers—especially those caring for older persons: "Strategies for preservation of muscle mass with advancing age and for increasing muscle mass and strength in the previously sedentary elderly may be an important way to increase functional independence and decrease the prevalence of many age-associated chronic diseases." Savings in health care costs are also a factor, as noted. Physicians and patients alike can contribute to a more general "rejuvenation" of our aging society through exercise.

7. CANCER

Earlier discussions have dealt with aging versus disease, and with geriatric syndromes as "conditions" rather than overt diseases—using frailty or sarcopenia as examples of these conditions. I might have thought that no such distinction was needed for cancer, by every account a disease in terms of its often relentless pathology, but then I came across an essay by Folkman and Kalluri (2004) entitled "Concept: Cancer without Disease." They write, "many of us have tiny tumors without knowing it," and the article goes on to point out the in situ presence of carcinomas discovered at autopsy, "not significant in the person's lifetime." The further discussion of cancer without disease examines factors that will concern us in this section: for example, the "body's inherent capacity to suppress cancer occurrence," a process intimately tied to tumor suppressors and the cellular alternative to cancer, namely, senescence (Chapter 3, section 3); also, the tumor's dependence on recruiting its own blood supply—angiogenesis—and the defenses based on angiogenesis inhibitors, aspects reflecting Folkman's pioneering research. Although the authors consider the ambitious idea of "new, nontoxic drugs capable of converting cancer into a chronic manageable disease," the more desirable goal might be "to prevent disease in those individuals whom the genetics favor progression of harmless in situ tumors into a lethal form of cancer."

Age would be a key aspect to consider. Geriatricians reading this book must be aware by now that the diseases they encounter in practice reflect enormous prevalence because of their high frequency in older persons. Of course, this is the constituency that consults health care providers most often. Virtually every review of cancer begins with a sentence relating to age. A small sample: "Advancing age is the most potent of all carcinogens" (DePinho, 2000); "cancer incidence rises exponentially with age" (Campisi, 2003); "age is the most important demographic risk factor for most of our life-threatening malignancies" (Irminger-Finger, 2004); "the age of the population is the most important factor affecting the overall incidence of cancer; persons 65 years and older bear the brunt of the cancer burden" (Yancik, 2004); "cancer is largely a disease of older people" (Balducci and Ershler, 2005). Therefore, I must present an account of the biology of cancer that discusses its link with aging. This link will reflect aspects already discussed in Chapter 3, section 3 concerning the aging cell's "choice" between senescence and neoplasia.

Note that the word *crossroads* has been used with reference to cancer in the context of aging and of inflammation (Irminger-Finger, 2004). (I also noted this in "the metabolic syndrome: the crossroads of diet and genetics" [Roche et al., 2005].) *Crossroads* implies a cell's point of decision about a path taken, or not. In terms of aging, Irminger-Finger (2004) summarized many pathways "implicated in the development and progression of cancer." These are worth quoting in full as an overview, and many of these have been discussed in this text (especially Chapter 3, section 3): "Telomeres and telomerase, apoptosis and its regulators, cancer predisposition genes, premature aging syndromes, genomic instability, epigenetic changes, intracellular damage from oxidative stress, altered extracellular matrix and stromal-epithelial interactions, immune dysfunction, and changes in stem cell homeostasis."

Before discussing what Campisi (2003) calls "rival demons"—the link between cancer and aging—I want to say something about inflammation, for it
is also an underlying theme of this book. Chronic inflammation and cancer
are closely associated in the intestine, witness the reduction in intestinal neoplasia with anti-inflammatory medication (e.g., inhibition of COX-2 activity
in the infiltrating macrophages) (Greten et al., 2004) and the increased incidence of cancer in ulcerative colitis (Clevers, 2004). Ulcerative colitis-associated colorectal cancer—compared with its sporadic counterpart ("the best
understood solid malignancy in man" [Clevers, 2004])—follows a different
sequence. The sporadic cancer develops initially as an aberrant crypt focus or
microadenoma, and proceeds to grow and evolve into a carcinoma in situ, and
then into invasive adenocarcinoma. The ulcerative colitis model, if you will,
evolves from inflammation promoting (rather than initiating) carcinogenesis
(Clevers, 2004). We are back on familiar territory in which "cells of the innate
immune system secrete proinflammatory cytokines and chemokines (TNF-α,
IL-1, IL-6) as well as matrix-degrading enzymes, growth factors, and reactive
oxygen species which can promote DNA damage" (Greten et al., 2004). (See
the extended summary of innate immunity in Chapter 5, section 5.) CSF-1, a
cytokine decisive in macrophage recruitment, plays a particularly important
role in human breast cancer, in which there is a positive correlation between
poor prognosis and the density of tumor-associated macrophages. Condeelis
and Pollard (2006), "in deference to Hanahan and Weinberg (2000)" (see below), propose six traits or trophic roles in which tumor cells co-opt macrophage functions and thereby "turn a wheel" on which are pinned inflammation, matrix remodeling, angiogenesis, seeding at distant sites, intravasation
(or the entry of tumor cells into blood vessels), and tumor cell invasion.

One of the major players in the inflammatory response of tumor-associated
macrophages is *NF-kB,* in general, recognized as a key cellular mediator acting
at the "*crossroads* (my italics) of life and death" (Zhou et al., 2005), discussed
earlier in atherosclerosis and osteoporosis. NF-kB now reappears with respect
to neoplastic diseases, including aspects I will discuss here—colitis-associated cancer and hormone-independent breast cancer. There are two distinct
NF-kB activation pathways: the classical or cannonical pathway and the alternate pathway. Most of our current knowledge of the molecular link between
inflammation and carcinogenesis derives from the knowledge about the pro-
carcinogenic functions of the classical pathway (Karin and Greten, 2005). An
IKK-β complex, activated by bacterial and viral infections and also by cytokines, especially TNF-α, phosphorylates NF-kB-bound Ikβs and targets these

inhibitory proteins for ubiquitin-dependent proteosomal degradation. "The 2004 Nobel Prize in Chemistry was awarded to the scientists who discovered the ubiquitin proteosome system (UPS), the major pathway for regulated degradation of intracellular proteins" (Reinstein and Ciechanover, 2006). This degradative process is complex, tightly regulated, and highly specific, with binding of a protein tag (ubiquitin) to the substrate and the degradation of the substrate by the proteosome complex. (See review by Reinstein and Ciechanover, 2006.) More specifically for us, NF-kB is liberated from its inhibitory complex and then translocates from the cytosol to the nucleus, where it binds promotor-specific kβ consensus elements and regulates transcription of target genes "at least 150" (Zhou et al., 2005)—mostly proinflammatory cytokines (Greten et al., 2004; Andreakos et al., 2005)—but also those genes involved in control of cell growth and transformation. NF-kB contributes to cell cycle progression into the S phase (Fig. 3.2) and cell survival by transcriptionally upregulating cyclin D1: the corresponding hyperphosphorylation of the tumor suppressor protein Rb permits cell replication to go forward (Baldwin, 2001a). (Hence, the "rival demons of cell senescence vs cell transformation.") Further, the "activation of NF-kB in a broad array of pathophysiologic disorders supports a belief that the NF-kB pathway is clinically relevant and a mechanistically important target for inhibition of new drugs" (although NF-kB may also be essential for normal organ development) (Zhou et al., 2005). The clinician must be aware of the biological basis for the development of such drugs that may be used in clinical practice, and I will discuss this further as part of anti-cytokine therapy in Chapter 5, section 5.

The section on osteoporosis presented an extensive discussion of CSF-1 and RANK L derived from osteoblasts that influenced osteoclast development. Those mediators and others also play a key role in two aspects of cancer relevant to the present chapter: the first is metastases to bone, and the second is in relation to mammary gland development and neoplasia. This discussion illustrates the "interconnectedness" of processes that affect bone and the breast in development and in malignancy.

In *osteolytic* metastases, the destruction of bone is mediated by osteoclasts rather than by tumor cells (Roodman, 2004). The leading osteoclastogenic factors are those I mentioned earlier, namely IL-6, IL-1, MIF, and RANK L, enhanced by PTHrP from lung and breast cancer cells. Bone resorption by osteoclasts in osteolytic metastases from the breast releases a variety of *bone-residing* components—including cytokines and growth factors—that in turn stimulate tumor growth and enhance bone destruction, a "vicious cycle" (Ross and Christiano, 2006). Parathyroid hormone-related peptide (PTH-rP) induces the expression of RANK L on stromal cells/osteoblasts that enhance osteoclastogenesis, as discussed earlier. In a murine model of cancer-induced bone disease, *tumor cells* that made OPG may have had a survival advantage (Holen and Shipman, 2006), perhaps by OPG inhibiting caspase-activating TRAIL (TNF-related apoptosis-inducing ligand) (Xing and Boyce, 2005), making these tumor cells less subject to apoptosis. Yet in mice with bone cancer, OPG also inhibited *osteoclastic* bone destruction and "blocked *behaviors* (my italics) indicative of pain" (Honore et al., 2000). Yoneda and Hiraga (2005) refer to the creation of a "fertile soil (for metastases) in the bone microenvironment" through the triggering and release of bone-stored growth factors, including IGF-1, TGF-β, as well as bone mineral components (calcium and phosphate). It must be appreciated how important these bone matrix components are clinically (e.g., hypercalcemia), and for our purposes, biologically. Bone-derived IGF-1 binds to its receptors in the cancer cell and activates a cascade of cytoplasmic signaling molecules, including insulin receptor substrate-1 (IRS-1), phosphatidylinositol-3-kinase (PI3K), protein kinase B/Akt, and the transcription factor NF-kB (Fig. 4.14). Then NF-kB enhances target gene transcription, thereby stimulating cell proliferation, inhibiting apoptosis, and promoting proliferation of tumor cells in the bone. Note the widespread nature of these *pathways* we have encountered: interaction of IGF-1 with its receptor in longevity paths across widely separate species (see Chapter 3, section 4); the activation of the IRS-1 and PI3K pathway in the discussion on insulin resistance (Chapter 4, section 3); the rather universal roles of NF-kB as part of the response to inflammatory mediators in atherosclerosis, cytokines and stress, osteoporosis, and cancer (Chapter 5, section 5).

TGF-β is also stored in bone (Janssens et al., 2005) and its release promotes PTHrP in the cancer cells; this in turn stimulates up-regulation of RANK L in osteoblasts, a further vicious cycle. Little wonder then, that Yoneda and Hiraga (2005) refer to "crosstalk between cancer cells and the bone microenvironment in bone metastases." But Danielpour and Song (2006) also refer to "crosstalk" between the IGF-1- and TGF-β-signaling pathways. TGF-β signals are propagated through serine/threonine kinase receptors which activate secondary signals, such as SMAD 2 and 3, conveying signals to the nucleus, and activating other signals noted with IGF-1, including PI3K and the mitogen-activated protein kinase (MAPK) family. Evidence exists that the transforma-

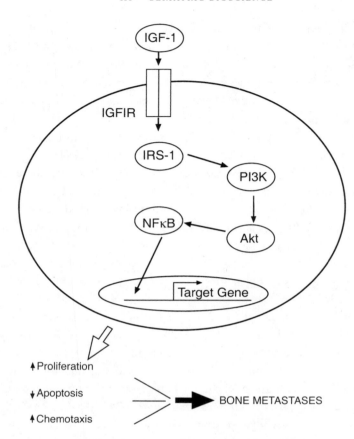

Figure 4.14. Role of bone-derived insulin-like growth factor-1 (IGF-1) in cancer cells. IGF-1 released from bone by osteoclastic bone resorption binds to type I IGF receptors (IGFIR) on the cancer cell membrane, causing homodimerization of IGFIR, followed by activation of insulin receptor substrate-1 (IRS-1), phosphatidylinositol-3-kinase (PI3K), protein kinase B (Akt), and the transcription factor NF-kB. NF-kB then stimulates target gene transcription, thereby stimulating cell proliferation and chemotaxis, and inhibiting apoptosis, leading to the promotion of bone metastases. Reprinted from Yoneda T, Hiraga T. 2005. Crosstalk between cancer cells and bone microenvironment in bone metastasis. *Biochem Biophys Res Commun* 328: 679–87, © 2005, with permission from the authors and Elsevier.

tion of TGF-β to tumor promoter from tumor suppressor is under the control of IGF-1 signaling pathway, creating for TGF-β a pleiotropic role—a word we have been familiar with in relation to aging, and will be in relation to statins. In fact, I have not written much until now about TGF-β and the variety of its biological roles, as reviewed by Sporn (2006) in a historical context. With respect to cancers, IGF-1 has been notable for the expression of prostate neoplasia, and TGF-β for epithelial-to-mesenchymal transitions in mammary cancers, with tumor progression (Cher et al., 2006).

It is interesting that, in the same review on osteolytic bone metastasis, Roodman (2004) could write that "the mechanisms of *osteoblastic* metastases and the factors involved are unknown." Mundy (2003) suggests this deficient knowledge is due to "the virtual non-existence of animal models that consistently produce prostate cancer-inducing osteoblastic metastases." In a model of osteoblastic metastasis caused by human breast cancer cells, endothelin-1 appeared to be the responsible mediator, but its role in bone remodeling is generally unfamiliar. Sharma (2006) mentions endothelin-1 as one of the endothelial secretory products of adipocytes that is a vasoconstrictor and inhibits further adipogenesis. Osteoblasts also produce endothelin-1 and perhaps this has mitogenic effects on bone growth with the addition of vascular growth factors from the tumor cells (Veillette and von Schroeder, 2004). Platelet-derived growth factor (PDGF) may also be involved in osteoblastic metastasis. Note also that, even in the active state of bone formation, bone resorption takes place, and markers are increased; hence, drugs like the bisphosphonates are effective in reducing bone pain and, because of osteoclastic inhibition, curtailing one part of the vicious cycle.

CSF-1 plays a prominent role in breast development and neoplasia. When CSF-1 binds to its cell surface receptor encoded by the *cellular* homologue of an oncogene called c-fms (derived from retroviral oncogene v-fms contained in a feline sarcoma virus), a series of events are triggered in macrophages that include chemotaxis, phagocytosis, synthesis, and secretion of proteolytic enzymes (urokinase-type plasminogen activator, thromboplastin), and other cytokines that have figured prominently in this book—IL-1, IL-6, and TNF-α (Sapi, 2004). Pertinent to this section is a role for CSF-1 and its receptor in the mammary gland during lactation, when it initiates ductal sprouting. It is also now known that abnormal activation of CSF-1 and its c-fms receptor occur in breast carcinoma with a poor prognosis (Pollard, 1997; Sapi and Kacinski, 1999; Lin et al., 2001), showing that pathways in normal development may extend to dramatic stimulation, growth, and malignant invasiveness, often accompanied by infiltration of macrophages (Lin et al., 2001). A link here may be with macrophage production of growth factors and proteases.

Activation by RANK L of its RANK receptor also occurs in mammary gland epithelial cells for lobular-alveolar development late in pregnancy, and for RNA message for the major milk protein β-casein. Thus, as Fata et al. (2000)

points out, there are "links with the development of the lactating mammary gland—the organ required to provide nourishment and calcium to offsprings in mammalian species—and bone, the principal source—through skeletal de-mineralization—of calcium demands of the newborn." Then, follows a sentence pregnant with ideas: "thus, mammals took a gene product that is the master regulator of bone metabolism and calcium turnover in the whole organism and subverted it to stimulation of mammary gland development during pregnancy."

Estrogen is implicated in the development of breast cancer by binding to the estrogen receptor and stimulating proliferation of mammary cells, and also perhaps by products of estrogen metabolism that directly damage DNA (Deroo and Korach, 2006). Chapter 5, Section 4 will say more about estrogen receptors and their modulation by selective estrogen receptor modulators (SERMs).

A final word about NF-kB in linking chronic inflammation and cancer, beyond the gut and the breast. Clevers (2004) notes that hepatocytes require an intact NF-kB-Ikkβ pathway to be protected against apoptosis to progress to hepatocellular carcinoma in a chronic inflammatory setting. "An intact apoptotic machinery is often unavailable in established malignancies" (Kahlem et al., 2004). The ability to evade apoptosis is one of the "hallmarks of cancer," and "induction of apoptosis during tumor promotion represents a critical step for chemoprevention" (Hanahan and Weinberg, 2000; Clevers, 2004). Indeed, Kahlem et al. (2004) point out that "it is a research priority to invent small compounds—or gene therapy-based approaches that may resensitize cancer cells to death signals. Likewise, one can envision lesion-based strategies to restore a defective senescence response." Thus, it is clear how useful the "two sides of the coin" thinking becomes, as discussed in Chapter 3, section 3 (Fig. 3.3): cell growth-arrest mechanisms via senescence/apoptosis, or unrestricted cell proliferation with malignancy.

Here we can see a glimpse of the ongoing struggle—on a biological basis, at least—between cells going gently into that good night of permanent sleep, called senescence, or to immortal transformation. The latter prevails when a multistep process of genetic and epigenetic changes occur over time (Lin et al., 2001). "Growth factors and their receptors represent signaling pathways that are involved in the transduction of mitogenic stimuli, such as mitogen-activated protein (MAP) kinase and the phosphatidylinositol (PI3K) kinase pathways. These are often constitutively activated in tumor cells, while upregulated expression of antiapoptotic genes also occurs frequently. All of these

alterations enable tumor cells to elude the commitment to terminal differen-
tation and quiescence that normally regulates tissue homeostasis." DePinho
(2000) presents a dramatic picture of the evolution of epithelial carcinomas,
beginning with the "age-dependent telomere attrition, somatic mutations during aberrant epithelial proliferation, neutralization of the Rb/INK/p53—dependent senescent checkpoint, culminating in a cellular crisis—a period of
severe telomere dysfunction, genomic instability and cell death." He notes
"dysfunctional *telo*mere-induced gen*omic* instability (or telonomic instability) combined with reactivation of telomerase, might enable a subset of post-
crisis cells to emerge with a genetic profile permissive for malignant progression."

Campisi (2004) views the dominant biological event related to *aging* as
cellular senescence—a broader concept than *replicative* senescence induced
primarily by short, dysfunctional telomeres. A host of determining factors
exist for this broader concept of cellular senescence, as discussed in Chapter 3, section 3, including DNA damage, oncogene expression, stressful mitogenic signals and changes in chromatin, and especially activation of tumor
suppressors p53 and Rb. The "senescence response is, very likely, a fail-safe
mechanism to prevent the growth of potentially malignant cells" (Campisi,
2004). Although senescence of cells tends to be toward the end of the story,
some cells, in particular, epithelial cells, may also be on the verge of further
emergence, as we have seen, to become partly or overtly neoplastic. Again,
"degradative enzymes, cytokines, and growth factors secreted by senescent
cells (highly accumulated in older age) create a tissue microenvironment that
is permissive for the expression of malignant phenotypes." With age, it is also
likely that an "oncogenic mutation has occurred with increasing frequency in
a cell" in close proximity to a senescent cell and its surrounding permissive
tissue environment. DePinho (2000), also citing Campisi, notes that "the accumulation of senescent stromal cells with ageing, coupled with altered expression and elaboration of proteases and cytokines, have been hypothesized
to create a tissue microenvironment that is more permissive for the growth of
oncogenically transformed epithelial cells." Thus, as I began, it is evident why
cancer is more widely prevalent in older age.

So "when all that story's finished, what's the news?"

In 2000 Hanahan and Weinberg put forth a general view of carcinogenesis
in a remarkable article synthesizing knowledge of the biology of the cancer
cell. Reflecting on the achievements and insights gained toward the end of
the twentieth century, the authors added a conceptual dimension that they

predicted would be the approach of the future. These are the authors' six fundamental and "essential alterations in cell-physiology that collectively dictate malignant growth":

- Self-sufficiency in growth signals. Many oncogenes in cancer also act by mimicking normal growth signals.
- Insensitivity to growth-inhibitory (and growth) signals. If the progression through the cell cycle phase (G_1-S) is to occur in a cancer cell, the tumor suppressor Rb function must be blocked.
- Evasion of programmed cell death (apoptosis): a hallmark of most and perhaps all cancer cells.
- Limitless replicative potential. This is the aspect that defies senescence by way of telomeres maintained by the enzyme telomerase, and thus failing to shorten.
- Sustained angiogenesis, derived by virtue of up-regulated access to vascular growth factors and down-regulated inhibitors to acquiring oxygen and nutrients.
- Tissue invasion and metastases.

Perhaps it is worthwhile to quote (somewhat selectively) the conclusion of Hanahan and Weinberg (2000), which is poetic in its prose and optimistic in its vision for the future, so that those readers particularly knowledgeable in this field may judge whether the authors' foresight in the year 2000 holds up today:

> With holistic clarity of mechanism, cancer progression and treatment will become a rational science, unrecognizable by current practitioners. It will be possible to recognize with precision how and why treatment regimens and specific anti-tumor drugs succeed or fail. We envision anti-cancer drugs targeted to each of the hallmark capabilities of cancer . . . to detect and identify all stages of disease progression, prevent incipient cancers from developing, and cure pre-exiting cancers, elusive goals at present. One day we imagine that cancer biology and treatment—at present a patchwork quilt of cell biology, genetics, histopathology, biochemistry, immunology, and pharmacology—will become a science with a conceptual structure and logical coherence that rivals chemistry and physics.

Few expressions so eloquently project the emergence of bioscience into the practice of clinical medicine.

8. ALZHEIMER DISEASE
Peter Davies, Ph.D.

Only in the past 25 years has Alzheimer disease been recognized as the largest single cause of impairment of cognitive function in elderly people. Before this time, terms such as organic brain syndrome, senile dementia, and senility were commonly applied to elderly individuals with dementia. Largely through careful neuropathologic studies of elderly people, the specific pathology of Alzheimer disease was found in about 70 percent of individuals with dementia. In 2006, it was estimated that about 4.5 million people in the United States have Alzheimer disease.

Neuropathologic Diagnosis of Alzheimer Disease

Formal criteria for diagnosis of Alzheimer disease require the demonstration of the presence of two lesions: the senile or neuritic plaque, and the neurofibrillary tangle in multiple regions of the cerebral cortex, including the hippocampus, a region thought to be of great importance for memory function. Small numbers of neurofibrillary tangles are frequently found in the hippocampus of healthy elderly patients, but the presence of tangles in temporal, parietal, and frontal cortex as well is almost invariably associated with dementia. Research has unraveled the molecular nature of both plaques and tangles, and because this work has implications for the development of therapies for Alzheimer disease, it is reviewed in brief below.

Neuritic (Senile) Plaques

The neuritic plaque is a complex, extracellular lesion. Protein deposits at the center of the plaque are composed largely of Abeta (also called β-amyloid), aggregated peptides of 38–43 amino acids formed by proteolytic cleavage of a ubiquitous precursor, usually referred to as amyloid precursor protein, or APP. Amyloid deposits usually contain several other proteins, including α-1-antichymotrypsin and ApoE, and in Alzheimer disease are surrounded by a halo of degenerating neuronal processes (neuritis), as well as reactive glial cells (microglia and astrocytes). Abeta deposits are also commonly found in the frontal cortex of healthy elderly people, but it is unusual in these cases to see the same degree of neuritic involvement and gliosis as is common in Alzheimer disease. Most workers in the field currently believe that plaques are formed in response to Abeta deposition, although it is still possible to imagine that amyloid deposition is the result of neuronal degeneration. The most popular hypothesis attempting to explain the pathophysiology of Alzheimer disease proposes that Abeta is the critical determinant in the development of both the neuritic plaques and neurofibrillary tangles, and has been the focus of numerous studies designed to lead to the development of treatments that might affect the course of Alzheimer disease. The "amyloid cascade hypothesis" proposed and supported by John Hardy and Dennis Selkoe (Hardy and Higgins, 1992; Hardy and Selkoe, 2003; Hardy, 2004; Selkoe, 2004), among others, proposes that the formation and deposition of Abeta initiates a cascade of pathologic changes in brain. The two most important pieces of evidence in favor of this hypothesis are discussed below.

Genetic Evidence

Although most cases of Alzheimer disease appear to be sporadic (occurring in patients with no clear family history of the disease), there are families in which the disease is inherited as an autosomal dominant disease. Efforts to uncover the genetic basis for these inherited cases resulted in the discovery of a few families in which mutations in the APP gene (on chromosome 21) segregated with the disease (Hardy, 1997; St George-Hyslop, 2000). Several APP mutations have been discovered, and these cause amino acid changes that cluster in and around the region cut from the precursor protein to yield Abeta (Fig. 4.15) (Tanzi et al., 1996). The fact that these mutations cause Alzheimer disease clearly indicates that abnormalities in the proteolytic processing of APP can be central to the disease process. Although APP mutations are rare even in families with familial Alzheimer disease, the search for additional genetic defects continued, and resulted in the discovery of mutations in two other genes, presenilin 1 on chromosome 14, and presenilin 2 on chromosome 1 (Tanzi et al., 1996; St George-Hyslop, 2000). Both genes appear to produce proteins that are intimately involved with the proteolytic processing of APP, in the cleavages that liberate the C terminus of the Abeta peptides (Fig. 4.15) (Hardy, 1997; Haass and De Strooper, 1999; Wolfe and Haass, 2001). Mutations in the presenilin 1 gene are the most common cause of familial Alzheimer disease, and more than 100 distinct mutations have been identified in this gene. This again argues that APP processing is a central event in Alzheimer disease,

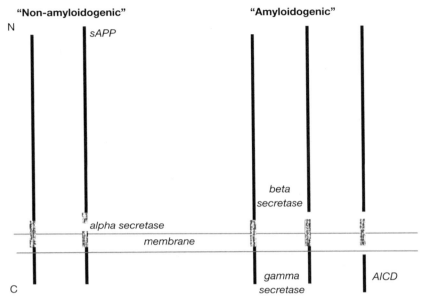

Figure 4.15. The amyloid precursor protein is thought to be a transmembrane protein, with the N terminus (N) outside the cell, and the C terminus (C) within. Cleavage of amyloid precursor protein (APP) follows one of two paths. The "nonamyloidogenic" path involves alpha secretase cleavage within the Abeta domain (pattern) to release a secreted part of the protein (sAPP). In most cell types, about 80 percent of the APP is processed through this pathway. The "amyloidogenic" pathway involves sequential cleavage of APP by beta secretase (also called BACE), and then gamma secretase cleavage to release the Abeta fragment. The "amyloidogenic" pathway operates in most cell types, and the Abeta released is not necessarily deposited. Further details regarding the gamma secretase cleavage are presented in Figure 4.16.

and the simplest explanation, adopted by the amyloid cascade hypothesis, is that abnormalities in the generation of Abeta are causal in the familial cases.

In Figure 4.15, two proteolytic activities are necessary to liberate the Abeta peptides from the precursor APP molecule. The presenilins are part of a complex multiprotein protease system called gamma secretase that liberates the C terminus: the N terminus is liberated by an enzyme called beta-secretase or BACE (Vassar and Citron, 2000; Citron, 2001). A vast amount of work on APP processing has revealed that two basic pathways for APP processing exist: the beta and gamma secretase processing that liberates Abeta, and a second pathway, involving a relatively poorly characterized enzyme called alpha secretase, which cuts the precursor within the Abeta peptide sequence. It is common to see these pathways referred to as the amyloidogenic and nonamyloidogenic pathways, although this is not strictly accurate. Both pathways are used, as far as we know, by all tissues, with perhaps 80–90 percent of APP being processed by alpha secretase. Thus the Abeta peptides are produced under normal conditions, usually without amyloid deposition.

Further genetic evidence apparently supporting the amyloid cascade hypothesis comes from two other areas. Individuals with Down syndrome (trisomy 21) have three copies of the APP gene (and of course of any other gene on chromosome 21). These individuals appear to develop the pathology of Alzheimer disease with a much higher frequency than the general population. Perhaps as many as 75 percent of individuals with Down syndrome who die at age 40 or over have the plaques and tangles typical of Alzheimer disease. The frequency of occurrence of dementia in this population is still under debate: it is difficult to apply standard neuropsychological tests in populations with severely limited intellectual capacity. It is reported that individuals with Down syndrome have detectable amyloid deposition as early as their teens, although how consistently this is true is currently unknown (Isacson et al., 2002). There are also recent reports of the development of Alzheimer disease in patients with duplications of small regions of chromosome 21 that contain the APP gene. These individuals do not have the Down syndrome phenotype, but develop Alzheimer disease in their early fifties (Rovelet-Lecrux et al., 2006).

Other Evidence

As mentioned above, examination of the brains of healthy (intellectually intact) elderly people often reveals amyloid deposits in the cerebral cortex, in the absence of neuritic or glial reactivity and in the absence of neurofibrillary tangles (Braak et al., 1989). One way to interpret this result is to postulate that amyloid deposition occurs before any of the other signs of Alzheimer disease. This would imply that had these healthy elderly people survived longer, they would have gone on to develop full-blown Alzheimer pathology. If this is true, then about 80 percent of the population would develop Alzheimer disease if they lived long enough: about 80 percent of the people 80–85 years old have significant amyloid deposition in the cerebral cortex. It is reported that individuals with Down syndrome have detectable amyloid deposition as early as their teens, although how consistently this is true is currently unknown.

Therapeutics Based on the Amyloid Cascade Hypothesis

If one accepts that the evidence for the amyloid cascade hypothesis is compelling, then the therapeutic opportunities are obvious. Because cleavage of the amyloid precursor at two sites is necessary to liberate Abeta peptides, it should be possible to develop specific inhibitors of one or both of the protease systems (beta and gamma secretases) and limit or prevent formation of Abeta (Dewachter and Van Leuven, 2002; Hardy and Selkoe, 2002). This indeed is an approach that the pharmaceutical industry has embraced. Recall the success in developing inhibitors for the HIV protease, and the success of such inhibitors in clinical practice. Most of the major pharmaceutical companies have programs to discover inhibitors of either beta secretase or gamma secretase or both. At the time of writing (spring 2006), the first clinical trials with secretase inhibitors are beginning or are in a late planning phase. Some potential problems have emerged. It is clear that the gamma secretase complex that liberates the C terminus of Abeta is also involved in the processing of other proteins, including some cell surface receptors. In this group is Notch, a receptor that is important in the developing brain. Mice with the presenilin 1 gene deleted die around birth with major skeletal and CNS malformations and pharmacologic inhibition of Notch processing can also cause major developmental abnormalities (Selkoe, 2001; Selkoe and Kopan, 2003). What role the Notch receptor system might play in the adult human is still not clear, but the potential for side effects from gamma secretase inhibitors seems clear, although there are indications that it is possible to develop inhibitors with some selective effects on APP processing. Mice in which the gene encoding beta secretase is deleted are born normally, and do not have major structural abnormalities. Whether or not there are more subtle defects is still controversial. The results from clinical trials with secretase inhibitors are eagerly awaited.

Other approaches to treating the "amyloid cascade" are also in clinical trials. Inhibitors of amyloid aggregation are being tested, with the assumption that the aggregation of Abeta is the critical factor in Alzheimer disease (Gestwicki et al., 2004; Rzepecki et al., 2004). Again, results are eagerly awaited. These pharmacologic studies are currently seen by many as direct tests of the validity of the amyloid cascade hypothesis.

Problems with the Amyloid Cascade Hypothesis

Much of the preceding discussion appears to be consistent with the amyloid cascade hypothesis, but from careful examination of brains from healthy elderly and early Alzheimer disease cases one of the major problems with the amyloid cascade hypothesis emerges. Neuropathologists and neuroanatomists who examine these cases agree that neurofibrillary tangle pathology begins in the hippocampal formation (Braak and Braak, 1995, 1997), and this makes sense in terms of the early appearance of memory problems in persons with Alzheimer disease. However, in the early stages of tangle pathology, there is little evidence of amyloid deposition in the hippocampus. Amyloid deposition is commonly found first in the frontal cortex, which is the region in which tangles are found only in relatively late stage cases. In general, it is agreed that a major spatial and temporal mismatch exists between amyloid deposition and neurofibrillary tangle formation (Braak et al., 1989; Armstrong et al., 1993; Davies, 1994). A simple reading of the amyloid cascade hypothesis would imply that close spatial and temporal association should exist, but it is hard to find evidence that this is the case. This is especially true in transgenic mouse models, made in part to test the amyloid cascade hypothesis. Transgenic mice have been constructed to express mutant human APP genes in the brain, sometimes along with mutant human presenilin 1 genes (Duff, 2001; Kurt et al., 2001). The rational here is that these mutations cause the disease in humans, and overexpression of the mutant human proteins should mimic Alzheimer disease in mice. To date at least, these models, although interesting and informative for research, have failed to yield good models of Alzheimer disease in humans. The basic problem is that many of these mice develop deposits of Abeta, and in some cases, enormous numbers of amyloid deposits can be found. However, these mice do not develop the neurofibrillary tangles so characteristic of the condition in humans, nor do they show much evidence of cell death and degeneration. These mice appear to be great models for the study of Abeta amyloid deposition, but not of the rest of the pathology.

The dissociation between amyloid deposition and neurofibrillary tangle formation and cell death has led to the suggestion that, in fact, the Abeta deposits are not causal agents of the neurodegeneration, but some other form of Abeta peptides (Selkoe, 2002). Two issues here may prove to be important. First, an assumption behind much of the work on Abeta in Alzheimer disease is that mutations that cause Alzheimer disease *increase* the amounts of Abeta

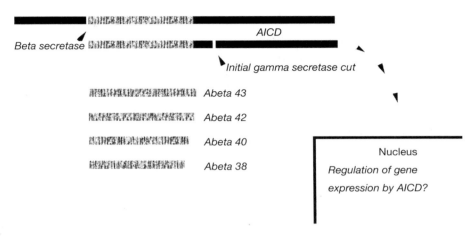

Figure 4.16. Although reference is often made to amyloid or Abeta (also called beta-amyloid) production and deposition, it is actually a mixture of peptides that is produced. The work of Kakuda et al. (2006) suggests that the initial gamma secretase cut generates an Abeta-containing peptide of 48–50 amino acids that is subsequently processed down to the shorter peptides. In brain, the predominant species are Abeta 40 and 42, but shorter and longer peptides are detectable. Most of the mutations in the presenilin 1 gene that cause Alzheimer disease reduce the total amount of amyloid precursor protein (APP) that is cleaved by gamma secretase, and reduce the amount of Abeta 40 produced. However, the amount of Abeta 42 produced appears to be less affected by the mutations, such that there is an increase in the ratio of Abeta 42 to 40. The significance of this change is still unclear. Mutations in the APP gene around the gamma secretase site cause Alzheimer disease and have a similar effect on Abeta production.

produced from APP. This may not be true of at least some of the mutations in the APP gene (Sato et al., 2003; Kakuda et al., 2006), nor does it seem to be true, in general, of the presenilin mutations. As one might expect, if presenilin 1 is an important part of a protease that cleaves Abeta, mutations in presenilin 1 would be expected to impair function of the protein, rather than increase its efficiency. This appears to be the case: careful analysis of the effects of prese-nilin mutations that cause Alzheimer disease shows that, in general, they do *decrease* the total amount of Abeta peptides produced (Bentahir et al., 2006). In doing so, they also seem to shift somewhat the site at which APP is cleaved, so that relatively more of a 42-amino-acid peptide is made, in place of the normally more abundant 40-amino-acid peptide (Fig. 4.16). Because the 42-amino-acid peptide aggregates more readily than the 40-amino-acid peptide, and is more toxic in some in vitro assays, the hypothesis can be modified to say that a relative increase in proportion of Abeta 42 produced is responsible for the amyloid cascade.

The second point is that the aggregation state of Abeta peptides appears to be a critical determinant of toxicity, at least in some in vitro assays. In most systems, monomeric Abeta peptides are not toxic at concentrations that could be expected to be present in vivo. The highly aggregated peptides found as deposits in mouse or human brain do not appear to be toxic either, based on the lack of neuronal death in mice with large numbers of amyloid deposits. Intermediate forms, currently called "oligomers," can be isolated under some conditions and appear to be toxic in vitro (Selkoe, 2002). Perhaps relatively small changes in either the rate of production of Abeta, or a change in the ratio of the 40- to 42-amino-acid peptide forms has a major impact on the formation of toxic oligomers, and sequestration of these species in amyloid deposits actually limits their toxicity.

In either case, it is clear that the simplest version of the amyloid cascade hypothesis—that amyloid deposition leads to the rest of the pathology of Alzheimer disease—will have to be discarded. Whether .the generation of some species of Abeta is a critical issue remains to be determined. It has been suggested that mutations in APP and presenilin genes do indeed implicate APP processing as important for Alzheimer disease, but generation of Abeta is just one consequence of APP processing. The C-terminal domain of APP, liberated inside cells on gamma secretase cleavage, may be an important regulator of either cytosolic calcium concentration, or of nuclear gene expression (Fig. 4.16) (Neve et al., 2000). Both functions have been suggested, and evidence shows that presenilin mutations alter regulation of calcium movement in vivo. Whatever the case, clinical trials of agents designed to inhibit Abeta formation or aggregation are currently under way, and the results of these studies may provide the next clues as to the relative importance of amyloid in Alzheimer disease.

ApoE and Other Risk Factors

There are three genes in which mutations have been found that cause Alzheimer disease: the amyloid precursor protein, presenilin 1, and presenilin 2. As discussed earlier, defects in these three genes appear to alter processing of the amyloid precursor protein, and cause early-onset (at less than 60 years of age in most cases) Alzheimer disease. An active search continues for genetic factors that are important in the more common late-onset disease, and the best established of these are in the gene encoding ApoE. There are three major variants of the ApoE gene in the human population: ApoE2, ApoE3, and ApoE4. It is now clear that individuals who carry an ApoE4 allele are at increased risk for the development of Alzheimer disease, and that having two ApoE4 alleles is associated with a further increase in risk (Strittmatter et al., 1993; Roses, 1998). In the general population, 15–20 percent of individuals carry an ApoE4 allele, but in surveys of patients with Alzheimer disease this figure is closer to 40 percent. It is still not clear exactly how the ApoE4 allele increases risk. ApoE is a major lipid transport protein (Mahley et al., 2006), and it has been suggested that less efficient cholesterol transport by ApoE4 (relative to ApoE2 or ApoE3) is an important issue (Wolozin et al., 2004).

The presence of one or even two ApoE4 alleles does not make the development of Alzheimer disease inevitable. ApoE4 is neither necessary nor sufficient for the development of this condition. ApoE4 is probably the strongest genetic risk factor identified to date, but probably many more genetic factors influence risk. Numerous genetic association studies have suggested associations between gene variants and the development of Alzheimer disease. A listing of these can be found at www.alzforum.org/res/com/gen/alzgene/default.asp.

Neurofibrillary Tangles

The numbers of neurofibrillary tangles in the cerebral cortex of patients with Alzheimer disease have been shown to be correlated with the extent of cognitive impairment (Dickson et al., 1995), and similar correlations have been difficult to find for amyloid deposits (Naslund et al., 2000). This has led many to conclude that the formation of tangles, rather than amyloid deposition, impairs neuronal function. Neurofibrillary tangles are formed largely from the microtubule-associated protein tau. Tau is a soluble cytoplasmic protein that is thought to be normally involved in the stabilization of neuronal microtubules. A single tau gene in humans is located on chromosome 17, and alternate splicing of the mRNA produces six isoforms of tau in the adult CNS. Microtubule binding involves imperfect repeat sequences coded in four exons, one of which (exon 10) is either included or excluded to yield tau with three (called 3R tau) or four (4R tau) microtubule binding domains (Fig. 4.17). Microtubule binding is believed to be regulated by phosphorylation of sequences in and around the microtubule binding domains. Neurofibrillary tangles are masses of filamentous tau aggregates, often made up of filaments with a paired helical appearance (Fig. 4.17), but often mixed with straight filaments. Tau isolated from tangles is hyperphosphorylated, relative to tau isolated from normal brain, and there is evidence for major conformational changes in the protein, which perhaps occurred before filament formation. Careful examination of cases of early-stage Alzheimer disease reveals certain tau phosphorylations and conformational changes are the earliest detectable neuronal abnormalities occurring in this condition.

For several years, changes in tau and the subsequent formation of neurofibrillary tangles were seen by many as secondary changes in Alzheimer disease, despite the evidence of their early occurrence. In 1998, mutations in the human tau gene were discovered to be the cause of some cases of "frontotemporal dementia" (Hutton et al., 1998, Spillantini et al., 1998). These are usually early-onset (in the forties and fifties) dementia cases that are characterized by massive degeneration of the frontal and temporal cortex, often with tangle-type pathology and signs of Parkinson disease. To date, more than thirty different tau mutations have been found as causes of frontotemporal dementia, and it thus appears that neuronal degeneration can be caused by a variety of abnormalities in tau, at least in the frontal and temporal cortex. It has been a surprise to learn that single-amino-acid changes in tau can be a cause of massive neuronal degeneration, and even more surprising that similar pathology can result from mutations that do not change the amino acid sequence of tau, but alter the splicing of exon 10, changing the ratio of 3R to 4R tau. Although tau mutations have not been found in cases of Alzheimer disease, the neuronal degeneration found in cases of tau mutation raises the possibility that changes in tau phosphorylation and conformation that do occur in Alzheimer disease may in fact be a cause of neuronal death. This possibility is supported by at least some of the transgenic mice that have been made, in which various forms of human tau are expressed in the mouse brain. Some of these mice clearly do show neuronal degeneration (Lewis et al., 2000; Andorfer et al., 2005). Games et al. (2006) review "mice as models: transgenic approaches and Alzheimer's disease." In these models, mutations in proteins such as amyloid-beta precursor protein and presenilin 1, other high-risk factors (ApoE4), as well as tau protein, can be expressed and the pathology studied.

Therapeutics for Tau Pathology

If phosphorylation and conformational changes in tau cause neuronal degeneration, then it will be necessary to develop drugs to try to prevent these

A

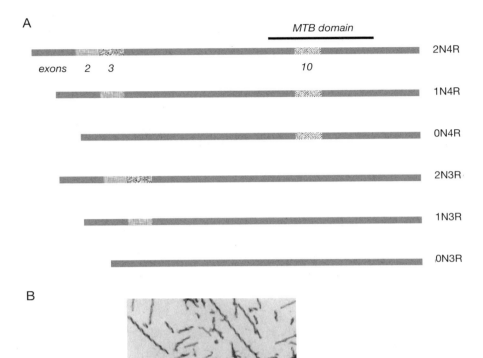

MTB domain

2N4R

exons   2   3   10

1N4R

0N4R

2N3R

1N3R

0N3R

B

Figure 4.17. (A) The tau protein is coded in a single gene, which through splicing of
the mRNA actually produces 6 different isoforms of the protein in the adult human
brain. Exons (protein coding sequences in the gene) 2 and 3 can be included or
excluded from the mRNA, producing protein isoforms that differ in the N terminus
(to the left in the figure). What difference these sequences make to the function of
tau is unclear. The microtubule binding (MTB) domain of tau is coded in four exons:
9, 10, 11, and 12. Exon 10 sequences can be included or excluded from tau, and the
longer microtubule binding domain that results from inclusion of exon 10 sequences
is thought to allow tighter tau binding and more stable microtubules. The common
abbreviation used for each protein isoform is written to the right of the figure.
(B) Paired helical filaments composed largely of tau are the basic building blocks of
the neurofibrillary tangles of Alzheimer disease. Masses of these filaments fill the
cytoplasm of neurons in this disease, although here they are shown dispersed after
purification.

changes. At present, most of the focus in this field has been to try to identify inhibitors of specific protein kinases that may be responsible for the abnormal phosphorylation. The presumption, still unproven, is that activation of a specific kinase is responsible for the increased phosphorylation, and that this leads to the conformational changes and tau aggregation into paired helical filament structures and tangles. The two candidate kinases cited most often are GSK3beta (Ferreira et al., 1997; Tomidokoro et al., 2001) and cdk5 (Noble et al., 2003), but several others have been suggested (Anderton et al., 2001; Zhu et al., 2001). Programs are underway in the pharmaceutical industry to develop inhibitors of these two kinases, but no clinical trials have yet been reported. Some companies are also trying to develop inhibitors of tau aggregation, but again there have been no clinical results.

Amyloid or Tau: Both or Neither?

For some time, a division has existed in the research on Alzheimer disease, with the majority focusing on amyloid and a vocal minority working on the tau area. This division is increasingly being seen as artificial, in that Alzheimer disease involves both plaque *and* tangle pathology. Although the amyloid cascade hypothesis in various modified forms remains dominant, interest is increasing in finding mechanisms that may lead to both amyloid and tau pathology. Most persons with Alzheimer disease do not have mutations in the amyloid precursor protein or presenilin genes, and presumably develop the pathology for other reasons. One promising area of research concerns activation of cell cycle or mitotic mechanisms.

It is widely believed that differentiated neurons do not reenter the cell cycle and that, in the adult brain, the only cells expressing cell cycle markers are glia, or the small number of precursor (stem) cells, that in some limited circumstances can differentiate into neurons. Thus, the first reports of expression of cell cycle markers in neurons in the brains of patients with Alzheimer disease (Vincent et al., 1996) were greeted with some skepticism. Several groups subsequently confirmed and expanded the observations, and these now include demonstration of G_1-, G_2/M-, and S-phase markers in neurons (Vincent et al., 1997; Yang et al., 2001; Herrup and Arendt, 2002; Yang et al., 2003). (See Fig. 3.2 for a diagram of the cell cycle.) Some mouse models of tangle formation also show clear signs of cell cycle-mediated neuronal death (Andorfer et al., 2005), and recent work by two independent groups provides strong evidence for activation of the cell cycle in experimental models of stroke (Kuan et al., 2004; Wen et al., 2004). This work has led to the suggestion that activation of protein kinases critical for cell cycle regulation might be what is responsible for the increased phosphorylation of tau in Alzheimer disease. This increased phosphorylation might lead to tangle formation. Note also that the amyloid precursor protein is heavily phosphorylated in cells in the G_2/M phase of mitosis (Suzuki et al., 1997). Perhaps the cell cycle kinases might also be responsible for alteration of processing of the amyloid precursors, with amyloid deposition as one result.

Further work on cell cycle or mitotic mechanisms in Alzheimer disease (and other neurodegenerative diseases) is clearly needed. The preceding discussion is intended simply to illustrate one new approach to the study of Alzheimer disease, an approach that seeks to unify the amyloid and tau fields. Several other "unification" theories are under development, and this is an area to watch over the next few years.

Other Therapeutics

In the late 1970s, evidence for a deficiency of the neurotransmitter acetylcholine was reported by several groups. Current treatments for Alzheimer disease include the cholinesterase inhibitors Aricept, Exelon, and Reminyl, designed to try to remedy this deficiency. Although these compounds have sufficient efficacy to merit FDA approval, they do not appear to influence the course of the disease, and they provide only modest, temporary improvements in patient symptoms (Lleo et al., 2006). This is also true of Namenda (memantine), which may provide modest benefits for patients in relatively late stages. A great deal of attention has been paid to the idea that there is an inflammatory component to the pathology of Alzheimer disease (Wyss-Coray, 2006); this idea stems from observations of increased microglial and astrocytic reactivity in the brains of persons with Alzheimer disease. There are also reports of decreased occurrence of Alzheimer disease in patients who used nonsteroidal anti-inflammatory drugs (NSAIDs) for long periods. At least to date, attempts to treat existing Alzheimer disease (rather than try to reduce the incidence of the disease) with NSAIDs have been disappointing (Lleo et al., 2006). These attempts do continue, and recent basic science data have suggested links between certain NSAIDs and amyloid production that do not necessarily involve inhibition of inflammation. Some NSAIDs may have direct effects on the production of amyloid peptides, and these compounds are the subject of intense basic scientific and clinical research.

In addition to the great interest in anti-inflammatory agents, a host of new compounds are in various stages of clinical trials for Alzheimer disease. These are too numerous and diverse to allow detailed discussion, and updated lists of formal clinical trials can be found by searching on "Alzheimer's disease" at: www.clinicaltrials.gov/. At the time of writing, 98 studies were listed as recruiting patients, an indication of the high level of activity in this field. Further developments in this area are likely to come quickly, as an increased understanding of the biology of Alzheimer disease leads to new approaches to therapy.

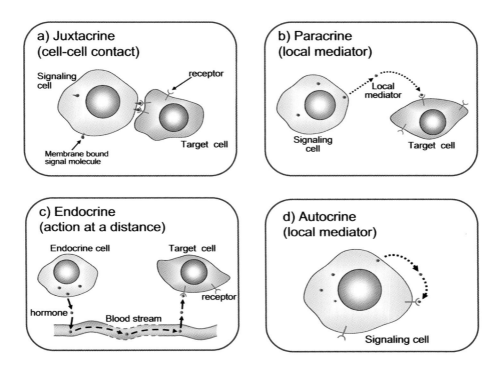

Figure 2.1. Modes of regulation by cellular signals, including cytokines, growth factors, and hormones. See text for details.

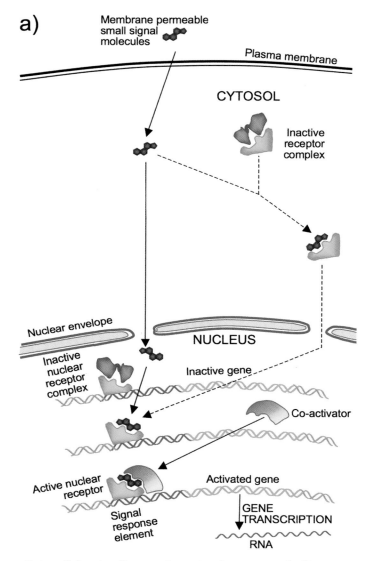

Figure 2.2. Types of intracellular signaling proteins acting downstream of cell surface and intracellular receptors. (a) The nuclear receptor subfamily. Binding of signal to a receptor leads to conformation change in the receptor, disassociation of an inhibitory complex, binding of coactivator proteins, and activation of target gene transcription. All nuclear receptors bind to DNA as homodimers or heterodimers. Monomeric forms are illustrated here for simplicity. Apart from the signals mentioned in the text, the prostaglandins and leukotrienes act via nuclear receptors termed PPARs that heterodimerize with RXR (see Chapter 5, section 4). (b) Cell surface receptors, showing how the signal of receptor activation is amplified, transduced, and distributed to regulate gene transcription. The functions of the labeled proteins are as follows: *Relay proteins* pass the message to the next signaling component. *Messenger proteins* carry the signal from one part of the cell to another without directly

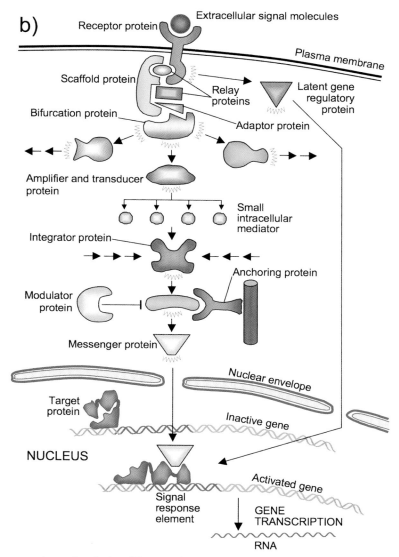

b)

Extracellular signal molecules

Receptor protein

Plasma membrane

Scaffold protein

Relay proteins

Latent gene regulatory protein

Bifurcation protein

Adaptor protein

Amplifier and transducer protein

Small intracellular mediator

Integrator protein

Anchoring protein

Modulator protein

Messenger protein

Nuclear envelope

Target protein

Inactive gene

NUCLEUS

Activated gene

Signal response element

GENE TRANSCRIPTION

RNA

conveying a signal. *Amplifier proteins* are enzymes that magnify the signal by producing large numbers of intracellular mediators (e.g., cyclic AMP) or by activating large numbers of downstream signaling proteins. *Transducer proteins* convert the signal to a different form. *Bifurcation proteins* distribute the signal to multiple signaling pathways. *Integrator proteins* receive signals from other signaling pathways and integrate them. *Latent gene regulatory proteins* are activated at the cell surface and translocated to the nucleus to regulate gene transcription. Other proteins with important roles in signaling include *scaffold proteins* that bind multiple signaling molecules in a functional complex, often at a specific location, *modulator proteins* that modify the activity of signaling proteins, and *anchoring proteins* that keep signaling molecules at specific cellular locations through tethers to a membrane or cytoskeleton. Adapted from Alberts et al. (2002).

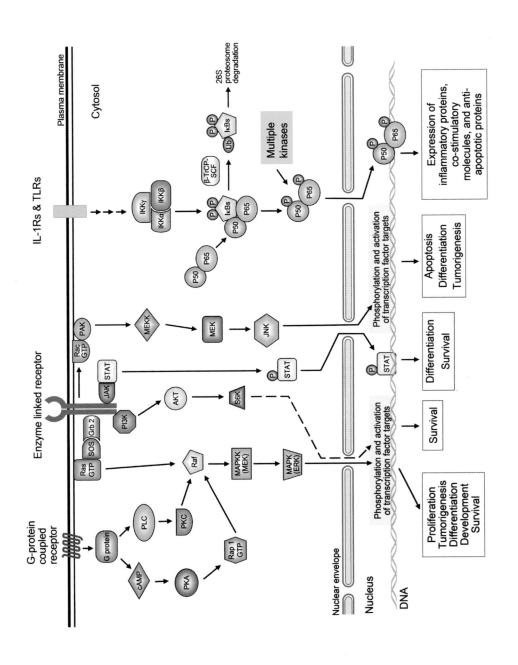

Figure 2.3. Some specific signaling pathways and the cellular functions they regulate. The pathways shown (left to right) are the G-protein-coupled pathways, the Ras/MAPK pathway, the PI3K pathway, the Jak/Stat pathway, the Jun kinase pathway, and the NF-kB pathway. Abbreviations: cAMP, cyclic AMP; PKA, protein kinase A; PLC, phospholipase C; PKC, protein kinase C; Grb2, Grb2 adapter; SOS, SOS guanine nucleotide exchange factor; RasGTP, activated small GTPase Ras; Raf, Raf serine/threonine kinase; MAPKK, mitogen-activated protein kinase kinase, also known as MEK; MAPK/ERK kinase; JNK, Jun N-terminal kinase; MAPK, mitogen-activated protein kinase, also known as ERK, extracellular signal-regulated kinase; PI3K, phosphatidylinositol-3-kinase; AKT, Akt kinase, also known as protein kinase B; S6K, p70/ribosomal protein S6kinase; JAK, Janus kinase; STAT, signal transducer and activator of transcription; P, phosphorylation (not indicated for most kinase cascades such as the Raf/MAPKK/MAPK cascade); PAK, p21-activated kinase; MEKK, MAPK/ERK kinase; JNK, Jun N-terminal kinase; IL-1R, interleukin-1 receptor; TLR, toll-like receptor (mediate innate immune responses); IKK, IkB kinase; Ub, ubiquitin.

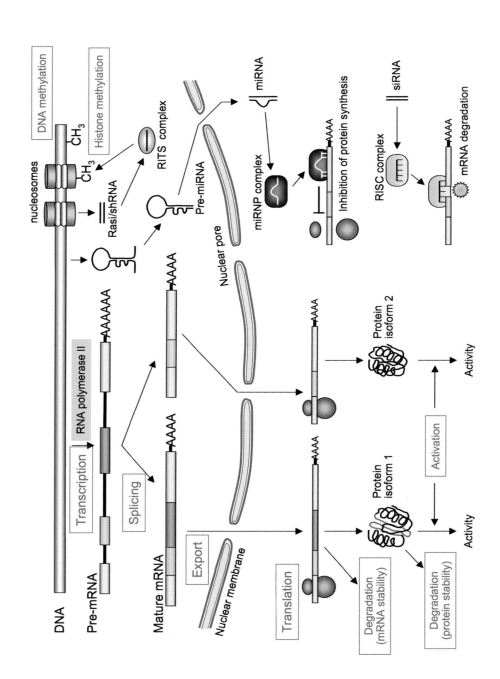

Figure 2.4. Regulation of gene expression. Transcription of the DNA by RNA polymerase II yields a primary transcript or mRNA precursor that contains *exons* (rectangles), including regions of the mature mRNA important for its regulation; and coding regions and noncoding regions known as *introns* (lines). Exonic sequences can also be removed as part of an intron to generate *alternative mRNAs* that can direct the synthesis of distinct *protein isoforms*. However, RNAs without coding capacity may also be similarly generated, including those that function as scaffolds for ribonuclear proteins that regulate processing of ribosomal RNA and other transcripts. An important example, represented here, is precursors of micro RNAs (miRNAs). Cleavage of these precursors by Drosha-type RNases in the nucleus and by Dicer-type enzymes in the cytoplasm generates 20–28 double-stranded miRNAs that assemble into RNP complexes that repress translation by complementary binding to the 3'-untranslated regions of mRNAs and/or cause mRNA degradation. Also shown is short inhibitory RNA (siRNA) that acts similarly and short hairpin RNA (shRNA), generated by bidirectional transcription and cleavage and often derived from repetitive DNA (rasiRNA), that can induce transcriptional silencing through histone and DNA methylation. Apart from this recently discovered and very important regulation by small double-stranded RNAs, gene expression may be regulated by many other mechanisms both normally and as a result of mutation. The steps at which regulation may occur include transcription, splicing, export of mRNA, mRNA stability, mRNA translation, and protein stability; these are indicated in red. Adapted from Mendes Soares and Varcárcel (2006).

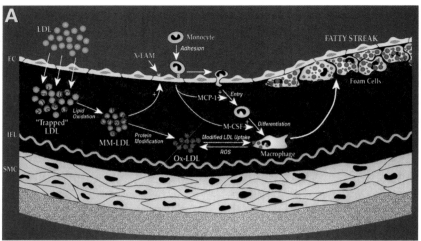

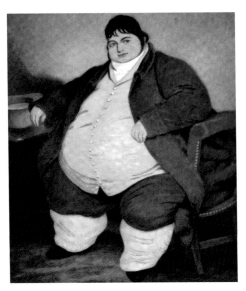

Figure 4.4. Daniel Lambert, reputedly the largest man in England, died in Stamford, Lincolnshire, in 1809, aged 39 years. At the time of his death, he weighed 739 pounds, and it is said that it took 20 men to lower his coffin into the grave in St. Martin's churchyard. His portrait now hangs in the mayor's office of Stamford Town Hall. This picture was displayed on the cover of the *Quarterly Journal of Medicine*, December 2004, vol. 97, and the legend is taken from there. Reproduced with the kind permission of Stamford Town Council, Lincolnshire, England.

Figure 4.3. Molecular and cellular mechanisms involved in the development and progression of atherosclerosis. (A) The fatty streak. Low-density lipoprotein cholesterol (LDL) in the lumen of the vessel penetrates an impaired endothelial cell (EC) layer and is trapped in the intima. Monocytes in the lumen attach to adhesion molecules (X-LAM) that arise from the effects of mildly oxidized LDL (MM-LDL), and under the influence of the chemokine monocyte chemoattractant protein (MCP-1) adhere to the ECs and enter the intima. Macrophage colony-stimulating factor (M-CSF) enhances macrophage development and these cells take up highly oxidized LDL (Ox-LDL) and release reactive oxygen species (ROS). These lipid-laden cells accumulate as foam cells, and constitute the fatty streak. (B) In the intermediate stage, smooth muscle cells (SMC) migrate through the internal elastic lamina (iel); these cells secrete polysaccharides and collagen, which form a fibrous cap on the necrotic core of the plaque. (The pericyte-like cells are mesenchymal stem cells that are also referred to as calcifying vascular cells.) (C) In the advanced lesion, the necrotic core of the fatty streak may calcify. Matrix metalloproteinases released by cells within the plaque contribute to the impending rupture of the fibrous cap. The plaque contents break through the endothelium into the lumen, and being highly thrombogenic, trap platelets; a clot forms that partially to completely occludes the lumen. Reprinted from Berliner et al. (1995), © Lippincott Williams & Wilkins, with permission.

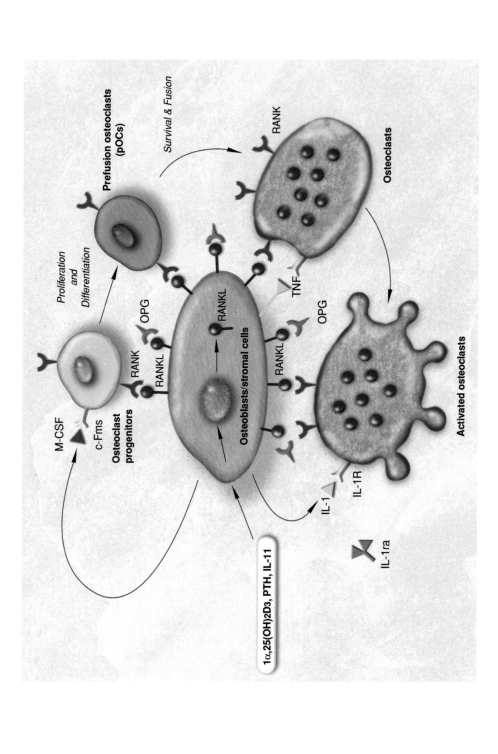

Prefusion osteoclasts (pOCs)

Survival & Fusion

RANK

Osteoclasts

Proliferation and Differentiation

OPG

RANKL

Osteoblasts/stromal cells

TNF

OPG

M-CSF

c-Fms

RANK

RANKL

Osteoclast progenitors

RANKL

IL-1

IL-1R

Activated osteoclasts

IL-1ra

$1\alpha,25(OH)_2D_3$, PTH, IL-11

Figure 4.9. Osteoblast/stromal cell involvement in osteoclastogenesis through cell–cell interactions with osteoclast progenitor cells. Osteoclast progenitors derived from hematopoietic cells of the monocyte-macrophage lineage evolve under their genetic direction and from factors arising from the osteoblast: macrophage colony-stimulating factor (M-CSF) and receptor activator of nuclear factor k-B ligand (RANK-L); these bind to c-Fms and to RANK receptors on the osteoclast, respectively. The osteoblast also releases osteoprotegerin (OPG), which acts as a decoy to block RANK-L-RANK binding, thus potentially limiting osteoclastogenesis. Additional cytokines arising from the osteoblasts that promote multinucleate osteoclasts include tumor necrosis factor α (TNF-α), and interleukin-1α, (IL-1α) that attach to their respective receptors on the osteoclast. Osteoclasts induced by TNF-α form resorption pits on bone surfaces in the presence of IL-1α, demonstrating the key independent role of these cytokines in bone resorption due to inflammation. Interleukin-1 receptor antagonist (IL-1ra) can inhibit IL-1 signals (Suda et al., 2001). Reprinted from Suda T, Takahashi N, Udagawa N, Jimi E, Gillespie M, Martin TJ. 1999. Modulation of osteoclast differentiation and function by the new members of the tumor necrosis factor receptor and ligand families. *Endocrine Reviews* 20: 345–57. Copyright 1999, The Endocrine Society.

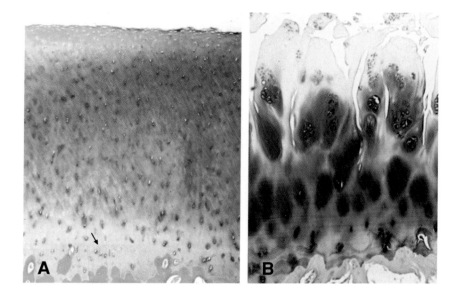

Figure 4.11. Comparison of normal and osteoarthritic cartilage stained with the
red dye Safranin O. (A) Normal. The surface is stained green by the counter stain
(Fast Green), while the cartilage below stains rather uniformly red as a result of
the binding of Safranin O to proteoglycans in the matrix. Note uniform distribution
of chondrocytes in the matrix. The arrow indicates the zone of calcified cartilage
(tidemark) separating the cartilage from the underlying subchondral bone, which
stains green. (B) Osteoarthritis. The surface, depleted of proteoglycans, shows severe
disruption with fissures and fibrillations extending deep into the remaining cartilage
where the chondrocytes are gathered into clusters, synthesizing proteoglycans in an
attempt to replace the depleted matrix (×100). Reprinted from Hamerman (1993),
with permission from Blackwell Publishing.

Interactive Therapies Significant for an Aging Population

My aim in this chapter is not to discuss the pharmacology and clinical use of a variety of medications to treat older patients; standard texts and reviews are more appropriate for that. Rather, my selection of therapies reflects a personal interest (my prior literature reviews and inclusion in my publications) and these criteria: (1) widespread use, or potential use, in an older population; (2) therapies designed to reduce the impact of diseases and conditions that were discussed in Chapter 4; and (3) overlapping effects of the medications on multiple-organ systems—an idea that I first considered in the paper on osteoporosis and atherosclerosis (Hamerman, 2005), where the therapies reviewed had an impact on aspects of both diseases (bone and the vasculature). Table 5.1 briefly outlines what I will discuss.

1. CYCLOOXYGENASES

Warner and Mitchell (2004), in their fine review, note that "cyclooxygenase (COX) must be the most common therapeutic drug target in human history. Inhibitors of this enzyme have been used for more than 3500 years—and we now consume tens of thousands of tons of these compounds each year." Brune and Hinz (2004) take us back about 200 years and trace the history of "extracts" with soothing powers—the range of anti-inflammatory precursors to aspirin; the use of animal models for skin and paw tests; the first use of cortisone; the origins of COX; and, highly relevant in today's world, the emergence of giant pharmaceutical industries manufacturing a host of related medications to suppress inflammation and pain. Dubois et al. (1998) also trace today's use of non-

TABLE 5.1
Interactive therapies discussed in Chapter 5

Therapy	Action	Effect
Cyclooxygenase inhibitors	Enzyme inhibition	Anti-inflammatory
Statins	Enzyme inhibition	Cardiovascular
		Bone density
Bisphosphonates	Osteoclast inhibition	Limit bone resorption
		Limit arterial calcification
Thiazolidinediones (TZDs)	PPARγ agonist	Metabolic
		Cardiovascular
Cytokine-based therapies	Inhibit cytokine mediators	Anti-inflammatory
Hormonal therapies	Supplementation	Increase muscle mass
		Increase bone mass

steroidal anti-inflammatory drugs (NSAIDs) from Felix Hoffman's discovery of aspirin a little more than 100 years ago (Brune and Hinz, 2004). Now, "in one year in the U.S. alone, approximately 50 million people, spending some 5–10 billion dollars, consume NSAIDs for the treatment of a wide spectrum of pathophysiological conditions." Perhaps I will be forgiven a bit of a personal, sentimental, and historical account. At a recent conference on medical and orthopedic perspectives in osteoarthritis, I noted for the geriatric fellows my awareness of three revolutionary medical breakthroughs dealing with the joint. The first, hearing Phillip Hench describe the use of cortisone in a patient with rheumatoid arthritis, for which he and his colleagues received the Nobel prize. The second was my meeting with John Charnley when he spoke about his creation of a hip prosthesis for joint replacement. Dr. Charnley was knighted. The third was also hearing John Vane describe aspirin-like drugs inhibiting COX. Dr. Vane was knighted *and* received the Nobel prize. I mention this historical note for most of it happened in my own professional lifetime.

The early 1990s were dominated by investigations of the two COX enzymes, COX-1 and COX-2, and more recently, the exploration of *selective* COX-2 inhibition. Although their genes are clearly different, COX-1 and COX-2 share 60 percent homology (similarity) at the amino acid level. When arachidonic acid is liberated from membrane-bound phospholipids by the action of phospholipase A_2, both COX enzymes form a variety of prostaglandins (PGs), initially PGG_2, and then PGH_2, which are subsequently converted to a variety of metabolites called eicosanoids that include PGE_2, PGD_2, $PGF_{2\alpha}$, PGI_2, and thromboxane. These are metabolized by different enzymatic pathways to a range of prostanoids which act on their respective receptors located on target tissues and cells. Therapeutic *antagonists* to those distal receptors may (1) reduce visceral pain, (2) suppress colitis, (3) reduce osteoclastogenesis, or (4) activate

PPARγ with multiple roles, as we shall see. DuBois et al. (1998) in their review put the diversity of PG reactions this way: "PGs are involved in blood clotting, ovulation, initiation of labor, bone metabolism, nerve growth, wound healing, kidney function, gastric mucosal protection, blood vessel tone, and immune responses. In contrast to hormones such as cortisone or thyroxine, which have broad systemic effects despite being released from a single site in the body, PGs are synthesized in a broad range of tissue types and serve as autocrine (same cell) or paracrine (cell-cell) mediators." Prostaglandin receptors transduce signals on binding to the G-coupled cytoplasmic receptor class and the nuclear PPAR receptor class that acts directly as a transcription factor (Chapter 2). I already mentioned PPARγ in the discussion of the metabolic syndrome, and will say more about it in considering the thiazolidinediones below.

An important distinction emerged between COX-1 and COX-2 by virtue of the evidence that the former is constitutively expressed in nearly all tissues (i.e., under basal conditions), producing PGs in "housekeeping functions" that regulate key physiologic processes under normal resting conditions. COX-1 in the kidney and stomach acts as a vasodilator; in the kidney it sustains renal plasma flow and glomerular filtration, and in the stomach it maintains mucosal defenses. In platelets, COX-1 generates thromboxane, which plays a key role in mediating platelet aggregation (Lipsky, 1999).

Alternatively, COX-2 was thought to be primarily expressed during inflammation, especially in those cells activated by cytokine mediators (interleukin 1 [IL-1]), including synoviocytes in the joint; macrophages in adipose tissue, vessels, and bone; and endothelial cells in the vasculature. Hence it was thought feasible to pursue rather more *selective* COX-2 inhibition that would achieve primarily anti-inflammatory effects with presumably less gastric and other side effects. Yet selective COX-2 inhibitors are capable of causing renal failure, hypertension, and exacerbation of cardiac failure in older persons (Savage, 2005). Thus, earlier views have been modified with the recognition that COX-2 is present in the kidney and other organs where selective COX-2 *inhibition* may have diverse effects, desirable or not: (1) control acute pain (Lee et al., 2005); (2) impair fracture healing (Simon et al., 2002; Seidenberg and An, 2004; Zhang et al., 2002); (3) shift mesenchymal cell differentiation in bone marrow into the adipocyte lineage rather than osteoblastogenesis—with the potential for reduced bone formation (Zhang, 2002; Warner and Mitchell, 2004; Gimble et al., 2006). Indeed, Li et al. (2006) note that COX-2 is "a key regulator of bone biology, capable of rapid and transient expression of bone cells and mediating osteoclastogenesis, mechanotransduction, bone formation, and fracture repair." (4) Inhibit a variety of cancers, especially gastrointestinal, where COX-2 is overexpressed (Brown and DuBois, 2005; Harris et al., 2005), including colonic (familial adenomatous polyposis and nonpolyposis) cancers (Brown and DuBois, 2005; Rodriguez-Moranta and Castells, 2005)—even considerations of chemoprevention (Evans and Kargman, 2004; Ristimaki, 2004; Pereg and Lishner, 2005), albeit with some reservations (Thun et al., 2004; Rigas and Kashfi, 2005); and (5) experimentally suppress neurodegenerative diseases, including Alzheimer disease (Gasparini et al., 2004; Klegeris and McGeer, 2005; Patrignani et al., 2005), and see Dr. Davies's discussion on this last point.

A range of PGs, thromboxanes, and leukotrienes also play a role in cardiovascular physiology and pathophysiology, as reviewed by Vila (2004). The "protective" or "deleterious" role of each COX isoform in cardiovascular disease needs to be sorted out as a result of the crisis precipitated by the use of selective COX-2 inhibitors (coxibs) (Fitzgerald, 2004) and the associated thrombotic and cardiovascular events. This outcome raises important medical, public health, and ethical issues, which I will not discuss (for a 2005 update, the reader may consult Graham et al. [2005]; Solomon [2005]; and Topol [2005], among many sources). As a result, coxibs have been withdrawn or had a black box warning. Marcus et al. (2002) considered that COX-2 may promote thrombotic events by virtue of selective COX-2 prostacyclin inhibition without affecting the COX-1-mediated generation of thromboxane by platelets resulting in a trend for platelet aggregation which a COX-1 inhibitor would also have suppressed. "While this view is biologically plausible, as it is compatible with evidence that introduction of COX-2 derived prostacyclin removes a protective constraint on thrombogenesis, hypertension and atherogenesis in vivo," the situation is likely to be more complex and many questions remain (Grosser et al., 2006).

The future of COX inhibitors continues to evolve, or, to put it another way, the past in this instance is prelude to new developments to come. What will these be? The therapeutic target of overall COX (1 or 2 or both) inhibition may be too broad. The aim may be to develop more selective action on the prostanoid receptors, of which Warner and Mitchell (2004) list five major subdivisions that have been defined pharmacologically. Alternatively, Meyer et al. (2005) explored studies of phospholipase A_2 inhibition, which would be a step before prostanoid production and "in theory, be a more effective anti-inflammatory approach. However, developing an inhibitor that would be selective for inflammatory metabolites and not inhibit the beneficial properties of phospholipase A_2 has so far proved elusive." But the key for the purpose of this text is the wide, multiple-system effects of cyclooxygenases that relate to geriatric bioscience in the broad categories of effects on pain and inflammation, malignancy, bone, vasculature, and the brain, which touch on many of the conditions discussed in Chapter 4.

2. STATINS

The statins are a class of drugs that inhibit the enzyme 3-hydroxy-3-methylglutaryl coenzyme A reductase (HMG-CoA reductase)—the first step in the cholesterol synthetic pathway where HMG-CoA (the "substrate") is converted to mevalonic acid. Inhibition of cholesterol synthesis in hepatocytes up-regulates the expression of hepatic low-density lipoprotein (LDL) receptors, and LDL and its precursors are cleared from the circulation (Werner et al., 2002). The historical record of statin development spans a much shorter time than that for COX, and Thompson presents a riveting account of the "foundations for blockbuster drugs" (Thompson, 2001) to show the extent of federally sponsored research with respect to the development of statins.

The isolation of cholesterol as the start of the "story" occurred in the 1800s, and biosynthesis of cholesterol began as recently as the late 1940s by groups in Munich, Boston, and London who shared the Nobel Prize. Statins were developed in a fermentation broth of a pharmaceutical company in Japan by Ando and colleagues, and later in London by Brown and coworkers, the latter naming their product (which they thought was antifungal) "compactin." These were the beginnings, and the rest, once again, is the multipharmaceutical historical development of the statin family with sales measured in the multiple billions. It is my impression that scientific grounding and broad medical understanding, combined with their relatively good safety record, provide a much happier culmination of the statin story than the COX story. (Or perhaps it is also partly "humble acknowledgment of the role of serendipity," as Brune and Hinz [2004] wrote of the development of anti-inflammatory drugs.)

Traditionally, statins "were directed away from individuals at low risk" (Durrington, 2004), to those with familial hypercholesterolemia, LDL-cholesterol levels of 190 mg/dl, and those with diabetes. A reappraisal is now in order for wider use. Durrington (2004) writes, "statins are pulling sharply ahead of other cardiovascular preventive therapies as the scale of the relative risk reduction which they can produce is finally becoming apparent. Their benefit appears to relate almost exclusively to the degree of LDL lowering achieved." To reduce LDL to levels well below previously accepted "norms" (100 mg/dl), statins have been used in "high-intensity" doses to bring LDL to an average of 70 mg/dl or lower, with concomitant slight rise in high-density lipoprotein (HDL). In one such study, Nissen et al. (2006) showed regression of atherosclerotic plaque by intravascular ultrasound studies in community dwellers who were undergoing angiography for angina or a positive stress test. That such LDL levels may be sustained over the long term was shown in a larger population study on the effect of DNA-sequence variations in a serum protease gene (PCSK9) that correlated with reduced plasma levels of LDL cholesterol over a *15-year interval.* These mutations were associated with moderate lifelong reduction in plasma LDL cholesterol and fewer coronary events (Cohen et al., 2006). O'Keefe et al. (2006) suggest that our dietary habits have evolved over millennia from those of "hunter-gatherers"—with the likelihood of LDL cholesterol levels of 50–70 mg/dl—to the present. Overweight, sedentary American adults have much higher LDL levels now and are at risk for, or have, overt atherosclerotic disease. The authors, in reviewing the evidence, also advise that "lower is better" in those at risk or with established cardiovascular disease to achieve LDL levels of 70 mg/dl *or less.* Finally they suggest that lowering LDL cholesterol "may improve not just atherosclerosis and its complications, but also other diseases commonly attributed to the aging process." This idea accords with the influence of proinflammatory cytokines arising from cardiovascular disease and other inflammatory conditions on the aging process (Ferrucci et al., 2005) and on related diseases discussed in Chapter 4. The editor of the *American Journal of Cardiology,* W. C. Roberts (2006), has also made an impassioned case for statin use to lower the LDL to <70 mg/dl *before* (his italics) a heart attack!

Topol (2004) suggested that the statins provide a "sea change in cardiovascular prevention, reducing the incidence of the major adverse outcomes of death, heart attack and stroke." It is remarkable, indeed, that now statins are often considered to be *underused;* only a fraction of the patients who should be treated with a statin are actually receiving such therapy, according to Topol (2004). Thus, in summary, an emphasis has now come to be placed on more aggressive therapy to lower LDL (and raise HDL) (Ballantyne, 2004). New treatment panels recommend minimal statin doses that result in a 30 to 40 percent reduction in LDL, with a goal of LDL below 70 mg/dl (Stone, 2004; Briel et al., 2005). But these recommendations were for patients with clinical coronary syndromes or postmyocardial infarction, and the issue before health care providers now is how "universal" statin therapy should be.

This aspect of statin use was vividly expressed by Thomas (2005) when he asked, "Should we screen asymptomatic individuals for coronary artery disease or implement universal lipid-lowering therapy? That is, for men age 30 and at the time of menopause for women." Thomas goes on: "while the cost of such a program would be substantial, although decreasing with the increasing availability of generic agents (and my editorial comment is that we are entering this era), this must be weighed against the direct and indirect costs of cardiovascular disease, estimated to be $368 billion in 2004. If such a strategy were implemented the goal of screening would shift from coronary artery disease (CAD) detection to detection of disease burden, such that therapies would decrease events." This reinforces two of the principal aims of this book (irrespective of the pros and cons of universal statin use): the value of risk factor assessment for vascular disease prevention (discussed in Chapter, 4, section 2) and the lifestyle approaches for preventive gerontology to be discussed in Chapter 6. Indeed, Denke (2005) stated: "how rapidly benefits accrue from nonstatin, lipid-lowering therapies is a 21st century question posed to data collected in the 20th century." It is interesting to contemplate a medication, perhaps uniquely, which is entering the realm of universal prescription (albeit with some age restriction at its inception). Indeed, "after simvastatin becomes a generic medication in the United States later in the year 2006, it is likely to be the preferred initial lipid-lowering therapy in many managed care formularies" (Blumenthal and Kapur, 2006) based on potency and cost. Kajinami et al. (2005) raise the "pharmacogenomic approach to identify individuals with the genetic variations that may modulate (or really improve) statin-induced cholesterol lowering."

For the sake of completeness, it is worth noting that an additional therapeutic target to reduce the risk of cardiovascular disease is to raise a low level (e.g., <40 mg/dl) of HDL cholesterol. Brousseau et al. (2004) explored an inhibitor of cholesteryl ester transfer protein (CETP) that facilitates the transfer of cholesteryl esters from HDL to apolipoprotein-B-containing lipoproteins. A trial of such an inhibitor—torcetrapib-increased HDL by more than 60 percent. However, an article in the *New York Times* of December 4, 2006 described the announcement by Pfizer of torcetrapib's "journey to failure." Early in its use, some elevations in systolic blood pressure were observed, but further in the clinical trial, when 82 persons taking torcetrapib died compared with 51 not taking the drug, it was withdrawn from the market.

Now, a further "story" of the statins highlights the cholesterol biosynthetic pathway, where inhibition of the first step involving the enzyme HMG-CoA

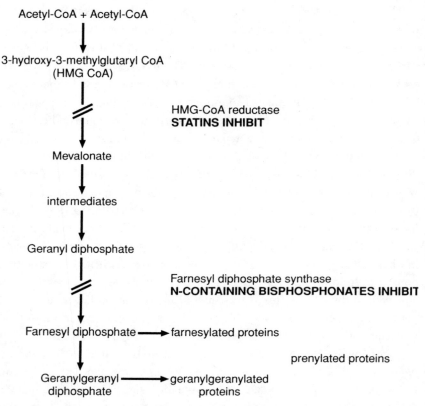

Figure 5.1. Abbreviated scheme of the enzymatic steps in the cholesterol synthetic pathway. Statins inhibit the first, rate-limiting step, thus contributing to extensive cellular modifications or pleiotropic effects (see Chapter 5, section 2). The N-containing bisphosphonates act further down this pathway, inhibiting formation of prenylated proteins, affecting more specifically osteoclast bone resorptive functions (see Chapter 5, section 3). Data from Dunford et al. (2006), Edwards et al. (2001), Libby and Aikawa (2003), Sowers (2003), and Werner et al. (2002).

reductase also affects the subsequent diverse metabolic events involving protein prenylation and causes "pleiotropic" effects because of their widespread actions on cells and tissues (Fig. 5.1). New aspects of biology have thus been revealed. It is relevant, then, to consider the biology of the mevalonate pathway and then discuss the pleiotropic effects of statins "beyond cholesterol lowering" (Liao, 2005; Liao and Laufs, 2005). (Until recently, all cholesterol-independent or pleiotropic effects of statins were believed to be mediated by inhibition of mevalonate synthesis. There is now evidence for statins binding to an integrin-associated antigen, independent of the mevalonate pathway, that plays a role in leukocyte trafficking and T-cell activation.) Yet while pleiotropic effects of statins have received widespread attention, "it is still difficult to prove," in the words of Liao and Laufs (2005), that the pleiotropic effects are real (i.e., independent of cholesterol lowering). Fritz (2005) presents a complete list of the pleiotropic effects attributed to statins in terms of their molecular and biological effects.

The pathway beyond mevalonate leads to the formation of isoprenoid intermediates, important for the modification of proteins after their synthesis (posttranslational). The players here are complex and relatively unfamiliar in terms of geriatric bioscience. Hall (1998) wrote, "the story begins back in the early 1990s with analysis of Rho, then a newly described member of the Ras superfamily of small guanosine triphosphatases (GTPases)." Members of the Ras and Rho GTPase family are major substrates for a process called prenylation. A prenylated protein is converted into a more lipophilic state and more readily interacts with cellular membranes. The key step in the activation of Rho is the attachment of isoprenoid geranylgeranyl, which translocates inactive Rho from the cytoplasm to its active form on cell membranes, with attachment of contractile actin-myosin filaments to the cytoskeleton. This provides a driving force for cell movement and division, triggering progression of the G_1 phase of the cell cycle, the potential for Ras-induced cell transformation (Walker and Olson, 2005), and coordinated control of other cellular activities, including gene transcription (Hall, 1998), and regulation of inflammatory pathways such as the JNK and p38 MAP kinase cascades (Abeles and Pillinger, 2006). Figure 5.1 also shows that nitrogen-containing bisphosphonates impair protein prenylation at a site *distal* to the action of statins, as will be discussed in Chapter 5, section 3 (Edwards et al., 2001).

Vasculature

The consequences of statins blocking prenylation of Rho, Ras, and other proteins are to modify a variety of cellular activities that particularly affect the vasculature. This is not surprising because statins have "made their mark" in reducing atherosclerotic disease complications. The improvement of endothelial functions as a result of the pleiotropic effects of statins are described in detail by Liao and Laufs (2005) and are briefly noted here: (1) inhibit reactive oxygen species (an antioxidant effect); (2) reduce vascular smooth muscle cell proliferation; (3) decrease macrophage accumulation in the plaque, tending to reduce lesions and plaque rupture, what Libby and Aikawa (2003) term "plaque stabilization;" (4) reduce levels of inflammatory phase reactants, including C-reactive protein (CRP), and plasminogen activator inhibitor-1 (PAI-1); (5) perhaps most notably, statins activate protein kinase Akt, which upregulates endothelium-derived nitric oxide synthase (eNOS) and production of nitric oxide (NO). Endothelial NO has been shown to inhibit several components of the atherogenic process (perhaps effects 1–4, listed previously), as well as promote endothelial relaxation and inhibit platelet aggregation.

Beyond statin effects on the vasculature are other, perhaps less clear-cut but suggestive benefits that still require further clarification in clinical studies.

Alzheimer Disease

Although elevated serum cholesterol levels are a risk factor for Alzheimer disease, some studies indicate that statins reduce risk of developing dementia independent of lipid effects (Hajjar et al., 2002; Liao, 2002; Werner et al., 2002), possibly by way of suppressing "inflammatory pathways" (Stuchbury and Munch, 2005). Nevertheless, the "molecular mechanisms are poorly understood" (Liao, 2002). Animal studies show reduced neuronal levels of β-amyloid peptides with statin use (Fassbender et al., 2001). "There is still some hope that statins may reduce or slow the progression of Alzheimer's disease" and large-scale, randomized controlled trials are underway (Miida et al., 2005).

Bone and Osteoporosis

About six years ago, Gregory Mundy and associates at the University of Texas Health Science Center screened a wide range of compounds for anabolic effects on bone, monitoring enhancement of bone morphogenetic protein-2 (BMP-2) (review, Garrett et al., 2001). Statins were found to augment BMP-2 in cell culture and enhance bone formation in animal studies. This work set the stage for a host of clinical studies on the ability of statins to limit the development of osteoporosis or to reduce the risk of fracture in statin users (Gonyeau, 2005). Although earlier clinical results on the effect of statins on bone density were variable, a recent review by Solomon et al. (2005b), taking into account a host of "potential confounders" that could affect the outcome, found that both lumbar spine and total hip bone mineral densities were higher in women who reported current or prior use of statins than in women who did not use statins.

However, an awareness of an association between high serum cholesterol levels and high BMD limits acceptance of a causal relationship between statins and BMD enhancement at this time. Fracture incidence was not explored in that study. In general, the literature is conflicting with respect to fracture incidence in statin users. As of August 2005, an analysis of 31 studies (24 observational and 7 randomized control trials) found that statin use was associated with fewer hip fractures, improved hip BMD, a nonsignificant reduction in vertebral fractures, and no effect on vertebral BMD. The authors conclude that "the improvement in hip fracture risk was seen only in case-control trials, not in either the eight prospective trials or two randomized controlled trials" (Hatzigeorgiou and Jackson, 2005). In a study of older, predominantly male U.S. veterans, statin use was associated with a reduction in fracture risk when compared with a non-lipid-lowering therapy, and also when compared with the nonstatin lipid-lowering therapy. Yet the authors indicate "more studies need to be performed to confirm or refute our findings" (Scranton et al., 2005). Although many studies lend little support to a reduction in fracture risk with statin therapy (Coons, 2002; Bauer et al., 2004; Reid et al., 2005), a recent large study from Denmark (Rejnmark et al., 2006) did find a reduced risk of hip fracture, but *not* when the statin was pravastatin. If, indeed, statin users at least improve their BMD (Cushenberry et al., 2002; Erdemli et al., 2005; Tanriverdi et al., 2005), then widespread use would have a duality of benefits (heart and bone) with important public health implications. A comment should also be made about statin use in limited clinical trials in patients with rheumatic and autoimmune diseases (Abeles and Pillinger, 2006), as mentioned earlier with respect to rheumatoid arthritis (Chapter 4, section 4): it does not have a likely future in rheumatologic therapy, at least as of this writing.

Cancer

Cancer is a recent addition to the literature on pleiotropic effects of statins (review, Stamm and Ornstein, 2005). Indeed, Fritz (2005) noted that "part of the effects provided by statins are due to inhibition of prenylation of low molecular weight GTPases, in particular Ras and Rho, which play key roles in signalling evoked by stimulation of cell surface receptors." Thus, Ras and Rho enhance the development of the malignant phenotype by promoting cell cycle progression, resistance to apoptotic stimuli, and tumor cell motility, and they are "appealing targets for cancer chemotherapeutic agents" (Walker and Olson, 2005) and/or offer preventive strategies for certain human cancers (Lam and Cao, 2005; Stamm and Ornstein, 2005). Nevertheless, a recent meta-analysis of many large series reported a "neutral effect" on cancer and cancer death risks in randomized control trials. There appeared to be no type of cancer that statins benefited or subtype of statin that reduced the risk of cancer (Dale et al., 2006).

3. BISPHOSPHONATES

It may not be apparent to the practicing geriatrician why bisphosphonates, used so selectively for the treatment of osteoporosis in postmenopausal women, should be included here with other therapies with more apparent multiple effects. Although bisphosphonates perhaps do not have the diversity of COX or statins, they are particularly interesting by virtue of their inhibition of the mevalonate pathway distal to the site where statins act and because of their potential role, albeit limited, in the treatment of atherosclerosis.

Bisphosphonates are analogs of inorganic pyrophosphate that contain stable P–C–P bonds rather than labile P–O–P bonds (Rogers et al., 1999; Russell and Rogers, 1999).

$$
\begin{array}{cc}
\underset{\underset{\text{OH}}{|}}{\overset{\overset{\text{OH}}{|}}{\text{O}=\text{P}}} - \underset{\underset{\text{OH}}{|}}{\overset{\overset{\text{OH}}{|}}{\text{O}}} - \underset{\underset{\text{OH}}{|}}{\overset{\overset{\text{OH}}{|}}{\text{P}}} = \text{O}
&
\underset{\underset{\text{OH}}{|}}{\overset{\overset{\text{OH}}{|}}{\text{O}=\text{P}}} - \underset{\underset{\text{R}''}{|}}{\overset{\overset{\text{R}'}{|}}{\text{C}}} - \underset{\underset{\text{OH}}{|}}{\overset{\overset{\text{OH}}{|}}{\text{P}}} = \text{O} \\
\text{Pyrophosphate} & \text{Bisphosphonate}
\end{array}
$$

Fleisch and his group (1998) in Berne studied inorganic pyrophosphate as a compound to inhibit in vitro ectopic calcium deposits. Because pyrophosphate was readily hydrolyzed when given parenterally, or inactive when given orally, a search for compounds that were stable, not degradable, took place, and the bisphosphonates were developed. Like pyrophosphate, bisphosphonates have a high affinity for bone mineral and prevent calcification both in vitro and in vivo. The "first generation" of bisphosphonates had a hydroxyl group attached to the C atom (etidronate) and were active in preventing calcification after absorption to bone surfaces. But compounds like clodronate (Cl attached to C) were even more active in inhibiting bone resorption. Then in the 1970s and 1980s, second-generation compounds were developed with an OH side chain left on, but a new chain added containing a nitrogen atom (pamidronate, ibandronate, risedronate, alendronate), and these latter compounds are much more active in inhibiting bone resorption. It is appropriate that Russell and Rogers (1999), who contributed so much to our understanding of the molecular basis of action of bisphosphonates, should review the journey "from the laboratory to the clinic and back again."

Bisphosphonates adhere to bone mineral and in the low-pH environment at the sealed site of osteoclastic resorption become free bisphosphonates, achieve high local concentration, and are taken up (internalized) by osteoclasts. Multiple effects that have to do with enhanced osteoclast apoptosis and an altered ruffled border and morphology, render the osteoclast inactive. The molecular mechanism appears to be inhibition in the osteoclast of the mevalonate pathway distal to the first step (where statins act), by impairment of protein prenylation. The inhibition of farnesylation and geranylgeranylation by inhibiting farnesyl diphosphate synthase (Fig. 5.1), seems to be the mechanism of action of nitrogen-containing bisphosphonates. By preventing their prenylation, nitrogen-containing bisphosphonates interfere with regulation of the activity of small GTPases required for osteoclast function: disruption of the osteoclast cytoskeleton and the bone-resorptive activities of these cells, and induction of osteoclast apoptosis (Dunford et al., 2006). When the question arises why statins are not as effective as bisphosphonates in inhibiting bone resorption by osteoclastic "poisoning," the answer seems to be that statins are largely taken up by the liver whereas bisphosphonates reach high levels in bone.

It perhaps followed that *combined* statin and bisphosphonate therapy needed to be studied, and indeed their therapeutic combination did produce significantly greater increases in BMD of the lumbar spine, but no differences in BMD of the total hip (Tanriverdi et al., 2005). Watts (2002), in his paper "Bisphosphonates, statins, osteoporosis, and atherosclerosis," found duality of therapeutic effects "interesting, but not something that can be applied in the clinic tomorrow; this is an area that begs for more research."

Long-term safety of bisphosphonates has come under scrutiny in view of what, until now, has been accepted as extended use. Liberman (2006) expressed reassurance for long-term administration "with no evidence of an adverse effect on bone health." Other reports are less reassuring. Ott (2005) notes that "unlike most medications, bisphosphonates remain in the body for decades"; their "open-ended" effects may be judged by markers of bone resorption and formation that remain suppressed for at least five years after bisphosphonate discontinuation. Ott cites studies on patients who took a bisphosphonate for up to eight years with "absence of bone formation" and concern about "adynamic bone disease" (as in renal failure) and delayed healing of nonspinal fractures. Women who discontinued alendronate after 5 years showed a moderate decline in BMD and a gradual rise in bone biochemical markers. There was no higher fracture risk than for further vertebral fractures in women at high risk for these. This group may benefit by continuing therapy beyond 5 years (Black et al., 2006). Of some concern is a literature search from a dental perspective by Woo et al. (2006) of bisphosphonate-associated osteonecrosis of the jaw. A total of 368 cases were found, alerting the medical community that the majority of cases occurred after a tooth extraction or other dentoalveolar surgery, especially after use of intravenous bisphosphonates (zoledronic acid or pamidronate) for multiple myeloma or metastatic cancer, or hypercalcemia associated with these conditions. Yet the rarity of jaw osteonecrosis with traditional use of oral bisphosphonates (alendronate, risedronate) to treat osteoporosis has raised some uncertainty about this presumed adverse skeletal effect (Grey and Cundy, 2006; Watts et al., 2006). Risks and benefits need to be weighed, dental consultation obtained if indicated, and the likelihood accepted that use of bisphosphonates "will do more good than withholding them" (Bilezikian, 2006).

In an earlier paper on biological linkages of osteoporosis and atherosclerosis (Hamerman, 2005), I made a strong case for the commonality of calcification mechanisms, bone formation, and their molecular processes and tissue factors in the bone and arterial wall. Several risk factors for osteoporosis and cardiovascular disease can coexist in apparently healthy postmenopausal women (Tanko et al., 2005a). A recurrent theme with respect to older patients is a need for "timely screening that would stress both heart and bone risk factors" (Masse et al., 2005). This certainly could be part of the clinical practice of geriatricians who may uniquely encounter older persons with both diseases. Indeed, it is not surprising that there is a literature on the effects of bisphosphonates in atherosclerosis, although still primarily from animal studies (Ylitalo, 2000). A practical means to deliver bisphosphonates in high concentration to the macrophages in the arterial wall has been developed, to encapsulate the bisphosphonates in liposomes for phagocytosis. This approach was taken by Danenberg et al. (2003), who showed that liposomal clodronate or liposomal alendronate depleted circulating monocytes and tissue macrophages, and inhibited in-stent neointimal hyperplasia in hypercholesterolemic rabbits. The potential was considered for therapeutic treatment in patients undergoing coronary catheterization and stent placement. In animal studies, bisphosphonates in doses that inhibited bone resorption also reduced arterial calcification. Yet in clinical studies in postmenopausal women, bisphosphonates apparently did not limit progression of aortic calcification (Tanko et al., 2005b).

4. THIAZOLIDINEDIONES AND PEROXISOME PROLIFERATOR-ACTIVATED RECEPTOR γ

Chapter 4, section 3 briefly mentioned this subject in connection with the metabolic syndrome and diabetes. In fact, the discussion to follow precisely illustrates the integration of a complex aspect of *molecular biology*—the family of peroxisome proliferator-activated receptors (PPARs)—and their *therapeutic* activation by thiazolidinediones (TZDs). The TZDs—a "new class of antidiabetic drugs that ameloriate insulin resistance systemically, improve insulin-mediated repression of gluconeogenesis, and stimulate glucose uptake in skeletal muscle" (Bouskila et al., 2005)—are in widespread use by the clinician. Thus, the "not-so-odd couple" of my preface—the molecular biologists, can write almost exclusively about the actions of the PPARs (Argmann et al., 2005; Li and Palinski 2006), while the clinicians specializing in diabetes can write predominantly about the TZDs (Einhorn et al., 2004; Yki-Järvinen, 2004). Yet the bioscience–clinical linkages are profoundly important in terms of understanding the basis for TZDs actions, and the potential for further drug development. Advances here will reduce the ravages of the most widespread yet interrelated of current human afflictions: insulin resistance, diabetes, atherosclerosis (the metabolic syndrome), and their myriad complications.

The PPARs are part of at least a 48-member superfamily of nuclear receptors (see Fig. 2.2a). The molecular binding components, or ligands, of the nuclear receptors may be endogenous (various polyunsaturated fatty acids, such as linoleic acid), or pharmacologic, like the TZDs. An earlier review of these receptors by Francis et al. (2003) described them as "master regulators integrating the homeostatic control of a range of metabolic functions, including the differentiation and energy storage by adipocytes." More recently, Christopher Glass (2006) edited a series on nuclear receptors that reviewed the three classes of PPARs: α, γ, and δ. Some mention will be made of each, but the appropriate extent to which this involved subject should be discussed here depends on what is in line with a theme of this book: to provide the clinician-reader with a background of the relevant bioscience. So, considerations of the nuclear receptors based on identification of their known ligands (activating molecules), their target genes and how they bind to DNA, and their physiologic function (Li and Palinski, 2006) are part of this chapter. One class of receptors will be familiar, and one will, I believe, be rather new terminology.

Endocrine receptors include estrogen, progesterone, androgen, glucocorticoid, and mineralocorticoid receptors. These play a role in the development of sexual maturity, reproduction, carbohydrate metabolism, immunity, and water and electrolyte balance. Ligands activating these receptors are steroid hormone-regulated through the hypothalamic-pituitary-adrenal axis, as discussed in Chapter 4, section 1.

Adopted orphan nuclear receptors include, among others, the PPARs, and this class regulates lipid and glucose homeostasis in the liver, adipocytes, skeletal muscle, and macrophages. PPARα, the first PPAR to be identified, is predominantly expressed in the liver and plays a critical role in liver and skeletal muscle lipid metabolism and glucose homeostasis by regulating the transport and catabolism of fatty acids (Schneider and Semenkovich, 2005; Lefebvre et al., 2006). The receptor is activated endogenously by lipids and pharmacologically by a class of drugs called fibrates (Bloomfield, 2006). Agents in this class (gemfibrozil and fenofibrate in the United States) are used to lower levels of circulating triglycerides and elevate HDL-cholesterol. Schneider and Semenkovich (2005) review a study by Gizard et al. (2005) pertaining to PPARα interfering with cell cycle progression in smooth muscle cells (SMCs). Recall from Chapter 4, section 2 that SMCs proliferate in atherosclerosis and secrete an extracellular matrix that forms a cap on the thrombogenic fatty streak, or may also proliferate as part of the restenotic process after stent replacement. In SMC proliferation, phosphorylation of Rb (discussed in Chapter 3, section 3, Fig. 3.2) by components of the cell cycle (cyclin-dependent kinases) promotes cell proliferation via cell cycle G_1/S *transition*. The cell cycle involves transitions from quiescence (G_0) to a pre-DNA replication phase (G_1), a DNA replication phase (S), a premitotic phase (G_2), and mitosis (M) (see Fig. 3.2). These transitions are stimulated and regulated by appropriate growth factors that bind to the cell membrane receptors which in turn transduce signals through the cell cytoplasm to the nucleus, as discussed in Chapter 2. PPARα induces transcription of p16 [INK4a] (a tumor suppressor, see Chapter 3, section 3), promoting cell cycle *arrest* at the G_1 to S phase by preventing cyclin-dependent kinases from phosphorylating Rb, thus maintaining Rb bound in its "pocket" region to the E2F family of transcription factors in its unphosphorylated or growth-suppressive state (Du and Pogoriler, 2006; Genovese et al., 2006; Giacinti and Giordano, 2006; Knudsen and Knudsen, 2006). Schneider and Semenkovich (2005) ask "whether PPARα is thus a savior, or a savage," the former as a growth suppressor, the latter because hepatic tumors were induced in rodent livers! Thus again, we may be seeing pleiotropic effects, discussed with statins, but perhaps prompting uncertainty about future use of PPARα agonists. Bloomfield (2006) raises the question of "fibrates in a statin world." She notes that gemfibrozil has "clearly stood the test of time," yet the evidence for use of statins is "more robust owing to the huge number of patients that have been studied in statin trials," and the gemfibrozil trials have not shown a convincing effect in reducing all-cause mortality. For further discussion of PPARα, consult other reviews (Vamecq and Latruffe, 1999; Knouff and Auwerx, 2004; Lefebvre et al., 2006; Li and Palinski, 2006).

Barish et al. (2006) described PPARδ as "a dagger in the heart of the metabolic syndrome." Its transcriptional program enhances the catabolism of fatty acid, decreases triglyceride stores, and "robustly" raises HDL cholesterol. Decreased hepatic glucose output also occurs. What remains to be better defined is the "predictive power of mouse models to determine the human carcinogenicity of PPAR-δ-targeted drugs," and "additional toxicology work will be required before PPARδ compounds can be marketed for chronic clinical use."

According to Semple et al. (2006), "current models of PPARα and PPARδ function (and hence their potential relevance to human disease) rely heavily on data from cell culture and mouse studies." PPARγ has more immediate relevance to human disease in terms of antidiabetogenic therapy. A "simplified" overview of the three PPAR isoforms (variations) is presented in Table 5.2 (Semple et al., 2006). PPARγ has a modular structure that harbors a ligand-independent transcriptional activation function domain (AF-1); the core DNA-binding domain, which contains two zinc finger motifs that bind the receptor to specific DNA sequences; and the region which contains the ligand-binding domain and ligand-dependent activation function (AF-2) (Gurnell, 2005). PPARγ forms a partner (or heterodimer) with retinoic X receptor and binds to PPAR response elements (PPRE) in the regulatory region of target genes—a process called transactivation. Ligand binding leads to preferential recruitment of coactivator complexes and favors dismissal of the corepressor complex (Semple et al., 2006). The activation of PPARγ is governed by the binding of ligands, which, as I've said, may be endogenous—and include polyunsaturated fatty acids and their eicosanoids, derived from nutrition or metabolic pathways; or from extrinsic pharmacologic sources, such as the TZDs or NSAIDs. In fact, there is a second gene-activating mechanism for these pharmacologic ligands: negative regulaton of gene expression, called transrepression, which antagonizes other classes of transcription factors, such as NF-kB and AP-1; this may underlie the anti-inflammatory actions of these

TABLE 5.2

Simplified overview of the current understanding of the metabolic roles of the three PPAR isoforms

Function	PPARα	PPARγ	PPARδ
Sites of highest expression	Liver, kidney, heart	Adipose tissue, macrophages	Adipose tissue, skin, brain, but widespread
Cellular processes activated	Fatty acid β-oxidation, lipoprotein synthesis, amino acid catabolism	Adipocyte differentiation, triglyceride synthesis	Fatty acid β-oxidation
Physiologic function	Coordination of metabolic response to fasting	Differentiation of adipocytes Fatty acid trapping	Muscle fiber type determination?
Examples of target genes	Carnitine palmitoyltransferase 1, HMG-CoA synthase 2, ApoA-1	Fatty acid-binding protein 4, lipoprotein lipase, adiponectin	Acyl-CoA oxidase, carnitine palmitoyltransferase 1
Metabolic phenotype of knockout mice	Fasting hypoglycemia, hypothermia, hypoketonemia, and hepatic steatosis	−/− lethal, −/+ more insulin sensitive at baseline	Reduced baseline adiposity; increased obesity on high-fat feeding

Source: Reprinted from Semple, Chatterjee, and O'Rahilly (2006), with permission from the authors and the American Society for Clinical Investigation.

nuclear receptors. Ligand binding alters the confirmation of PPARγ, with various activation and repressors released, resulting in increased transcriptional activation of target genes. As Knouff and Auwerx (2004) put in (with my additions): "a Westernized life style, characterized by high caloric intake and a lack of physical exercise (environmental and choice) exposes people to chronically high levels of free fatty acids, the endogenous ligands for PPARγ (metabolic), which causes the feed-forward activation of genetic programs leading to a metabolic state favorable for the development of obesity (molecular)."

Further variations of the PPARγ family itself determine tissue distribution, which is often selective. PPARγ1 is widely expressed, and PPARγ2 is adipose tissue restricted. The presence of PPARγ in adipocytes, skeletal muscle, and liver indicates tissue interactions (crosstalk) controlling lipid metabolism and glucose homeostasis, in particular, in the aspect of insulin sensitivity; but the wider distribution of PPARγ in endothelial cells and macrophages suggests how interrelated are the metabolic aspects concerning atherosclerosis, and in a wider sense, inflammation as well.

Several mutations are also documented in the human PPARγ gene "that have contributed significantly to our understanding of the biology of this receptor" (Gurnell, 2005). A loss-of-function mutation in PPARγ (PPARγ ligand resistance syndrome) has been observed rarely, in particular, in a few young women with partial lipodystrophy (see Chapter 4, section 3), dyslipidemia, insulin resistance with type 2 diabetes, hypertension, and polycystic ovarian syndrome.

My justification for discussing PPARγ at such length derives from its key role as a regulator of adipocyte differentiation and metabolic homeostasis, as I mentioned. As such, "elucidation of the function of PPARγ is crucial to an understanding of the diseases associated with the metabolic syndrome, including obesity, diabetes, hypertension, and atherosclerosis" (Knouff and Auwerx, 2004), which formed a major part of the discussion in Chapter 4, section 3. Again, the authors write of the "pleiotropic" effects of PPARγ (as discussed with the statins), and in this case refer to the actions of the PPARγ pharmacologic agonists, especially the thiazolidinediones (TZDs), or glitazones, which are most important for the geriatrician in the treatment of diabetes. These are "the only compounds currently available that specifically target tissue insulin resistance, and are highly selective and potent synthetic agonists for PPARγ" (Einhorn et al., 2004). Troglitazone, the first on the market, was removed because of hepatic toxicity; the two current PPARγ agonists (second generation), rosiglitazone and pioglitazone, were approved in the United States as recently as 1999.

What follows is an attempt to express the moderating effects of the glitazones on aspects of the metabolic syndrome rather than a therapeutic discussion, as such. Pleiotropic effects again come into play because of the diversity of tissue and cell actions (adipose tissue, liver, muscle, vasculature, macrophage), as well as a word about bone, all of which is consistent with therapeutic diversity.

Metabolic Syndrome Complex

van Wijk and Rabelink (2004) and Semple et al. (2006) present a broad schema of TZD actions that capture at the transcription level the many aspects of improved metabolic balance.

With respect to activation of PPARγ-retinoic acid receptor transcription, TZDs decrease hepatic gluconeogenesis and increase skeletal muscle glucose uptake, improving insulin sensitivity. Perhaps most important is the remodeling and redistribution of fat mass to selective accumulation of subcutaneous adipose tissue. If experimental studies are confirmed in humans, TZDs are likely to enhance adipose tissue's capacity as a "sump" for dietary fatty acids, safely sequestering them away from other insulin-sensitive tissue such as skeletal muscle (Semple et al., 2006). This has been termed the "lipid steal" hypothesis or "fatty acid steal hypothesis" (Yki-Järvinen, 2004), and consistent with this, TZDs have almost universally lowered fasting free fatty acid (FFA) levels in clinical trials, with improved insulin sensitivity (as discussed in Chapter 4, section 3). The enhanced lipoprotein hydrolysis by endothelial lipoprotein lipase and uptake of FFAs into adipocytes occurs by way of fatty acid transport protein and CD36 (see below) (Semple et al., 2006). Makowski and Hotamisligil (2004) discuss the intracellular lipid-binding proteins called fatty-acid-binding proteins; these appear to "fine-tune by way of c-jun amino-terminal kinase (JNK) and PPARγ, the balance between the availability of metabolic resources, keeping in check a too-robust inflammatory response" induced by stressors. These authors ask whether fatty-acid-binding proteins are "friend or foe—that remains to be determined," reminiscent of the query about PPARα as "savior or savage," cited earlier.

With respect to the alternate transcription pathway concerned with trans-repression of inflammatory mediators, there would be widespread beneficial effects on several systems and cells: adipogenesis—decreased tumor necrosis factor α (TNFα); vascular cells—decreased vascular smooth muscle cell proliferation, and matrix metalloproteinases; monocytes—decreased proinflammatory cytokines. Despite the increased fat mass, TZDs augment adiponectin blood levels, which in part may explain improved insulin sensitivity (Bouskila et al., 2005). (In obese patients, recall that adiponectin blood levels are low.) TZDs have also been shown to decrease CRP (Li and Palinski, 2006).

Clearly, the multiple effects of TZDs as a PPARγ agonist are more complex than all this, but improved metabolic balance, anti-inflammatory, and antiatherogenic effects in the vessel wall are well established (Roberts et al., 2003; Chilton and Chiquette, 2005). Two further pleiotropic actions of TZDs on PPARγ merit discussion.

Macrophages

TZD activation of PPARγ in macrophages could suppress their inflammatory response. Recall that adherence (ligation) of receptors initiates a signaling cascade in macrophages that culminates in activation of both NF-kB and activator protein (AP-1). It took development of in vitro cellular and mouse models to clarify the role of PPARγ itself and its potential activation with respect to macrophage functions and mediator release. It does appear that activation of PPARγ couples a pathway for the *uptake* of oxidized LDL via CD36, to a pathway via the liver X receptor α (LXRα) for cholesterol *efflux*. Transcriptional coupling is achieved by two components—9- and 13-hydroxyoctadecadienoic acids (HODE)—within the oxidized LDL particle which activate PPARγ and LXRγ, respectively. Treatment with TZDs enhances flux through both the uptake and the efflux arms of this cycle, thereby reducing the deposition of pathogenic oxidized LDL in the vascular wall (Zhang and Chawla, 2004). Yet Zhang and Chawla (2004) point out that the actions of PPARγ in macrophage lipid metabolism and inflammatory response that have been elucidated in mouse models might limit the application of these findings to humans, and they present a list of species differences.

I wish to pursue a bit further the paths of cholesterol transport from cells and tissues. Free or unesterified cholesterol is toxic to macrophages, and as Cuchel and Rader (2006) state in a recent review "the first line of defense for macrophages" is the esterification of cholesterol to cholesteryl ester by the enzyme acyl:coenzyme A cholesterol *O*-acetyltransferase-1. Although this may "spare" the macrophage, the lipid droplets accumulate and lead to the formation of the foam cell. But then a "second line of defense" is important for the possible regression of atherosclerosis—cholesterol efflux or "reverse cholesterol transport" (Cuchel and Rader, 2006). Mechanisms include the mitochondrial cytochrome P-450 enzyme cholesterol 27-hydroxylase that converts cholesterol into 27-hydroxycholesterol; and the adenosine triphosphate-binding cassette transporter (ABCA1) in macrophages, the liver, and other tissues, which transports cholesterol to apolipoprotein A-1, the major apolipoprotein of HDL. Cholesterol associated with HDL in the plasma may be esterified by lecithin:cholesterol acetyltransferase (LCAT) with the formation of the mature HDL. A cholesterol ester transfer protein (CETP) can transfer cholesteryl ester from cholesterol-rich lipoproteins in exchange for triglycerides. This process leads to triglyceride-enriched HDL particles and LDL particles "which are a substrate for hepatic lipase and lipoprotein lipase, resulting in lower HDL concentrations and small dense LDL" (Sandhofer et al., 2006)—an "atherogenic profile" indeed. HDL particles function as an integral component of the reverse cholesterol transport system, and the cholesterol brought to the liver is excreted from the body as bile acids. Maintenance of high levels of HDL is a key antiatherogenic therapeutic goal. "Besides its role in reverse cholesterol transport to limit cholesterol accumulation in the vascular wall, HDL appears to offer protection (at least in part) against the development of atherosclerosis by inhibiting oxidative modification of LDL," as Sampietro et al. (2006) note in a review of HDL as a "new" target of cardiovascular medicine. Therapy to *inhibit* CETP (noted in Chapter 5, section 2—Brousseau et al., 2004), or mutations in the CETP gene with low plasma levels, is associated with marked elevations of HDL values (Cuchel and Rader, 2006) and evidence for longevity (Atzmon et al., 2006). These aspects of cholesterol transport are thus relevant to health and disease, with insights gained from genetic alterations in humans that influence longevity, as I will mention in the epilogue.

Bone

Rosen and MacDougald (2006) in a review of adipocyte differentiation note that adipocytes derive from multipotent mesenchymal stem cells. What concerns us are the "conditions that tip the balance leading to a cascade that promotes one cell fate while repressing other cell fates." Thus, ligand activation of PPARγ drives bone marrow mesenchymal stem cells toward adipocytes rather than osteoblasts (Gimble et al., 2006), with lower bone factors (Runx2/Cbfa1) and limited osteoblastogenesis (Rzonca et al., 2003). Genes in the residual osteoblasts appear to manifest reduced Runx2/Cbfa1, and osteoblastogenesis is limited (Rzonca et al., 2003). Increase in marrow adipose tissue accompanies the decreased bone mass observed in age-related osteoporosis (Pei and Tontonoz, 2004). Where cells and animals are deficient in PPARγ expression (Akune et al., 2004), the progenitor cells expand to osteoblasts rather than to adipocytes, and Pei and Tontonoz (2004) note that "fat's loss is bone's gain." These experimental findings suggest that older persons with diabetes treated with TZDs are at risk for bone loss, and indeed this seems to be the case. Watts and D'Alessio (2006) in an editorial, "Type 2 diabetes, thiazolidinediones: bad to the bone," review studies on the complex aspects of bone density in diabetic persons per se, but go on to note that the use of TZDs was "associated with a significant 50% increase in the annualized rate of whole-body bone loss in women (but not in men)." Commenting on recent studies in rodent models (some cited above, and more in their editorial), Watts and D'Alessio point out that the concordance of these laboratory studies "makes the findings in the clinical study credible," and the common mechanism (a shift in the flow of mesenchymal precursor cells from osteoblastic to adipogenic lineages) consistent with the known biology of PPARγ. The use of TZDs is becoming more widespread, and their indication may expand with ongoing trials of their effects on other aspects of diabetes, including metabolic effects and vascular disease; more human studies on TZDs and bone are needed. "This is a question for which a randomized clinical trial is imperative."

Further, in appraising the effect of PPARγ in reducing BMD and contributing to osteoporosis, Pei and Tontonoz (2004) strike a note consistent with geriatric bioscience: from bench to clinic. They consider development of PPARγ modulators or antagonists that would serve to increase bone mass, yet without enhancing insulin resistance. That is, "a bone-selective PPAR modulator/antagonist that may exploit PPARγ as a target in osteoporosis." This thought mirrors selective estrogen receptor modulators (SERMs), like raloxifene, acting as an estrogen antagonist in the breast, but an estrogen agonist in bone (Yki-Järvinen, 2004). Binding a SERM to the estrogen receptor causes a specific conformational change in the receptor and the resulting three-dimensional structure determines which coactivators and/or corepressors are recruited to the promoter. For example, tamoxifen recruits a coactivator complex to estrogen-regulated genes in endometrial cells but a corepressor complex to the same gene in breast cancer cells. The estrogen receptor α or estrogen receptor β ratio varies between tissues, and which estrogen receptor form predominates also affects SERM activity. Thus, the estrogen receptor may be targeted by a SERM, or the production of estrogen may be reduced by aromatase inhibitors (Deroo and Korach, 2006).

To conclude, the range of actions of PPARγ goes beyond its role, as I said at the outset, of a "master regulator of adipocyte differentiation and metabolic homeostasis" (Francis et al., 2003; Knouff and Auwerx, 2004), and includes a full panoply of pleiotropic actions. Indeed, the elucidation of PPARγ functions has shed new light on the diseases and conditions associated with the metabolic syndrome—a veritable "explosion of research on the effects of PPARγ and the TZD drugs" (Knouff and Auwerx, 2004). New prospects of therapy are opening up where molecular insights on cell functions in the *absence* of PPARγ, or *reduced* PPARγ, or *activated* PPARγ, may be translated to differential therapies—at present and in the future. Argmann et al. (2005) raise a question—"PPARγ: the more the merrier?" They go on, "we are guided by the principle that more is better." Yet the initial development of full PPARγ agonists may give way to modulating levels of activation, as noted above. "Collectively, our current knowledge suggests that modulating (or inhibiting) PPARγ activity rather than enhancing it will be the preferred therapeutic strategy to treat metabolic disorders that encompass all of today's most prevalent diseases; this will improve glucose homeostasis and prevent adipogenesis—coordinating evolutionary benefit and adaptive responses." These concepts are innovative, and seem to go to the heart of the interrelations of geriatric bioscience, initiated by the experimentalist and used by the geriatrician-clinician aware of these new frontiers of science applicable to the treatment of disease in humans.

5. CYTOKINE INHIBITORS AS A LANDMARK FOR NEWER ANTI-INFLAMMATORY THERAPIES

I have chosen to introduce one aspect of newer therapies—cytokine inhibitors—recognizing that at this time their use is reserved almost exclusively for the rheumatologic specialist rather than the geriatrician. Yet, as Leishman and Bundick (2005) note, "recent biopharmaceuticals that enable the targeting of a specific cytokine, such as blocking, or neutralizing antibodies, decoy receptors, and more recently cytokine traps (e.g., IL-1 Trap), have arguably been one of the most significant recent advances in medicine." According to Dinarello (2005), "the IL-1 Trap is a new concept using soluble forms of cytokine receptors to bind and neutralize a specific cytokine. The Trap takes advantage of the high affinity of the two signaling chains of the cell surface IL-1 receptor linked by the Fc portion of IgG_1." A whole new era has emerged in the treatment of rheumatoid arthritis (RA) with an IL-1 receptor antagonist (IL-1 Ra), and especially with blocking antibodies to TNF-α, that are likely to become the forerunner for the treatment of other diseases as well when the targets and therapies in many inflammatory conditions are further defined. (See, for example, Fig. 4.9 relating to osteoblast-osteoclast interactions.)

Cytokines as part of stress reactions constitute key aspects of the geriatric bioscience literature in aging, longevity, and virtually all the conditions and diseases discussed in Chapter 4. It seemed that *anticytokine* therapy may reflect a prototype of future therapies *beyond* the treatment of RA—the disease for which such therapy was introduced—largely by the work of Maini and Feldmann, who shared a Lasker Award in 2003 for its development (Maini, 2005). The possibility that such therapy and related ones will emerge beyond their use in RA for generations of future physicians and their patients is, of course, entirely speculative on my part and must await the trend of the scientific development. The "newness" of this subject on the medical therapeutic scene is further attested to by publications in 2004–2006.

In their paper on "endocrinology of the stress response," Charmandari et al. (2005) define stress as "a state of threat to the maintenance of a complex dynamic equilibrium termed homeostasis." When I read that, I thought of the names of three early pioneers in medical research concerned with stress and homeostasis with whom I "grew up" in my medical education. The first was

Claude Bernard in France (1813–1878), who proposed the constancy of the milieu intérieur (internal environment). The second was Walter B. Cannon (1871–1945) at Harvard, who coined the term "homeostasis" (in his book *The Wisdom of the Body*) to describe the organism's tendency to maintain physiologic stability. He was also famous for proposing an animal's "fight or flight" response to stress. The third was Hans Selye, who, in the early 1950s, gained fame (even notoriety) for proposing the endocrine response to alarm followed by adaptation to stress.

At present, as noted in Chapter 4, section 1, the concept of stress and the nature of the response is likely to be conceived of in immunologic, biochemical, and molecular terms that focus primarily on cytokines. In a paper on stress-activated cytokines and the heart, Mann (2003) noted, "cytokines expressed within the myocardium in response to environmental injury, namely TNF-α, IL-1, and IL-6, play an important role in initiating and integrating homeostatic responses. However, at high concentrations, the cytokines all have the potential to produce cardiac decompensation." Again in the short term there is adaptation to stress, but maladaptation occurs over the long term.

Synovial joints in rheumatoid arthritis (RA) are the sites of acute and chronic inflammation and from such ongoing maladaptation, so to speak, a rationale for anticytokine therapy emerged. RA and its therapy would not ordinarily be a subject included in a text on geriatric bioscience for geriatricians. Indeed, I had not previously chosen to discuss RA, although there are interesting ramifications: statins have been used in RA to reduce inflammatory mediators and "accelerated atherogenesis" (Gonzalez-Gay et al., 2005) predisposing to cardiovascular mortality (Costenbader and Coblyn, 2005). Pasceri and Yeh (1999) describe atherosclerosis and RA as a "tale of two diseases," with overlapping biological linkages. Perhaps the association is not surprising, based on the aspects of "systemic inflammation" that are features of both diseases (Snow and Mikuls, 2005).

But the point I am getting at is that investigators were able to identify a specific pathologic system—the inflamed synovial membrane of RA patients—and demonstrate that the addition of neutralizing anti-TNF antibodies suppressed inflammation and strikingly reduced the production of a number of other inflammatory cytokines (IL-1, granulocyte macrophage–colony-stimulating factor [GM-CSF], IL-6). As Maini and Taylor (2000) put it, "these observations support the concept that TNF-α occupies a dominant position at the apex of a proinflammatory cytokine network. As a pleiotropic cytokine that can enhance synovial proliferation and production of prostaglandins and metalloproteinases, as well as regulate the cytokine network , TNF-α was seen as a potential therapeutic target in RA." Some of the earlier efforts to block the TNF receptor were described by Beutler (1999), and at present, three anti-TNF-α biologicals have emerged that limit TNF-α binding to its receptors: *etanercept,* a soluble TNF-α type II receptor—IgG1 fusion protein; and two monoclonal antibodies against TNF-α—*infliximab,* a chimeric (human-murine) IgG1 anti-TNF-α; and *adalimumab,* a recombinant humanized anti-TNF-α antibody (Scott and Kingsley, 2006). All these have differing pharmacokinetic and pharmacodynamic properties (Nesterov, 2005). With respect to his most recent views, Maini (2005) wrote, "There can be no doubt that anti-TNF drugs provide a reference point from which to move forward. The understanding of the mechanisms of action of TNF blockade that are being uncovered provides us with an overview of the big picture of loss of biological homeostasis in chronic rheumatic diseases." However, the issue of "pharmacoeconomics" arises with the use of TNF inhibitors for RA, and "evolving for use in other systemic inflammatory diseases" (i.e., problems relating to toxicity, quality of life, productivity, and costs) (Kavanaugh, 2005). Already there appears to be a high incidence of infections and tuberculosis in patients treated with TNF inhibitors (Crum et al., 2005; Winthrop, 2005), because TNF activates several immune and antimicrobial effects in the lung concerned with the clearance of mycobacterial pathogens (Stenger, 2005).

It is also recognized that focusing on TNF-α, important as this is in the hierarchy of cytokines (Beutler, 1999), may fail to engage IL-6 and IL-1, for example, elevated in many inflammatory conditions, including chronic heart failure (Mann, 2003; Bolton, 2005). Indeed, anti-inflammatory therapy in chronic heart failure has thus far proved disappointing (Gullestad and Ankrust, 2005). Bolton (2005) considered a broader approach—immune modulation, which would down-regulate proinflammatory cytokines and up-regulate anti-inflammatory cytokines (IL-4, IL-10, TGF-β). As he puts it, "this alteration in the balance between pro-inflammatory and anti-inflammatory cytokines may be more appropriate than neutralizing the activity of a single cytokine."

Yet the anti-TNF agents did initiate a trend, and one senses an accelerated pace of development that may still apply to RA therapy or, more likely, extend beyond in experimental studies designed to suppress other inflammatory diseases, as I mentioned (Hochberg et al., 2005). Indeed, a trial of etanercept has been used "to reduce C-reactive protein levels . . . and interrupt the inflammatory cascade that occurs in abdominal obesity" associated with the metabolic syndrome (Bernstein et al., 2006). Leishman and Bundick (2005) wrote, "one of the driving forces in the field of immunology is the ambition to translate experimental research into useful therapies." Chapter 4, sections 2 and 3, mentioned the *innate* immune system in relation to the inflammatory cascade that is part of atherosclerosis, obesity, and the metabolic syndrome. The response is through "pattern recognition receptors" (PRRs) on macrophages and dendritic cells that recognize pathogen-associated molecular patterns (PAMPs) (Arend, 2001). The binding of PAMPs to PRRs triggers the cells to an immediate effector response (unlike more specific B- and T-cell responses) by way of Toll-like receptors, mammalian homologs of the *Drosophila* protein Toll, which induces effective immune responses to *Aspergillus* (Uematsu and Akira, 2006). The Toll-like receptors also have homology to the IL-1R and the IL-18R families that transduce signals with kinase activation, resulting in phosphorylation, ubiquitination, and proteosomal degradation of Ikβ proteins. As discussed in Chapter 4, section 7, Ikβ proteins are inhibitors of NF-kB; the freed NF-kB then translocates from the cytosol to the nucleus, with multigene activation (Andreakos et al., 2005).

The point is that the NF-kB pathway is involved in the production of a host of inflammatory chemokines, cytokines, adhesion molecules, tissue-degrading enzymes, antiapoptotic proteins, and costimulatory molecules necessary for the induction of subsequent B- and T-cell responses, and the emergence of the *adaptive* immune system. It is evident why NF-kB has engendered so much interest in the molecular basis of inflammation in relation to human disease (Baldwin, 2001b). In the light of NF-kB's role as a transcription factor, the panorama of related diseases that unfold include rheumatoid arthritis, autoimmune diseases, atherosclerosis, sarcopenia, cancer cachexia, diabetes, osteoporosis, and cancer—the "whose who" of Chapter 4 (Arend, 2001; Baldwin, 2001a; Firestein, 2004; Hamerman, 2004). As Baldwin (2001b) writes: "NF-kB is a primary effector of human disease. For this reason efforts are underway to develop safe inhibitors of NF-kB."

So, with this background, here are some "snapshots" of evolving therapies to counter cytokine-induced inflammation:

- Modulating the IKK-β complex-driven NF-kB activation pathway to suppress NF-kB activation would be desirable (1) to treat malignancies (Aggrawal, 2005) and to convert inflammation-induced tumor growth to inflammation-induced tumor regression (Luo et al., 2005); (2) to diminish insulin resistance (Arkan et al., 2005; Sjoholm and Nystrom, 2006); (3) to reduce RANK activation of NF-kB and decrease osteoclastogenesis and inflammation-induced bone loss (Ruocco and Karin, 2005); (4) to suppress joint inflammation, cartilage degradation, and cell proliferation in OA, as well as angiogenesis and pannus formation in rheumatoid arthritis (Roman-Blas and Jiminez, 2006); (5) to prevent an inflammatory component of muscle degeneration and myofiber death in clinical conditions of dystrophy and atrophy (Karin, 2006).

- Antagonizing the cytokine macrophage migration inhibitory factor (MIF) has been suggested as a potential therapeutic strategy in inflammatory diseases (Morand, 2005). It is evident that the macrophage has "come into center stage," as Campion (1994) puts it, and is a dominant cell type in the inflammatory processes of adiposity and insulin resistance, atherosclerosis, and even osteoporosis.

- Inhibiting sirtuins—a family of histone deacetylases (discussed in Chapter 3, section 3)—could reactivate p53, a tumor suppressor protein (Fig. 3-3), and shift proliferating cells to senescent cells. Thus, a SIRT1 inhibitor (*Sirtinol*) has been used in cancer treatment (Longo and Kennedy, 2006), and histone deacetylase inhibitors "may be new drugs for treatment of inflammatory diseases" (Blanchard and Chipoy, 2005). However, such therapies get at the heart of conditions controlling aging and longevity as well, because of the crucial effects of SIRT1 on metabolic pathways (insulin/IGF-1). As Longo and Kennedy (2006) state, "the safest bet at this point is that Sir2 deacetylases both play pro -and anti-aging roles in different contexts," and antideacetylase therapies for diseases may raise truly complex issues.

- Selectively inhibiting adipokines that promote insulin resistance (e.g., TNF-α, possibly resistin) potentiating adiponectin (Kobayashi, 2005), or inhibiting of cytokine-activated protein kinase Jun N-terminal kinase (JNK) could be used to reduce insulin resistance. Indeed, I discussed in Chapter 4, section 3 that JNK mediates the inhibitory effect of TNF-α on insulin signaling by way of serine phosphorylation (Musi and Goodyear, 2006). In fact, Liu and Rondinone (2005) review advances in small-molecule inhibitors of JNK "that have also been targeted for other diseases with an inflammatory component, such as stroke, rheumatoid arthritis, Alzheimer's and Parkinson's diseases." Yet the "lingering concern remains about how much JNK isoform selectively is required for drug discovery," and this also relates to safety issues. These are concerns similar to those of PPARγ modulators discussed earlier. In the case of JNK, knockout mice (i.e., those with genetic deletion of JNK) had enhanced tumor burden. Further, in humans, molecular action can vary from organ to organ, as I discussed with the SERMs. "It will be of great interest to see how many JNK inhibitors progress to clinical trials, which specific diseases are to be targeted, and whether the desired therapeutic outcomes can be achieved without unwanted side effects."

- Blocking of IL-1 receptors (Dinarello, 2005) may inhibit inflammation. The subject is also well reviewed by Braddock and Quinn (2004), who discuss *anakinra*—"at present the only drug in clinical practice for the treatment of RA that inhibits IL-1 signalling" by virtue of its role as a receptor antagonist (IL-1Ra). Other drugs under development that inhibit IL-1 are reviewed, promising that what began as cytokine inhibition in RA may extend to other diseases as well, as I noted previously. However, to reiterate, too much inhibition may not be a good thing: the combination of anakinra and an anti-TNF-α inhibitor may increase infections by "compromising the host defense system such that the balance is tipped towards an unacceptable profile."

- Antagonizing chemokine receptors CCRI and CCR2 (Koch, 2005; Ribiero and Horuk, 2005) has clinical potential.

- Adding omega-3 fatty acids to the diet "to decrease the risk and severity of inflammatory conditions, such as cardiovascular disease," may limit inflammation. Stephensen and Kelley (2006) review aspects of dietary supplementation that may curtail the inflammatory components generated by the innate immune system, but may have risks in terms of suppression of host defenses.

To conclude this section, genetic variations may affect the individual's response to cytokines (Duff, 2006). Indeed, responses to inflammatory mediators—a discussion that has dominated so much of this book—are to an extent under genetic regulation. Thus, IL-1 and IL-18 are cytokines that are part of the same structural family (Dinarello, 2006). IL-1β and IL-18 in their respective forms are inactive until cleaved by the intracellular cysteine protease, caspase-1. The "NALP 3 inflammasome" is a protein complex that stimulates caspase-1 activation to promote the processing and secretion of proinflammatory cytokines (Ogura et al., 2006). Patients with gain-of-function mutations in the NALP 3 gene secrete more IL-1β and IL-18 and have "systemic inflammatory diseases" with high circulating levels of IL-6, serum amyloid A, and CRP (Dinarello, 2006). In these patients, blockade of the IL-1 receptor will reduce serum levels of inflammatory cytokines. Westendorp (2006) suggests that investment in the proinflammatory state was a developmental priority, perhaps to resist infection, but at the expense of an anti-inflammatory state. But now our species may be ready to favor genetic variations that promote anti-inflammatory cytokines (e.g., IL-10) or diminish inflammatory ones (IL-1, TNF-α), and so achieve more healthy aging (Chapter 6, section 3).

Recall that, in Chapter 4, section 3, in persons with manifestations of the metabolic syndrome, *low plasma levels* of adiponectin were noted to be a risk factor for cardiovascular events. Now, in a study based on laboratory analyses, certain *genetic variants* of adiponectin have appeared to predict *protection* from incident ischemic stroke in persons without diabetes (Hegener et al., 2006). In the future, studies of adiponectin gene variants may become clinically useful in identifying persons at risk for the metabolic syndrome, type 2 diabetes, and cardiovascular disease (Gable et al., 2006).

Chapter 4, section 7, in the discussion on osteolytic metastases, mentioned that osteoprotegerin (OPG) could inhibit TRAIL (TNF-related apoptosis inducing ligand), conferring on the tumor cell that produced OPG a survival advantage by reducing apoptosis. TRAIL is of interest because it induces apoptosis by cross-linking its receptors and forming a death-inducing signaling complex by way of caspase activation. Tsokos and Tsokos (2003) discuss "gene-cell therapy" in models with mice, whereby genetically modified dendritic cells transfer TRAIL to tumor cells, or to T cells, or to synovial cells, activating the apoptosis cascade, suppressing cancer, autoimmune disease, or rheumatoid arthritis, respectively.

Finally, Kornman (2006) discusses IL-1 genetics, pointing out the need to develop the technology (as in Hegener et al., 2006) to identify persons with genetic variations that alter or diminish the trajectory of the inflammatory response and clinical outcomes. Single nucleotide polymorphisms (SNPs) that are inherited in variations or blocks called haplotypes have been recognized in the IL-1 gene (Duff, 2006). In these clusters, IL-1 genetic variations have been identified for periodontal disease, cardiovascular disease, and gastric cancer. Kornman (2006) has important things to say about these studies as part of risk factor identification to prevent disease. Until now, the most widespread applications have been a *public health* model in which lifestyle factors—smoking cessation, prudent diet, exercise—are addressed (as discussed in Chapter 6, section 2). We are entering a new phase in which more *individual* patient management may also have widespread health benefits—such as reducing cholesterol or normalizing blood pressure—and "these approaches have been primarily driven by the availability of drugs to modify the risk." The statins in particular have been discussed (Chapter 5, section 2) from this perspective. Thus, an array of serologic and genetic tests is emerging that will be available to clinicians to anticipate "common chronic diseases well before their initiation is evident" in their patients (Kornman, 2006). Indeed, the earlier concept of pharmacogenetics involving one allele at a time may now be enhanced by "pharmacogenomics—where investigators are beginning to explore much larger sets of genes, including pathways up to the whole genome, and how variations in these pathways may affect drug response" (Roden et al., 2006). The "increasingly sophisticated understanding of the relationship between genetic variants and variability in drug response represents an approach that may enter and enrich clinical practice." It may not be too far fetched to consider that ways will be developed not only to safely modify genetic risks to limit the expression of disease but also to use genomic-based therapies—perhaps the ultimate step in translation—to reduce risk and promote health and prospects for longevity. Thus, the potential promise of this era of "genomic medicine" (Guttmacher et al., 2004).

6. HORMONAL THERAPIES

It seems reasonable for the clinician to consider replenishing hormonal mediators that decline rather generally in older persons, such as growth hormone (GH)—insulin-like growth factor (IGF-1) (for the somatopause) and testosterone (for the andropause). But certain problems arise about such a discussion based on the criteria for inclusion set forth in the introduction to this chapter: these therapies are not used to treat diseases and are certainly not multidisease suppressive. With respect to GH, von Werder (1999) notes, "GH secretion in the elderly—the somatopause—if it occurs at all, is unlike the menopause which occurs during midlife and often causes severe symptoms necessitating hormone replacement therapy. No acute symptoms can be attributed to GH deficiency." Indeed, whether GH decline is "homeostatic or pathologic and in need of treatment" remains debatable (Hoffman, 2005). Indications for hormonal (GH, testosterone) replacement remain uncertain and controversial, with recognized adverse side effects. Moreover, use of GH has often been presented as an appeal to the public to "prevent aging" (Vance, 2003). Vance (2003) quotes a lay publication, "Grow Young with HGH." Morley (1999) raised the question of "GH as the fountain of youth or death hormone?"

Yet I decided to include these two therapies. Their "physiology" is of great importance for the geriatrician, and this is a field of active investigation with a possibility of their revival or at least their reappraisal from a time when they seemed to have fallen into disfavor. For example, in a recent editorial, Rosen and Wüster (2003) asked, "Growth hormone rising: did we quit too quickly?" Reckelhoff et al. (2005) wrote, "treatment of aging men and women with testosterone supplementation is increasing." Basaria (2005) notes, "the number of T prescriptions written has soared and the industry is actively marketing androgen products."

Growth Hormone and Insulin-like Growth Factor

Cummings and Merriam (2003), in their review, point out that "it has long been axiomatic that GH is essential for normal linear growth, and thus GH deficiency arising in childhood should be treated." Only recently has it also become apparent that GH continues to serve many important functions throughout life. Yet it is also true, as Rosenfeld (2006) points out in an introduction to a symposium in the journal *Hormone Research,* "that GH is not the major mediator of skeletal growth; rather the IGF option has emerged as the true mediator of skeletal growth." Nevertheless, it is important to look at their overall interaction.

GH secretion from the anterior pituitary is under dual hypothalamic control via stimulation by GH-releasing (GHR) hormone and the inhibitory neurohormone somatostatin. Release of GH is pulsatile, with variations in serum levels ranging from virtually immeasurable to higher values after meals, exercise, sleep; this limits use of direct GH serum measurement and further encourages assays of factors that GH promotes, especially IGF-1 and IGF-binding protein 3. IGF-1 plays a role in feedback control, stimulating somatostatin release at the hypothalamic level and inhibiting GH gene expression at the pituitary level (von Werder, 1999). Mauras and Haymond (2005) review and contrast the metabolic effects of GH and IGF-1. GH binding to its receptor activates a complex signaling transduction cascade of events, including (a system I have mentioned)—Janus kinase (JAK)/signal transducers binding to Src domains (Rawlings et al., 2004; Waters et al., 2006) and generating production of IGF-1 in the liver (Cohen, 2006). The stimulatory effects of GH are mainly observed in linear growth and body composition. The inhibitory effects of GH are mediated in part through suppressors of cytokine signaling (SOCS) (Alexander and Hilton, 2004). The effect of GH to induce a state of mild insulin resistance in adults is due to GH activation of the insulin receptor substrate (IRS) independent of IGF-1 generation; in fact, IGF-1 suppresses GH, has potent hypoglycemic effects, and enhances insulin sensitivity (Clemmons, 2004). Although the IGF-1 receptor and insulin receptor are highly similar and activate overlapping downstream signaling components (insulin receptor substrate proteins and PI3K—discussed in Chapter 4, section 3), the IGF-1 receptor primarily regulates growth and development, and has only a minor function in metabolism (Engelman et al., 2006). IGFs circulate in plasma complexed to a family of binding proteins (IGFBPs), in particular, IGFBP-3 or IGFBP-5, and these seem to be "modulators of IGF-1 bioactivity" (Frystyk, 2005). IGF-2 is expressed in a limited number of tissues and is not regulated by growth hormone. Levels of IGF-2 in general are higher than IGF-1 and do not decline after puberty (Danielpour and Song, 2006), but will not be discussed further.

A recent development in our further understanding of GH release from the pituitary is the role in this process of small synthetic molecules called growth hormone secretagogues (GHSs). These are potent stimulators of GH release that work through the GH secretagogue receptor (GHS-R). Kojima and Kangawa (2005) "unexpectedly" purified and identified an endogenous (i.e., within the body) ligand for the GHS-R from rat stomach and named it "ghrelin," after a word root (*ghre*) in Proto-Indo-European languages meaning "grow." These authors note that ghrelin is a peptide hormone in which the third amino acid, usually a serine but in some species a threonine, is modified by a fatty acid (a unique *n*-octanoyl ester), a modification essential for ghrelin's activity. Thus, the discovery of ghrelin indicates that the release of GH from the pituitary might be regulated not only by GH-releasing hormone, which stimulates GH release through binding to the GH-releasing hormone receptor, but also by ghrelin, derived from the stomach. This latter discovery raises interesting associations. In the diagram on signals (Fig. 4.6) we can add ghrelin: it is orexigenic. After secretion from the stomach, it enters the circulation and acts on the arcuate nucleus in the hypothalamus to stimulate appetite. Thus, ghrelin plays important roles in maintaining GH release and energy homeostasis in vertebrates. In its role as an endogenous agonist of the GHS-R (independent of GHR hormone), it is not clear whether ghrelin production decreases with age (Smith et al., 2005). This is interesting in the light of GH decline as well as appetite diminution with aging, and Smith et al. (2005) go on to say "restoration of the GH-IGF axis in elderly subjects by ghrelin or a ghrelin mimetic reduced inflammatory cytokines, improved appetite by way of leptin antagonism, and improved body composition." Ghrelin is one of many "gut hormones" that influence energy balance and metabolism. For a review of this subject, the reader may wish to consult the paper by Perez-Tilve et al. (2006), but as Jørgensen (2006) notes in an editorial, "the convoluted tale of ghrelin" continues due to evolving "revelations of its diverse and complicated physiology." Although administered (exogenous) ghrelin strongly stimulates GH release in humans, it has been difficult to demonstrate that circulating (endogenous) ghrelin levels regulate pituitary GH secretion. Ghrelin administration does stimulate food intake and its orexigenic action may be by way of neuropeptide Y and agouti-related peptide-containing neurons in the arcuate nucleus (Fig. 4.6). Ghrelin infusion does significantly increase gastric motility, stimulate appetite, and decrease satiety. One point of uncertainly that awaits clarification is the extent to which ghrelin's effects are due to its role as a ligand binding to the GH secretagogue receptor.

Veldhuis et al. (1997) list "multiple confounding factors that attend aging that strongly influence neuroendocrine activity of the GH-IGF axis": an increase in visceral fat, varying levels of sex-steroid hormones (declining testosterone in older men and diminished estrogens in postmenopausal women), diminished exercise, sleep variations, and nutritional deficiencies. Individuals who are moderately to markedly obese and manifest other features of the metabolic syndrome (Chapter 4, section 3) have profound suppression of serum GH through all ages (i.e., there is little additional discernible diminution with age). Further, the usual positive association of serum testosterone in relation to daily GH secretion is lost with obesity. The implications are interesting, for according to Veldhuis et al. (1997, 2005a) a "disorderliness" of GH release that occurs with aging—an increasing irregularity—has also been demonstrated for ACTH-cortisol, luteinizing hormone, testosterone, and insulin, "a more general decline of neuroendocrine servomechanistic control." Indeed, Veldhuis et al. (2005b), in writing about the ensemble "mechanisms in aging," point out that a fourfold diminution of estrogen, GH, IGF, and testosterone is accompanied "by an increased prevalence of clinical, hormonal, biochemical, and structural features of frailty, disability, and reduced quality of life." All this seems to be well put and sounds like "loss of complexity with aging," according to Lipsitz and Goldberger (1992) and cited in Chapter 3, section 2. Yet a "paradox" (Giordano et al., 2005) remains in that high concentrations of GH in mice are associated with their reduced life expectancy; in dwarf mice, or in a range of species on caloric restriction, low GH (and hence IGF) results in longevity, along with increased insulin sensitivity, reduced reactive oxygen species, and improved resistance to stress (Bartke, 2005; Rincon et al., 2005). Would this argue against use of GH as "antiaging" treatment?

In the light of all this, what would alert the physician to use GH in older persons? Prescribing GH should be considered in those with a history of child-hood GH deficiency, as noted, or significant hypothalamic-pituitary damage (Cummings and Merriam, 2003). The hypothalamic–GH-IGF-1 axis is often the first hormonal system to be affected by disease of the hypothalamic-pituitary region (neoplasm, irradiation, infection, surgery). The "growth-promoting effect of recombinant human IGF-1 (rhIGF-1) in persons with the extensively studied GH receptor mutation who develop the GH insensitivity or Laron Syndrome, signifies an endocrine role for the GH-IGF system" (Ranke, 2005), an indication for IGF treatment, and the potential for use of the newly developed compound—recombinant IGF-1/IGF binding protein-3 (Aimaretti et al., 2005; David et al., 2005). The broader question relevant to geriatrics is whether spontaneous GH deficiency syndrome can be clinically defined in older patients (Merriam et al., 2004). A characteristic adult GH deficiency syndrome has been *hypothesized,* according to Hoffman (2005), consisting of "alterations in body composition" (increased visceral adiposity, decreased lean body mass, decreased bone mineral density), an atherogenic lipid profile, decreased muscle strength and aerobic capacity, and a poor quality of life, with low energy, lassitude, and social isolation." Aimaretti et al. (2005) and Cummings and Merriam (2003) have expressed similar views. This seems to cover a wide territory indeed (multisystem, aging, lifestyle behavior). Hoffman (2005) himself agrees that the components of this "syndrome" are nonspecific, to a large degree subjective, and compatible with a broad older population, especially of men at risk. According to Isley (2002), the "growth hormone deficiency syndrome exists but it is unclear whether growth hormone deficiency is the primary etiologic factor." The issues relate to its diagnostic substantiation, and the indications for growth hormone therapy, weighing the salutory effects of improving body composition and decreasing cardiovascular mortality as compared with "doing harm." Isley concludes "for most patients, growth hormone therapy is not quite ready for prime time."

Hoffman (2005) cites recent studies describing the benefits of GH replacement in persons with presumed adult GH deficiency. So to pose the question again, what indication would a geriatrician in clinical practice need to treat an older male patient with recombinant GH or IGF-1 (Lombardi et al., 2005)? The first requirement, of course, would be suspicion of adult GH deficiency, and the clinician would be hard pressed to distinguish this from expected body composition changes and metabolic events with aging itself, as I've noted. Because I chose not to discuss actual therapeutic interventions, it would seem appropriate that the next step would be endocrinologic specialty consultation.

If clinical guidelines and indications become more apparent, measurements of serum IGF would be indicated. Unlike levels of GH itself, serum levels of IGF-1 and IGF-binding protein-3 are stable throughout the day, show no diurnal variation, and need not be drawn fasting (Cummings and Merriam, 2003). I am not certain how available this laboratory test is, even in teaching hospitals. Moreover, as many as 40 percent of GH-deficient adults have normal age-adjusted IGF-1 levels, so this may not resolve the situation (Hoffman, 2005).

By way of a summary of this part of the discussion, a task force of the Endocrine Society has issued clinical practice guidelines (Molitch et al., 2006), noting that "confirmation through stimulation testing is usually required unless there is a proven genetic/structural lesion persistent from childhood." Essentially, "the final decision to treat adults with GH deficiency requires thoughtful clinical judgment and careful evaluation." The most recent conclusion about the use of GH in the "healthy elderly" (Liu et al., 2007) is that such use is associated with small changes in body composition and increased rates of adverse events. "On the basis of this evidence, GH cannot be recommended as an antiaging therapy."

At this point a brief look in retrospect at GH therapy is necessary to observe how the concept of treating GH "deficiency" arose: it did so based largely on a study by Daniel Rudman (1990) when he gave GH to older men with low serum levels of IGF compared with levels in healthy young men. Mary Lee Vance (2003) wrote a brief retrospective for the *New England Journal of Medicine* expressing the importance of the original studies by Rudman, and noting many of her reservations about this and subsequent studies, all on older, healthy men and for short-term GH use. Her conclusion applies to the last chapter of this text—that "going to the gym" is as beneficial in improving body composition and muscle strength—and "certainly cheaper than growth hormone." Yet as Cummings and Merriam (2003) note, "beneficial effects on body composition have been almost universally observed in adult GH deficiency receiving GH," including bone mineral content in postmenopausal women (Landin-Wilhelmsen et al., 2003). Consult recent papers for the adverse effects—especially fluid retention, arthralgia, and carpel tunnel syndrome.

The issue also exists of the effects of GH promoting enhanced BMD. Hypopituitary patients, ages 16–81, "severely deficient in GH" had significantly more vertebral fractures than controls or those growth hormone-deficient patients receiving rhGH (Mazziotti et al., 2006). A recent study showed that serum IGF-1 (and certain IGF-binding proteins) were significantly lower in women with vertebral fractures than in those without fractures (Yamaguchi et al., 2006). In a group of "healthy" men 60 to 70 years of age, serum IGF-1 was reduced in persons with low BMD, but GH secretion and reserve (measured by exercise and glucagon stimulation tests) were normal. Hepatic resistance to GH was considered as a possible explanation (Patel et al., 2005). Rosen and Wüster (2003) citing "GH rising," reviewed a paper in the *Journal of Bone and Mineral Research* on the use of GH as an anabolic agent for bone (Landin-Wilhelmsen et al., 2003). The background for the use of the anabolic agent teriparatide (PTH 1-34) (Tashjian and Gagel, 2006) was traced by Rosen and Wüster (2003), who noted in the study under review that GH treatment of women with postmenopausal osteoporosis showed "remarkable increases in bone mineral content and density at several skeletal sites after about a year and a half of treatment and an additional longer term follow-up." Yet to my knowledge, therapy with GH has not "risen" as a therapeutic consideration for osteoporosis.

Andropause

The decline of testosterone and its clinical significance in men—the andropause—raises issues similar to the somatopause (Isidori and Lenzi, 2005; Wang and Stocco, 2005). Vermeulen (2000) (from whom much of this discussion is taken) also compared the andropause with the menopause in women, as noted above, and found no similarity. In contrast to a "relatively abrupt and universal loss of ovarian function in middle-aged women, men develop a more gradual and incomplete age-associated decline in gonadal function," which shows a high degree of "interindividual variability." He goes on, "as fertility persists until very old age, the andropause as the equivalent of the menopause does not exist." (Indeed, he does cite paternity in men over 90 years!) So the andropause has come to be based on late-onset hypogonadism with a decline in serum testosterone, with the proviso that about 20 percent of men over 75 years still have plasma testosterone levels within the range of young men. Amory et al. (2004) described men with "low" testosterone levels as less than 12.1 nmol/liter, and Snyder et al. (2000) studied "hypogonadal" men (mostly of secondary nature) with serum testosterone levels of 8.7 nmol/liter. A general screening of "older" men for testosterone is not feasible (Schubert and Jockenhovel, 2005), and there is "no consensus for diagnosing elderly men with the andropause (Basaria, 2005). Testosterone levels are a "modest" predictor of BMD in sites of predominantly cortical bone (hip, arm), but not in sites of trabecular bone (lumbar spine), and lower levels are predictive of prevalent osteoporosis-related vertebral fractures (Mellström et al., 2006). A reader interested in pursuing the subject of "hormones for men," relating particularly to testosterone replacement therapy, would find the article by Jerome Groopman in the July 29, 2002, issue of the *New Yorker* particularly relevant.

Decline in testicular function diminishes serum testosterone, but the adrenal cortex is the source of other androgens, notably dehydroepiandrosterone-sulfate (DHEAS) and dehydroepiandrosterone (DHEA) itself, and both decline with age. As I wrote in Chapter 3, section 2, their role in physiologic aging may be by way of conversion to estrogens, androgens, or both, or as a steroid hormone, especially in the postmenopausal period (Labrie et al., 2005). DHEA may reduce visceral fat and heighten insulin sensitivity, and improve endothelial function by virtue of increased NO. Perrini et al. (2005) write that "physiological age-associated decline in circulating DHEA per se does not justify DHEA supplementation since the effects of this hormone on metabolic abnormalities, endothelial function in vivo, and cardiovascular events are contradictory."

The clinical and therapeutic implications of the andropause are far more uncertain than the somatopause. Issues of libido, erectile dysfunction, and impotence arise, but are not necessarily attributable to lower testosterone levels. Enhanced visceral fat mass and decline in muscle mass and bone density that may be found with low testosterone levels occur with aging as well, and as they do with apparent GH deficiency. Liu et al. (2005) pointed out a wider endocrine deficiency in older men with declining testosterone levels as a result of "simultaneous *disruption* of hypothalamic gonadotropin-releasing hormone-luteinizing hormone-stimulated Leydig cell–testosterone production axis."

Vermeulen (2000) reviewed the evidence for "restorative" effects with testosterone therapy. The extent to which the practicing geriatrician would even consider the "andropause," much less seek a serum measure of testosterone, must be scarce indeed. Hijazi and Cunningham (2005) discussed the andropause in the *Annual Review of Medicine*. An estimated 70 percent of men 70–80 years of age have low bioavailable or free testosterone levels, yet the "symptoms of testosterone deficiency are similar to those associated with aging: loss of energy, depressed mood, decreased libido and erectile dysfunction, decreased muscle mass and strength, increased fat mass, frailty, and osteopenia."

Kaufman and Vermeulen (2005) in *Endocrine Reviews* also consider the implications of androgen decline in elderly men and point out the uncertainty of diagnosis, inability to distinguish between "substitutive" and "pharmacologic" androgen administration, and the limitations to date of randomized trials. They conclude: "Until the long-term risk-benefit ratio for androgen administration to elderly men is established in adequately powered trials of longer duration, androgen administration should be reserved for the minority of those who have both clear clinical symptoms of hypogonadism and frankly low serum testosterone levels." Morley et al. (2005) emphasize the need for more placebo-controlled trials "in view of the substantial placebo effect on aging symptomatology." Perhaps an alternative clinical role for the geriatrician might be to provide counseling for older adults on sexual function in aging; prevent men from casually receiving testosterone from free-standing clinics; and inform male patients about the adverse side effects of testosterone use (Reckelhoff et al., 2005), especially in regard to cardiovascular events or development of prostatic cancer (Basaria, 2005; Hijazi and Cunningham, 2005).

But, as mentioned in Chapter 4, section 4 on osteoporosis, consideration of bone loss in *men* does raise interesting and unique biological issues and, of course, public health problems. Based on the prospect of longer life expectancy in men, and specifically with respect to hip fractures, 424,000 were estimated world wide in 2000, and 800,000 in 2025, representing an increase of 89 percent (Seeman et al., 2004). If we exclude secondary causes of osteoporosis (medical diseases, medications, and environmental factors) that pertain to women as well and will not be considered further, osteoporosis in men can be divided into that occurring in older men (e.g., over 60 years)—what Khosla (2004) in his fine review calls involutional—and that occurring in young and middle-aged men, less of a geriatric subject but of great interest nevertheless because these men age in place.

With respect to the involutional group, these men have a marked age-related increase in circulating sex hormone-binding globulin (SHBG). Although a high level of SHBG may have no direct effect on bone, indirectly "the proximate cause" does: declining IGF-1 and GH, as well as significant reductions in serum levels of bioavailable estrogen and testosterone. These hormonal deficiencies are associated with bone loss, impaired bone formation, in particular, periosteal apposition, low BMD, at predominantly cortical bone sites, and fracture risk, especially in those men with a low peak bone mass at an early age (Mellström et al., 2006).

In a small group of men younger than 60 years, similar serum findings may be apparent, but in those with an aromatase deficiency, testosterone levels may be normal and estrogen levels low, reflecting inadequate conversions of testosterone as the principal source of estrogen. It has come to be appreciated by virtue of rare cases of "experiments of nature"—albeit at a much younger age—in which genetic losses contribute to absence of the estrogen receptor, or to aromatase deficiency, that estrogen plays a dominant role in the mineral density of the male skeleton as well (Gennari et al., 2004).

Practical issues arise for the geriatrician. A DEXA study should be considered in men older than 65 years, although treatment options (traditional bisphosphonates versus physiologic testosterone replacement) may need to be weighed. Bisphosphonates are likely to be efficacious in men; evidence for antifracture effects of androgens in men with androgen deficiency and osteoporosis is lacking. There is no evidence to support the use of androgens in eugonadal men (Seeman et al., 2004). A high SHBG level may be helpful, but interpretation of testosterone levels may be more problematic. Care must be exercised, as noted above, that testosterone therapy does not enhance prostatism or prostate cancer, and finasteride has been used to limit untoward prostate effects based on its blocking the enzyme 5α-reductase which converts testosterone to dihydrotestosterone—the major form of the hormone associated with prostatism (Amory et al., 2004).

CHAPTER SIX

Preventive Gerontology

1. THE COMPRESSION OR EXPANSION OF MORBIDITY?

An issue that uniquely applies to geriatrics is the question posed by James Fries (1980) more than two decades ago: whether the "average age at first infirmity can be raised, making the morbidity curve more rectangular, thereby extending adult vigor and compressing the period of senescence [sic] near the end of life." The *compression* of morbidity hypothesis, as this came to be known, was based on evidence that primary prevention measures could be adopted to postpone the onset of chronic illness. That is, the age of the time of initial disability would increase more than the gain of longevity, resulting in fewer years of disability. Indeed, in extensive longitudinal follow-up, the studies of Fries and colleagues (Vita et al., 1998) identified smoking, body mass index, and exercise patterns at midlife and late adulthood as predictors of disability. "Not only do persons with better health habits survive longer, but in such persons disability is postponed and compressed into fewer years at the end of life." In this scenario, the ideal might be the 88-year-old man who dropped dead on the golf course. Indeed, evidence gathered by Manton and Gu in 2001 and described by Cutler (2001) in a commentary that accompanied their article noted "clear, overwhelming evidence that the average health of the elderly population is improving. In a 17 year span from 1982 to 1999 the number of elderly persons with disabilities (limited capacity to function independently with ease) declined from 26.2% to 19.7%."

On the other hand, Olshansky and Cassel (1997), and perhaps Verbrugge and Patrick (1995), viewed the accumulation of widely prevalent "chronic, non-fatal conditions" as surpassing fatal diseases. Accordingly, the *expansion* of morbidity view is consistent with the medical community continuing to *extend* the lives of older persons by reducing death from fatal illnesses, and thereby exposing survivors to additive diseases and conditions associated with advanced age. Jerry Avorn (1986) created a fictitious scenario of "too long life" centered on a character he called "Oliver Shay." This surely was a tongue-in-cheek selection by Avorn because "The Wonderful One Horse Shay," by Oliver Wendell Holmes, "provided the metaphor for the perfect life-span: the carriage that lasted exactly a hundred years and then collapsed into a mound of dust, going to pieces all at once" (Kennedy, 2004). Yet "Oliver Shay" starting at age *67,* successively encountered (1) bacterial pneumonia, (2) coronary bypass surgery, (3) hip replacement, (4) a stroke and cardiac electric defibrillation, (5) a pacemaker, (6) Alzheimer disease, (7) hemodialysis, and (8) at age 91, terminally, a ruptured abdominal aortic aneurysm. Based on this scenario, Avorn expressed skepticism about the compression of morbidity.

Both views may be compatible: in the next half-century morbidity in older persons will continue to be compressed, but the trend toward longer life span and absolute numbers of the oldest old will increase, as will morbidities in these survivors. It is true that these morbidities will be for fewer years, but more persons will exhibit them. Verbrugge and Patrick (1995) were correct to place chronic illness as a societal challenge by way of priorities: a need for an overall effort to address *chronic* conditions that call for research resource allocations often reserved for acute, fatal conditions and, further, a gerontologic orientation to health care provision. (This would enhance geriatric practice.) Yet the achievement of compression of morbidities (even though there may be more persons with them) needs to come from lifestyle practices of older persons, and scientific advances in the basic biology of aging by the experimentalist with potential application by physicians in clinical practice. I will discuss lifestyle aspects in a moment. I want to give one example of the merging of the bioscience and clinical application put forth, again by Olshansky and Cassel (1997), and drawing on theories of Kirkwood (2005a): "that senescence is not programmed directly as a biological clock designed to go off at a predetermined time—it is a by-product of an evolved reproductive pattern and survival into an age range that permits its expression. Thus, one of the most important biological properties of senescence is that its expression can be modified without manipulating the genome itself. This has important implications for medical and gerontological research focused on extending life—particularly the healthy years of life." Westendorp (2006) echoes the view that "aging is not programmed nor is it inevitable." Thus, in a sense, this chapter is the opposite side of the coin to views presented in Chapter 3, sections 3 and 4, where aging was discussed in terms of the evidence for molecular mechanisms: rather, now, we must seek to "optimize fitness to help to protect and preserve healthy aging" (Westendorp, 2006). Olshansky and Cassel (1997) speak of methods to manipulate the expression of senescence-related diseases and conditions by avoiding "senescence accelerators" (cigarettes, radiation—such as the sun, excessive alcohol, stress), and encouraging use of "senescence decelerators" (dietary antioxidants). I would like to go beyond this and add, from what I have discussed before: unraveling the principles of caloric restriction that might apply in the short term to humans; using pharmaceuticals to improve metabolic status (e.g., statins), and seeking to reduce inflammatory mediators with safety in well-defined instances beyond RA (Chapter 5, sections 2, 4, and 5); and, ultimately, promoting the personal aspects of patient–physician interactions to foster lifestyle modifications, discussed below.

If I have made a convincing argument, at any level of retarding (or compressing) morbidity development (which I prefer to use rather than senescence deceleration), a bioscientific foundation exists from which enlightened clinical application may be derived: molecular-genomic, pharmacologic, immune, nutritional-dietary, and choices for health practices. The issue is not to postpone aging (or "antiaging," as the ads say), but to improve individual (and so collective) health and well-being despite aging.

2. HEALTHY LIFESTYLE PRACTICES

Healthy lifestyle practices are the responsibility of the individual, but also the educational task of the geriatrician, indeed of all health care providers in clinical practice. The notion that the "risky components of aging cannot be modulated is no longer tenable" (Najjar et al., 2005). Andrews (2001), in a commentary focusing on "Care of older people," also notes that health and well-being at older ages are modifiable, and substantial gains can be made by investment in promoting health and fitness throughout life. Interventions can be practiced at any age—obviously the earlier the better. Many societal factors have an impact on health practices, a full discussion of which is beyond the scope of this book. Yet, there are some to consider.

First is the issue of *literacy* itself—whereby limited literacy in community-dwelling persons over age 70 was associated with a nearly twofold increase in mortality. It is evident that every societal aspect related to well-being is diminished, including poor personal health practices, reduced perception of chronic disease burden, and limited access to health care. "Ultimately, the public health issues can be projected on a global scale" (Sudore et al., 2006). Then there is the important aspect of what Wolf et al. (2005) call *health literacy:* individuals with limited health literacy have less health knowledge, worse self-management skills, lower use of preventive services, and higher hospitalization rates. Yet in one large survey of more than 150,000 persons, "widely known healthy lifestyle characteristics"—including nonsmoking, healthy weight (BMI, 18.5–25.0), consumption of fruits and vegetables, and regular physical activity—were observed in "very few adults in the U.S. No demographic subgroup followed this combination to a level remotely consistent with clinical or public health recommendations" (Reeves and Rafferty, 2005). "Low-risk status" identified by stringent control of blood pressure (BP), cholesterol levels less than 200 mg/dl, and nonsmoking "is uncommon among U.S. adults with a prevalence of only 3% to 20%" (Daviglus and Liu, 2004). Chapter 4, section 6 already noted how infrequently older persons engage in exercise. All these adverse health practices have led some critics to characterize the prevalence of a low-risk status as an unachievable, impractical goal. And yet a favorable outcome of one aspect—lowering diastolic BP from 38 percent to 70 percent did occur in the 30 years from 1970 to 2000. Indeed, reduction of cardiovascular incidents is an extremely important and achievable goal. In fact, *lifestyle* interventions recommended and proven effective to lower blood pressure, that include improved diet, aerobic exercise, alcohol and sodium restriction, and fish oil supplementation (Dickinson et al., 2006), clearly have health ramifications for almost all the conditions discussed in Chapter 4, including stress, atherosclerosis, the metabolic syndrome, and osteoporosis. Beyond lifestyle practices, *therapeutic* modalities play an essential role in cardiovascular prevention (Andrawes et al., 2005) in which the geriatrician is aware of risk factors (primary prevention) or treats established hypertension, dyslipidemia, insulin resistance (secondary prevention), all discussed earlier. Smoking reduction has also occurred, an important preventable cause of cardiovascular mortality. Ezzati et al. (2005) estimated 1.6 million cardiovascular deaths worldwide (11% of total global cardiovascular deaths) were due to smoking. Other major health issues related to smoking are, of course, lung cancer, and perhaps less well known, increased osteoporotic fracture risk in older men (Olofsson et al., 2005). A recent editorial (Davis, 2006) summarizes many facets of the toll of smoking: the "dose-response relationship" with mortality, the benefits of smoking cessation, the greater detrimental effects of starting at a younger age, and the response of health care providers in counseling their patients (about 65 percent received advice to quit). As Daviglus and Liu (2004) note, "achieving favorable levels of *all risk factors* simultaneously requires effective cooperation among voluntary, professional, and governmental organizations, and the unrelenting repetition of the fundamental message." I would add—the message must reach consumers as well, and the choice made by them to engage in favorable lifestyle practices.

Multiple agencies among the National Institutes of Health, many commercial organizations, and a wide representation among academic medical centers have vigorously embarked on a randomized controlled trial to determine "whether intentional weight loss reduces cardiovascular morbidity and mortality in overweight individuals (BMI about 36 kg/m²) with type 2 diabetes." This is a "Look AHEAD" (*a*ction for *hea*lth in *d*iabetes) Research Group (2006) that is now at about the midpoint in a study to induce a mean loss of about 7 percent of initial weight and to increase participants' modestly intense physical activity to about 175 minutes per week. Eventually, awareness of this study by the widest possible network of health care providers, and by all consumers older than 60 years, through journal articles and the lay press, could have a wide influence and induce a favorable momentum.

An important lesson also can be derived from a recent study by Manini et al. (2006) and is further analyzed in an accompanying editorial (Blair and Haskell, 2006). The Manini paper describes "free living energy expenditure," uniquely and quantitatively measured in active community dwellers aged 70–82 years by use of isotopically labeled water (^{2}H and ^{18}O) over a two-week period rather than employing questionnaires of self-reported physical activity—reports that are rather widely recognized as inaccurate. The point is that the activity was not specific (e.g., at a gym) but rather consisted of many *routine* daily tasks, such as household chores, climbing stairs, walking, working for pay or as a volunteer, and even "fidgeting." Figures for free-living energy expenditures were derived from these activities. It is important that these activities could be correlated in a quantitative way in follow-up to show that higher levels of energy expenditure were observed in those with a lower risk of mortality than for those who were largely sedentary with high risk for adverse events. From all these recent studies, the key issue for public health education is how to *translate* weight loss and exercise—including these nonstandardized favorable activities—to practical use by a wide cohort of free-living older adults to improve their health outcomes as part of healthy lifestyle practices.

3. "SUCCESSFUL AGING"

I became aware of the term *successful aging*—which might represent a summation of healthy lifestyle practices—with the publication in *Science* by Rowe and Kahn in 1987, but an editorial by Phelan and Larson in 2002 cites references dating back more than three decades. During that time a strong publication history by academic professionals has grown in a range of disciplines which Phelan and Larson (2002) draw together to ask, in effect, where this topic is going. Their article—"intended as a succinct reference for clinicians who provide health care for aging adults and researchers engaged in the study of aging"—presents in tabular form a diverse representation of successful aging. This subject is likely to be heavily weighted by input from psychologists and social scientists, for the major elements included in their table are life satisfaction, mastery/growth, active engagement with life, high/independent functioning, positive adaptation, as well as the more traditional freedom from disability. Of course, longevity is a necessity to enjoy the fruits of successful aging. The two authors make the interesting suggestion that the actual beliefs of aging individuals themselves were not solicited by the researchers, and these attitudes might be a way for their health care providers to choose from an array of health care recommendations that are most relevant to their patients' individual needs.

About ten years after their first article, Rowe and Kahn (1997) revisited successful aging, with this being in the center of three intersecting loops: avoidance [sic] of disease and disability, high cognitive and physical function, and engagement with life. I had more trouble with this article than I did before because it seemed to have a judgmental quality, a sort of "class system" compared with those who could only manifest (or achieve) "usual aging." In 2003 Newman and coworkers conceptualized successful aging as freedom from disease—especially cardiovascular disease. This reflected a compression of morbidity—"shifting the burden of vascular disease to later in life, attenuating by several years the age-related trajectory of decline." This has merit because cardiovascular diseases are the leading cause of morbidity and mortality in developed countries, and there has been great success in reducing them through a variety of interventions (Andrawes et al., 2005)—including public health education (smoking cessation, obesity control); pharmacologic approaches (lipid lowering, blood pressure reduction); encouragement of good health practices (diet, exercise); and risk factor assessment and intervention.

Many health practice modalities are specifically aimed at reducing the manifestations of the metabolic syndrome (Després, 2005; Foreyt, 2005; Stone and Saxon, 2005; Sullivan et al., 2005). But these multisystem approaches require physician *awareness* of risk factor assessment to target primary preventive interventions—themes of this book. Present and emerging therapies (Chapter 5) will be increasingly effective in controlling many of the aging-related diseases discussed in Chapter 4.

In a recent supplement to the *American Journal of Clinical Nutrition* entitled "Living Well to 100," concerned with the influence of nutrition on aging, several authors used the term "healthy aging." This seems to fit better the sense of a more personal choice in this process than the word "successful," which poses a competitive edge with the less-desirable "usual aging," as I've said. Moreover, there is more of a biological underpinning, as Kennedy (2006) notes, where "healthy aging involves the interaction between genes, the environment, and lifestyle factors, particularly diet and physical activity." In the preventive approach to health care, "nutrition research has shifted from focusing exclusively on alleviating nutrient deficiencies to also stressing chronic disease prevention." Dwyer (2006) echoes that a "prevention-oriented, life cycle approach is critical to establishing and maintaining health throughout life . . . to delay and compress morbidity and the social toll associated with chronic disease and disability for as long as possible into old age." This is a life-long initiative, "starting down the right path with appropriate nutrition, and staying on it as important components of healthy aging" (Dwyer, 2006).

The outcome of all these motivational and "health-wise" approaches will have unquestioned societal and cost benefits as well. With evidence of effectiveness and wider population use, they would cast the broadest possible "health" net. By virtue of modifying disease risks, these measures would have a "significant impact on late survival" (Benetos et al., 2005). One might be optimistic, then, about the value of additional steps widely used in comprehensive geriatric practices: immunizations, falls prevention, accepted cancer-screening procedures, and encouragement of social engagement—promoting the widest possible scope of "successful" aging that might be summed up in a single word—health. While not guaranteeing life's satisfaction, approaching this state would perhaps be most widely appreciated among its recipients—the growing numbers of older persons in our society.

Epilogue

Now, at the conclusion of this book, we need to take another look at "geriatric bioscience" and whether the chapters herein have helped us to better understand the links (or interrelations) between the biology of aging and the emergence of aging-related diseases—which was our mission.

The issue has been joined by Cynthia Kenyon, director of the University of California Hillblom Center for the Biology of Aging. In her Darwin lecture (2006) at Cambridge University, England, she asks, "What links the normal ageing process to age-related disease?" She goes on:

The insulin/IGF-1 hormone system appears to link ageing to many related diseases. Scientists have looked at a range of disease in the long-lived worm mutants. One laboratory has put the protein causing Huntington's Disease into worms—which resulted in the worms getting the equivalent of Huntington's disease as they age. However, in long-lived *daf-2* mutants, the response is delayed, as though they are protected from the toxic effects of the disease-causing protein. Normal worms develop a muscle condition during ageing that resembles a human muscle wasting condition known as sarcopenia, although long-lived mutants do not develop this condition until they are much older. We have also shown that long-lived *C. elegans* mutants are resistant to tumors.

Interesting results also have been discovered in mice. Long-lived mutants have been shown to be resistant to cancer as well as other diseases, and every few months there are new reports about other diseases that are postponed in one of these long-lived animals. This is an interesting revelation—that these long-

lived animals age more slowly in the most fundamental ways—because diseases
of ageing don't afflict them until later.

As discussed in Chapter 3, section 4, the exciting current ideas of the genetic basis for *longevity extension* in yeast, worms, flies, and mice are based
on interactive models of gene mutations and caloric restriction. Longevity is,
of course, linked to aging by virtue of the species growing older—surviving
longer—although Hayflick (2000) suggests aging and longevity are separable:
"ageing is a stochastic (random) process not programmed by genes" (a view,
mentioned by Olshansky and Cassel in Chapter 6, section 1), "whereas longevity is not random and is subject to genetic modification." A somewhat
different view is expressed by Guarente and Kenyon (2000). They write, "the
ageing process like most biological processes is subject to regulation and can
be studied using classical genetics. When single genes are changed, animals
that should be old stay young. In humans, these mutants would be analogous
to a ninety year old who looks and feels forty-four." "On this basis we begin to
think of ageing as a disease that can be cured, or at least postponed." However,
in Chapter 3, I set forth views, which I share, of the separability of aging and
disease, and this epilogue continues that discussion. The key issue is one that
Kenyon and others have expressed—that those models of longevity in worms,
flies, and mice appear to represent the postponement or possibly the exclu-
sion of disease.

What links aging and the potential for the inception of disease is the time
course that cell senescence and its wider ramifications take (Fig. 3.3), including the local effects of senescent cells on tissues and the occurrence of apoptotic cell dropout—all of which contribute to deterioration of organ function;
or, more rarely, the multiple factors that limit senescence and induce neoplastic cell transformation. *Traditional,* cumulative timing of cell senescence may
be the basis for the *usual* life span of worms, flies, and mice. In experimental
models in which mutations or genetic down-regulation modifies certain metabolic pathways (e.g., GH/IGF), or with caloric restriction, we might suppose
that these events and other factors *postpone* or *delay* cell senescence and its
consequences, and longevity occurs by virtue of delayed or absent disease expression. Fossel (2004, pp. 14–15) takes a strong stand on cell senescence as
being "consistent with our current understanding of age-related clinical disease." He goes on: "free radicals may be villains in this drama, but the script
is that of cell senescence, and it is here that we can intervene most effectively.

The importance of cell senescence lies in how it defines human pathology and then only if we can alter the script and prevent age-related disease."

Blagosklonny (2006) also views senescence as part of the "developmental program that is not turned off, hyper-functional and damaging, causing diseases of aging." He proposes a crucial regulator of cell growth that I have not previously mentioned: mammalian target of rapamycin (M-TOR)—raptor kinase complex, one of the downstream targets of the PI3K–protein kinase B/Akt signaling pathway—which features prominently in Chapter 4, section 3—and also regulates FOXO (see below) (Engelman et al., 2006). Blagosklonny suggests that "in animals from worms to mammals, caloric restriction . . . and numerous mutations that inhibit TOR increase longevity and slow down aging."

In humans the actual molecular/genetic links between longevity and delayed expression of disease are not at the level of understanding they are in lower species, which is understandable. To achieve longevity, humans cannot activate genes engaged in the silencing of chromatin, as yeast can, or enter a dauer state, as worms can, or trade a mutation leading to dwarfism, as flies can, or extend caloric restriction for years, as mice can; and it is doubtful whether low GH/IGF activity can be sustained with impunity (Guarente and Kenyon, 2000; Longo and Finch, 2003; Kenyon, 2005). Yet these and related pathways are conserved through all these species, and are likely to "spark molecular insights in causes of human ageing" (Guarente and Kenyon, 2000). Thus, a general decrease in IGF-1 receptor levels can increase life span in mammalian species. The genetic link between insulin-like signaling and longevity, originally discovered in nonvertebrates, also seems to exist in higher vertebrates (Holzenberger et al., 2003; Richardson et al., 2004). Kenyon, in a personal communication, cites other examples of pathways for longevity in mammals: pituitary mutants (Ames and Snell dwarf mice), growth hormone receptor mutants, insulin receptor FIRKO mutants, and mice with low IGF-1 levels. Indeed, pathways by which the pituitary dwarf mutations in mice lead to longevity may overlap only partly with the effects of caloric restriction (Miller et al., 2002), suggesting interactive but not identical determinants. "Stress resistance—the ability to mount an effective response to environmental and cellular stressors"—may be where linkages influence the onset and progression of late-life aging and diseases (Harper et al., 2006). In long-lived worms and mice, superoxide dismutase, a powerful antioxidant, is up-regulated (also shown in figure 2 of Longo and Finch [2003], crossing species' lines). In fact, Martin and Groteweil (2006) go so far as to suggest that "many studies (although not universally) in a variety of species implicate oxidative damage with age as a significant cause of functional senescence [sic] and an important determinant of shortened life span."

Martin (2006) also approaches the question of how we age by classifying "modalities of gene action suggested by the classical evolutionary biological theory of ageing": longevity assurance genes with "good" alleles exert beneficial effects during both early and late stages of life, whereas other genes may exhibit early benefits but with potential deleterious effects later in life, the so-called antagonistic pleiotropy discussed in Chapter 3.

Perhaps research on centenarians and their notably long-lived offspring will provide new insights on genetic factors that promote longevity and delay the expression of disease (Atzmon et al., 2004). As Perls et al. (2002) point out, and as discussed in Chapter 3, section 4 and by Martin above, "there may be two classes of genes at play: on the one hand, the probability of achieving exceptional old age is likely enhanced by lacking genetic polymorphisms that predispose to diseases that cause premature mortality" (i.e., a "disease gene" such as apolipoprotein E-4 allele). On the other hand, genes "that slow aging at its most basic levels and therefore also confer resistance to age-related diseases are longevity enabling genes." Nir Barzilai and colleagues at the Albert Einstein College of Medicine have pursued studies of genes that favorably influence certain phenotypes and increase the likelihood of longevity in a group of Ashkenazi Jewish centenarians (Atzmon et al., 2004). In unpublished studies, Barzilai, Suh, and Cohen found several nonsynonymous mutations (a change in amino acids) in the IGF-1 receptor gene in several centenarians, consistent with longevity. Polymorphisms in apolipoproteins and related genes are likely to be one of several potential pathways in which genetic variants influence longevity. Thus, polymorphism in the cholesterol ester transfer protein (CETP) gene, with low CETP levels and a favorable lipoprotein profile, as discussed in Chapter 5, section 4, promote longevity. Polymorphism in the APOC 3 gene—a major component of very-low-density lipoprotein and chylomicrons—implicates a genotype with low serum levels of APOC 3, and again, a favorable lipid phenotype related to longer life span (Atzmon et al., 2006). The point of great interest for this further discussion is that the activator of APOC3 is FOXO-1.

Chapter 3, section 4 introduced the transcription factor forkhead box-containing protein of the O subfamily (FOXO) as the mammalian counterpart of the nematode daf-16, with a major role in development, reproduction, and longevity (Greer and Brunet, 2005). In this epilogue I wish to discuss FOXO-1 in relation to the interplay of classical obesity-related signals—leptin and insulin—that influence secretion of hypothalamic neuropeptides (orexigens and anorexigens) and are so important in the regulation of energy homeostasis, which Morton et al. (2006) describe as "the adjustment of food intake over time so as to promote stability in the amount of body fuel stored as fat." The discussion of obesity signals in Chapter 4, section 3 did not include FOXO-1, and now I wish to incorporate it and present a more comprehensive integration and overview of many aspects in this book, especially metabolic pathways that concern control of food (energy) intake, caloric restriction, and obesity—all features that influence a healthy "phenotype" or, alternatively, predispose to morbid events. I do not address here the large array of FOXO's other crucial functions at the "crossroads" (Accili and Arden, 2004) of differentiation, transformation, and longevity (Engelman et al., 2006) discussed in Chapter 3, section 4.

Morton et al. (2006), in a fine review, point out convergent actions of leptin and insulin involving their downstream products—STAT3, and PI3K:protein kinase β/Akt—respectively, that regulate energy homeostasis and hypothalamic neurocircuits. The interplay of these pathways promotes release of anorexigens (proopiomelanocortins, POMCs), and inactivates FOXO-1 by phosphorylation and its translocation from the nucleus to the cytoplasm (Greer and Brunet, 2005; Kitamura et al., 2006; Morton et al., 2006; Plum et al., 2006). This means that FOXO-1 is unable to promote release of orexigens (Agouti-related protein, AgRP). Conversely, during fasting, dephosphorylation of FOXO-1 translocates it to the nucleus and this *blunts* the effects of leptin to decrease food intake and to suppress AgRP (Kitamura et al., 2006).

I would like to propose three conditions:

1. In times of food shortage and potential weight loss, caloric restriction (CR) is in effect, with low plasma levels of insulin and leptin, enhanced FOXO-1 activation, hypothalamic release of orexigens (AgRP), and reduced levels of anorexigens (POMC). While these neuronal mediators may lead to increased energy intake and weight gain, relative CR still prevails. This is a "favorable phenotype" that prevailed millennia ago when our forebears were "hunter-gatherers" (O'Keefe et al., 2006) (see Chapter 5, section 2).

2. In present times of food abundance, but with essentially normal body weight, levels of insulin and leptin activate their respective molecular

receptors in the hypothalamus; FOXO-1 is inactivated, and the overall
effect is release of anorexigens and reduced orexigens. The result is to
sustain limited food intake, retaining favorable energy homeostasis and
"relative CR." This maintains a favorable phenotype.

3. In present times of food abundance, a combination of genetic (Barsh
et al., 2000) and environmental factors (Berthoud, 2004) contribute to
certain persons consuming excess calories and failing to exercise, contributing to their becoming overweight or obese. A state of insulin and
leptin *resistance* prevails with FOXO-1 activation, release of orexigens,
diminished anorexigens, and continuing food consumption with further weight gain. In a sense, this is the opposite of CR, i.e., "ad lib" or
unrestricted food intake, part of an unfavorable phenotype: a metabolic
state promotes adverse events and the potential for diminished longevity exists. The validity of this proposal depends on whether the issue of
CNS resistance to leptin and insulin is ultimately resolved (Morton et
al., 2006).

Note that I am considering "caloric restriction" beyond merely short-term
restricted consumption of food, but rather in the fascinating "physiologic,"
more long-term sense that has been shown to reduce disease inception and
promote longevity in yeast, worms, flies, and mice. Furthermore, in this scenario, which I mentioned in Chapter 3, section 4, food scarcity millennia ago,
and food abundance (especially "fast foods") at present, promoted adaptations
that were initially favorable but now are maladaptive for an increasing population worldwide: the trend toward obesity. This is in line with the concept of
"antagonistic pleiotropy."

In conclusion, the genetically related conditions that influence longevity in humans are expressed phenotypically in multiple biological ways described in this book: antioxidant activities that reduce organelle and vascular
"stresses" imposed by ROS; diminished expression of inflammatory cytokines
and macrophage infiltration, especially in abdominal (visceral) fat depots;
maintenance of insulin sensitivity; preservation of muscle strength and bone
mass; sustained low LDL cholesterol and high HDL cholesterol; diminished
expression of cell transformation and cancer. These aspects are discussed in
Chapter 4. Therapies that promote these manifestations of "phenotypic health
or fitness" are reviewed in Chapter 5. Although these favorable outcomes can
derive from a vast assortment of genes, the lifestyle practices—the environ-

mental factors—discussed in Chapter 6 have a profound influence on their expression. To engage in these practices is under human choice: they are the mainstay of healthy aging and provide prospects for longevity and disease prevention, or delayed expression, that may compress morbidities and make growing very old tolerable, until perhaps, the "fountain of youth" is really discovered!

Glossary

In the text, the first use of an abbreviation is spelled out, often with a description.

For abbreviations referring to specific signaling pathways and the cellular functions they regulate, Figure 2.3 and its legend can also be consulted.

Some of the less-familiar terms in the glossary are followed by a brief definition.

ACE	angiotensin-converting enzyme
adipokines	secretory products of adipocytes, or fat cells, that act locally or at a distance to promote many metabolic effects, especially inflammation
agonist	any compound that activates a receptor
allele	a variant form of a given gene
AMPK	adenosine monophosphate-activated protein kinase
antagonist	any compound that inhibits a receptor
ApoE	apolipoprotein E
AP-1	activator protein
apoptosis	programmed cell death
ATP	adenosine triphosphate
BMD	bone mineral density
BMP	bone morphogenetic protein
chemokine	a superfamily of small proteins active in immune and inflammatory reactions
CNS	central nervous system
COX	cyclooxygenase 1, 2, enzymes converting arachidonic acid to prostaglandins
CR	caloric restriction
CRP	C-reactive protein
CSF	colony-stimulating factor
cytokines	proteins or glycoproteins that regulate cell responses, including proliferation

cytosol	cytoplasm of the cell
DEXA	dual energy X-ray absorptiometry
DNA	deoxyribose nucleic acid
FFAs	free fatty acids
FIRKO	fat-specific insulin receptor knockout mice
FOXO	forkhead box-containing protein, O subfamily, a transcription factor
GH	growth hormone
HDL	high-density lipoprotein cholesterol
HMG-CoA reductase	3-*hydroxyl*-3-*methylglutaryl coenzyme A reductase, the first enzyme in the cholesterol biosynthetic pathway, and inhibited by statins
HPA	hypothalamic-pituitary-adrenal axis
IFN	interferon
IGF-1	insulin-like growth factor-1
IKK-β	Inhibitor kappa kinase-beta—a coordinator of inflammatory responses through activation of NF-kB (see below)
IL	interleukin—a family of cytokines active in inflammatory and immune responses
IR	insulin receptor
IRS	insulin receptor substrate
JNK	c-Jun amino-terminal kinase
KO	knockout of a gene in animal studies
LDL	low-density lipoprotein cholesterol
LRP	low-density lipoprotein receptor-related protein
MAPK	mitogen-activated protein kinase
MCP	monocyte chemoattractant protein
M-CSF	macrophage colony stimulating factor, also known as colony-stimulating factor-1
MIF	macrophage inflammatory protein
MMP	matrix metalloproteinase
motif	a DNA sequence pattern within a gene that, because of its similarity to sequences in other known genes, suggests a possible function of the gene, its protein product, or both
NADPH	nicotinamide adenine dinucleotide phosphate (reduced)
NF-kB	nuclear factor-kappa B—a transcription factor with a prominent role in immunity, cancer, and inflammation

NO	nitric oxide—a mediator of vascular tone
NOS	nitric oxide synthase
OA	osteoarthritis
oncogenes	cancer-promoting genes
OPG	osteoprotegerin
PAI-1	plasminogen activator inhibitor-1
PGE	e E
PI3K	phosphatidylinositol-3-kinase
PPARγ	peroxisome proliferator-activated receptor-gamma
promoter	upstream region of gene regulating transcription
PTH	parathyroid hormone
PTHrP	parathyroid hormone related peptide
RA	rheumatoid arthritis
RANK	receptor activator nuclear factor kappa
RANK L	receptor activator nuclear factor kappa ligand
redox	balance of oxidation-reduction reactions in a cell
repressor	a protein that suppresses transcription
ROS	reactive oxygen species
SIR	silent information regular gene
SMC	smooth muscle cells in arterial wall
SNS	sympathetic nervous system
SOCS	suppressors of cytokine signaling
SOD	superoxide dismutase—an antioxidant
superfamily	groups of proteins (e.g., receptors) with significant sequence similarity suggestive of an evolutionary relationship
telomerase	ribonucleoprotein enzyme with reverse transcriptase activity able to reform telomere DNA
telomere	DNA at end of genes progressively less transcribed with each replication
TGF-β	transforming growth factor-beta
TLRs	toll-like receptors
TNF-α	tumor necrosis factor-alpha
transgenic	introduction of a gene into the genome
TZDs	thiazolidinediones—"glitazone" class of antidiabetic drugs, pharmacologic PPARγ ligands

References

Abeles AM, Pillinger M. 2006. Statins as antiinflammatory and immunomodulatory agents: a future in rheumatologic therapy? *Arthritis Rheum* 54: 393–407.

Accili D, Arden KC. 2004. FoxOs at the crossroads of cellular metabolism, differentiation, and transformation. *Cell* 117: 421–26.

Aggarwal BB. 2005. Nuclear factor k-β as target for chemoprevention. In: Surh Y-J, Packer L (eds.), *Oxidative stress, inflammation, and health.* New York: Taylor and Francis, pp. 107–26.

Aigner T, Bartnik E, Sohler F, Zimmer R. 2004. Functional genomics of osteoarthritis: on the way to evaluate disease hypothesis. *Clin Orthop Relat Res* 427 (Suppl.): S138–43

Aimaretti G, Baldelli R, Corneli G, Croce C, Rovere S, Baffoni C, et al. 2005. IGFs and IGFBPs in clinical management of adult growth hormone deficiency. *Endocr Dev* 9: 76–88.

Ahima RS, Qi Y, Singhal NS, Jackson MB, Scherer PE. 2006. Brain adipocytokine action and metabolic regulation. Diabetes 55: S145–54.

Akune T, Ohba S, Kamekura S, Yamaguchi M, Chung U, Kubota N, et al. 2004. PPARγ insufficiency enhances osteogenesis through osteoblast formation from bone marrow progenitors. *J Clin Invest* 113: 846–55.

Alexander WS, Hilton DJ. 2004. The role of suppressors of cytokine signaling (SOCS) proteins in regulation of the immune response. *Annu Rev Immunol* 22: 503–29.

Alwin DF, Wray LA. 2005. A life-span developmental perspective on social status and health. *J Gerontol B Psychol Sci Soc Sci* 60 Spec No. 2: S7–14.

American Geriatrics Society, Core Writing Group of the Task Force on the Future of Geriatric Medicine. 2005. Caring for older Americans: the future of geriatric medicine. *J Am Geriatr Soc* 53: 5245–56.

Ames BN. 2005. Increasing longevity by turning up metabolism. *EMBO Rep* 6: S20–24.

Amizuka N, Davidson D, Liu H, Valverde-Franco G, Chai S, Maeda T, Ozawa H, Hammond V, Ornitz DM, Goltzman D, Henderson JE. 2004. Signalling by fibroblast growth factor receptor 3 and parathyroid hormone-related peptide coordinate cartilage and bone development. *Bone* 34: 13–25.

Amory JK, Watts NB, Easley KA, Sutton PR, Anawalt BD, Matsumoto AM, Bremner WJ, Tenover JL. 2004. Exogenous testosterone or testosterone with finasteride increases bone mineral density in older men with low serum testosterone. *J Clin Endocrinol Metab* 89: 503–10.

Anderson RE, Blair SN, Cheskin LJ, Bartlett SJ. 1997. Encouraging patients to become more physically active: the physician's role. *Ann Intern Med* 127: 395–400.

Anderton BH, Betts J, Blackstock WP, Brion JP, Chapman S, Connell J, Dayanandan R, Gallo JM, Gibb G, Hanger DP, Hutton M, Kardalinou E, Leroy K, Lovestone S, Mack T, Reynolds CH, Van Slegtenhorst M. 2001. Sites of phosphorylation in tau and factors affecting their regulation. *Biochem Soc Symp* 67: 73–80.

Andorfer C, Acker CM, Kress Y, Hof PR, Duff K, Davies P. 2005. Cell-cycle reentry and cell death in transgenic mice expressing nonmutant human tau isoforms. *J Neurosci* 25(22): 5446–54.

Andrawes WF, Bussy C, Belmin J. 2005. Prevention of cardiovascular events in elderly people. *Drugs Aging* 22: 859–76.

Andreakos E, Sacre S, Foxwell BM, Feldmann M. 2005. The toll-like receptor-nuclear factor kB pathway in rheumatoid arthritis. *Front Biosci* 10: 2478–88.

Andrews GR. 2001. Care of older people. Promoting health and function in an ageing population. *BMJ* 322: 728–29.

Arabi A. 2005. Leptin effect on bone: Is it ethnic specific? *J Lab Clin Med* 146: 7–8.

Arend WP. 2001. The innate immune system in rheumatoid arthritis. *Arthritis Rheum* 44: 2224–34.

Argmann CA, Cock TA, Auwerx J. 2005. Peroxisome proliferator-activated receptor γ: the more the merrier? *Eur J Clin Invest* 35: 82–92.

Arkan MC, Hevener AL, Greten FR, Maeda S, Li ZW, Long JM, Wynshaw-Boris A, Poli G, Olefsky J, Karin M. 2005. IKK-beta links inflammation to obesity-induced insulin resistance. *Nat Med* 11: 191–98.

Armstrong RA, Myers D, Smith CU. 1993. The spatial patterns of plaques and tangles in Alzheimer's disease do not support the 'cascade hypothesis.' *Dementia* 4(1): 16–20.

Arner P. 2005a. Resistin: yet another adipokine tells us that men are not mice. *Diabetologia* 48: 2203–5.

Arner P. 2005b. Human fat cells lipolysis: biochemistry, regulation and clinical role. *Best Pract Res Clin Endocrinol Metab* 19: 471–82.

Artandi SE. 2006. Telomeres, telomerase, and human disease. *N Engl J Med* 355: 1195–97.

Ashcroft GS, Horan MA, Ferguson MWJ. 1998. Aging alters the inflammatory and endothelial cell adhesion molecule profiles during human cutaneous wound healing. *Lab Invest* 78: 47–58.

Aspden RM, Scheven BAA, Hutchison JD. 2001. Osteoarthritis as a systemic disorder including stromal cell differentiation and lipid metabolism. *Lancet* 357: 1118–20.

Attur MG, Dave M, Akamatsu M, Katoh M, Amin AR. 2002. Osteoarthritis or osteoarthrosis: the definition of inflammation becomes a semantic issue in the genomic era of molecular medicine. *Osteoarthritis Cartilage* 10: 1–4.

Atzmon G, Rincon M, Schechter CB, Shuldiner AR, Lipton RB, Bergman A, Barzilai N. 2006. Lipoprotein genotype and conserved pathway for exceptional longevity in humans. *PLoS Biol* 4: 56–69.

Atzmon G, Schechter C, Greiner W, Davidson D, Rennert G, Barzilai N. 2004. Clinical phenotype of families with longevity. *J Am Geriatr Soc* 52: 274–77.

Avlund K. 2004. Disability in old age. *Dan Med Bull* 51: 315–49.

Avorn JL. 1986. Medicine, health and the geriatric transformation. *Daedalus* 115: 211–25.

Babij P, Zhao W, Small C, Kharode Y, Yaworsky PJ, Bouxsein ML, Reddy PS, Bodine PV, Robinson JA, Bhat B, Marzolf J, Moran RA, Bex F. 2003. High bone mass in mice expressing a mutant LRP5 gene. *J Bone Miner Res* 18: 960–74.

Bachmann MF, Kopf M, Marsland BJ. 2006. Chemokines: more than just road signs. *Nat Rev Immunol* 6: 159–64.

Bailey AJ, Knott L. 1999. Molecular changes in bone collagen in osteoporosis and osteo-arthritis in the elderly. *Exp Gerontol* 34: 337–51.

Bajayo A, Goshen I, Feldman S, Csernus V, Iverfeldt K, Shohami E, Yirmiya R, Bab I. 2005. Central IL-1 receptor signaling regulates bone growth and mass. *Proc Natl Acad Sci USA* 102: 12956–61.

Balaban RS, Nemoto S, Finkel T. 2005. Mitochondria, oxidants, and aging. *Cell* 120: 483–95.

Balducci L, Ershler WB. 2005. Cancer and ageing: a nexus at several levels. *Nat Rev Cancer* 5: 655–62.

Baldwin AS. 2001a. Control of oncogenesis and cancer therapy resistance by the tran-scription factor NFkB. *J Clin Invest* 107: 241–46.

Baldwin AS, Jr. 2001b. The transcription factor NF-kB and human disease. *J Clin Invest* 107: 3–5.

Ballantyne CM. 2004. Achieving greater reductions in cardiovascular risk: Lessons from statin therapy on risk measures and risk reduction. *Am Heart J* 148: S3–8.

Bandeen-Roche K, Xue Q-L, Ferrucci L, Walston J, Guralnik JM, Chaves P, Zeger SL, Fried LP. 2006. Phenotype of frailty: characterization in the women's health and ag-ing studies. *J Gerontol Med Sci* 61A: 262–66.

Barish GD, Narker VA, Evans RM. 2006. PPARδ: a dagger in the heart of the metabolic syndrome. *J Clin Invest* 116: 590–97.

Barker DJ. 2005. The developmental origins of insulin resistance. *Horm Res* 64 (Suppl. 3): 2–7.

Barsh GS, Farroqi S, O'Rahilly S. 2000. Genetics of body-weight reduction. *Nature* 404: 644–51.

Bartke A. 2005 Minireview: role of the growth hormone/insulin-like growth factor sys-tem in mammalian aging. *Endocrinology* 146: 3718–23.

Bartke A. 2006. Long-lived Klotho mice: new insights into the roles of IGF-1 and insulin in aging. *Trends Endocrinol Metab* 17: 33–35.

Barton A, Mulley G. 2003. History of the development of geriatric medicine in the UK. *Postgrad Med J* 79: 229–34.

Barton BE. 1997. IL-6: insights into novel biological activities. *Clin Immunol Immuno-pathol* 85: 16–20.

Barzilai N, Atzmon G, Schechter C, Schaefer EJ, Cupples AL, Lipton R, Cheung S, Shuldiner AR. 2003. Unique lipoprotein phenotype and genotype associated with exceptional longevity. *JAMA* 290: 2030–40.

Barzilai N, Schuldiner AR. 2001. Searching for human longevity genes: the future history of gerontology in the post-genomic era. *J Gerontol Biol Sci Med Sci* 56: M83–87.

Basaria S. 2005. Andropause: need for concrete guidelines until more evidence becomes available. *J Endocrinol Invest* 28: 667–70.

Bauer DC, Mundy GR, Jamal SA, Black DM, Cauley JA, Ensrud KE, et al. 2004. Use of statins and fracture. Results of 4 prospective studies and cumulative meta-analysis of observational studies and controlled trials. *Arch Intern Med* 164: 146–52.

Bauer DC, Hunter DJ, Abramson MD, Attur M, Corr M, Felson D, Heinegård D, Jordan JM, Kepler TB, Lane NE, Saxne T, Tyree B, Kraus VB. 2006. Classification of osteoarthritis biomarkers: a proposed approach. *Osteoarthritis Cartilage* 14: 723–27.

Baumeister R, Schaffitzel E, Hertweck M. 2006. Endocrine signaling in *Caenorhabditis elegans* controls stress response and longevity. *J Endocrinol* 190: 191–202.

Baumgartner RN, Koehler KM, Gallagher D, Romero L, Heymsfield SB, Ross RR, Gary PJ, Lindeman RD. 1998. Epidemiology of sarcopenia among the elderly in New Mexico. *Am J Epidemiol* 147: 755–63.

Bautmans J, Njemini R, Vasseur S, Chabert H, Moens L, Demanet C, Mets T. 2005. Biochemical changes in response to intensive resistance exercise training in the elderly. *Gerontology* 51: 253–65.

Bekker PJ, Holloway D, Nakanishi A, Arrighi M, Leese PT, Dunstan CT. 2001. The effect of a single dose of osteoprotegerin in postmenipausal women. *J Bone Miner Res* 16: 348–60.

Benetos A, Thomas F, Bean KE, Pannier B, Guize L. 2005. Role of modifiable risk factors in life expectancy in the elderly. *J Hypertens* 23: 1803–8.

Ben-Porath I, Weinberg RA. 2005. The signals and pathways activating cellular senescence. *Int J Biochem Cell Biol* 37: 961–76.

Ben-Shlomo Y, Kuh D. 2002. A life course approach to chronic disease epidemiology: conceptual models, empirical challenges and interdisciplinary perspectives. *Int J Epidemiol* 31: 285–93.

Bentahir M, Nyabi O, Verhamme J, Tolia A, Horre K, Wiltfang J, Esselmann H, De Strooper B. 2006. Presenilin clinical mutations can affect gamma-secretase activity by different mechanisms. *J Neurochem* 96(3): 732–42.

Berenson RA. 2006. Does more health care spending produce better health and happier doctors? *Ann Intern Med* 144: 694–96.

Bergink AP, Uitterlinden AG, Van Leeuwen JP, Hofman A, Verhaar JA, Pols HA. 2005. Bone mineral density and vertebral fracture history are associated with incident and progressive radiographic knee osteoarthritis in elderly men and women: The Rotterdam Study. *Bone* 37: 446–56.

Bergman RN, Kim SP, Catalano KJ, Hsu IR, Chiu JD, Kabir M, Hucking K, Ader M. 2006. Why visceral fat is bad: mechanisms of the metabolic syndrome. *Obesity* 14 (Suppl.): 16–19S.

Berliner JA, Navab M, Fogelman AM, Frank JS, Demer LL, Edwards PA, Watson AD, Lusis AJ. 1995. Atherosclerosis: basic mechanisms. Oxidation, inflammation, and genetics. *Circulation* 91: 2488–96.

Bernstein DM, Laney C, Morris EK, Loftus EF. 2005. False beliefs about fattening foods can have healthy consequences. *Proc Natl Acad Sci USA* 102: 13724–31.

Bernstein LE, Berry J, Kim S, Canavan B, Grinspoon SK. 2006. Effects of etanercept in patients with the metabolic syndrome. *Arch Intern Med* 166: 902–8.

Berthoud HR. 2004. Mind versus metabolism in the control of food intake and energy balance. *Physiol Behav* 81: 781–93.

Beutler BA. 1999. The role of tumor necrosis factor in health and disease. *J Rheumatol* 26 (Suppl. 57): 16–21.

Bilezikian JP. 2006. Osteonecrosis of the jaw - Do bisphosphonates pose a risk? N Engl J Med 355: 2278–81.

Billings PR, Carlson RJ, Carlson J, Cain M, Wilson C, Shorett P, Everett W. 2005. Ready for genomic medicine? Perspectives of health care decision makers. *Arch Intern Med* 165: 1917–19.

Black, DM, Schwartz AV, Ensrud KE, Cauley JA, Levis S, Quandt SA, Satterfield S, Wallace RB, Bauer DC, Palermo L, Wehren LE, Lombardi A, Santora AC, Cummings SR. 2006. Effects of continuing or stopping alendronate after 5 years of treatment. The Fracture Intervention Trial Long-term Extension (FLEX): a randomized trial. JAMA 296: 2927–38.

Black PH. 2003. The inflammatory response is an integral part of the stress response: implications for atherosclerosis, insulin resistance, type II diabetes and metabolic syndrome X. *Brain Behav Immun* 17: 350–64.

Blackburn EH. 2005. Telomerase and cancer. *Mol Cancer Res* 3: 477–82.

Blagosklonny MV. 2006. Aging and immortality: Quasi-programmed senescence and its pharmacologic inhibition. *Cell Cycle* 5: 2087–2102.

Blair MC, Robinson LJ, Zaidi M. 2005. Osteoclast signalling pathways. *Biochem Biophys Res Comm* 328: 728–38.

Blair SN, Haskell WL. 2006. Objectively measured physical activity and mortality in older adults. *JAMA* 296: 216–18.

Blair SN, LaMonte MJ. 2005. How much and what type of physical activity is enough? What physicians should tell their patients. *Arch Intern Med* 165: 2324–25.

Blanchard F, Chipoy C. 2005. Histone deacetylase inhibitors: new drugs for the treatment of inflammatory diseases? *Drug Discov Today* 10: 197–204.

Blasco MA. 2005. Telomeres and human disease: ageing, cancer, and beyond. *Nat Rev Genet* 6: 611–22.

Blaum CS, Xue QL, Michelon E, Semba RD, Fried LP. 2005. The association between obesity and the frailty syndrome in older women: the Women's Health and Aging Studies. *J Am Geriatr Soc* 53: 927–34.

Bloomfield HE. 2006. The role of fibrates in a statin world. *Arch Intern Med* 166: 715–16.

Blumenthal RS, Kapur NK. 2006. Can a potent statin actually regress coronary atherosclerosis? *JAMA* 295: 1583–84.

Bodenheimer T. 2006. Primary care—will it survive? *N Engl J Med* 355: 861–64.

Bolton AE. 2005. Biologic effects and basic science of a novel immune-modulation therapy. *Am J Cardiol* 95: 24–29C.

Bonora E. 2006. The metabolic syndrome and cardiovascular disease. *Ann Med* 38: 64–80.

Bouloumié A, Curat CA, Sengenès C, Lolmèda K, Miranville A, Busse R. 2005. Role of macrophage tissue infiltration in metabolic diseases. *Curr Opin Clin Nutr Metab Care* 8: 349–54.

Bouras EP, Lange SM, Scolapio JS. 2001. Rational approach to patients with unintentional weight loss. *Mayo Clin Proc* 76: 923–29.

Bouskila M, Pajvani UB, Scherer PE. 2005. Adiponectin: a relevant player in PPARγ-agonist-mediated improvements in hepatic insulin sensitivity? *Int J Obes* 29: S17–23.

Boyd CM, Xue QL, Simpson CF, Guralnik JM, Fried LP. 2005. Frailty, hospitalization, and progression of disability in a cohort of disabled older women. *Am J Med* 118: 1225–31.

Boyden LM, Mao J, Belsky J, Mitzner L, Farhi A, Mitnik MA, Wu D, Insogna K, Lifton RP. 2002. High bone density due to a mutation in LDL-receptor-related protein 5. *N Engl J Med* 346: 1513–21.

Boyle JJ. 2005. Macrophage activation in atherosclerosis: pathogenesis and pharmacology of plaque rupture. *Curr Vasc Pharmacol* 3: 63–68.

Boyle WJ, Simone WS, Lacey DL. 2003. Osteoclast differentiation and activation. *Nature* 423: 337–42.

Braak H, Braak E. 1995. Staging of Alzheimer's disease-related neurofibrillary changes. *Neurobiol Aging* 16(3): 271–78; discussion 278–84.

Braak H, Braak E. 1997. Staging of Alzheimer-related cortical destruction. *Int Psychogeriatr* 9 (Suppl. 1): 257–61; discussion 269–72.

Braak H, Braak E, Ohm T, Bohl J. 1989. Alzheimer's disease: mismatch between amyloid plaques and neuritic plaques. *Neurosci Lett* 103(1): 24–28.

Braddock M, Quinn A. 2004. Targeting IL-1 in inflammatory disease: new opportunities for therapeutic intervention. *Nat Rev Drug Discov* 3: 1–10.

Brady MJ. 2004. IRS2 takes center stage in the development of type 2 diabetes. *J Clin Invest* 114: 886–88.

Braithwaite RS, Col NF, Wong JB. 2003. Estimating hip fracture morbidity, mortality, and costs. *J Am Geriatr Soc* 51: 364–70.

Bramlage P, Mühlen I, Randeva H, Spanswick D, Lehnert H. 2006. Cardiovascular risk management by blocking the endocannabinoid system. *Exp Clin Endocrinol Diabetes* 114: 77–81.

Brandt KD, Mazzuca SA, Katz BP, Lane KA, Buckwalter KA, Yocum DE, Wolfe F, Schnitzer TJ, Moreland LW, Manzi S, Bradley JD, Sharma L, Oddis CV, Hugenberg ST, Heck LW. 2005. Effects of doxycycline on progression of osteoarthritis: results of a randomized, placebo-controlled, double-blind trial. *Arthritis Rheum* 52: 2015–25.

Briel M, Nordmann AJ, Bucher JC. 2005. Statin therapy and treatment of acute and chronic cardiovascular diseases: update on recent trials and metaanalyses. *Curr Opin Lipidol* 16: 601–5.

Bristow MR. 1998. Editorial. Tumor necrosis factor-α and cardiomyopathy. *Circulation* 97: 1340–41.

Brockelhurst JC. 1997. Geriatric medicine in Britain: the growth of a specialty. *Age Ageing* 26-S4: 5–8.

Brousseau ME, Schaefer EJ, Wolfe ML, Bloedon LT, Digenio AG, Clark RW, Mancuso JP,

Rader DJ. 2004. Effects of an inhibitor of cholesteryl ester transfer protein on HDL cholesterol. *N Engl J Med* 350: 1505–15.

Brown JR, DuBois RN. 2005. COX-2: a molecular target for colorectal cancer prevention. *J Clin Oncol* 23: 2840–55.

Brownlee M. 2005. The pathobiology of diabetic complications: a unifying mechanism. *Diabetes* 54: 1615–25.

Bruder SP, Jaiswal N, Haynesworth SE. 1997. Growth kinetics, self-renewal, and the osteogenic potential of purified human mesenchymal stem cells during extensive subcultivation and following cryopreservation. *J Cell Biochem* 64: 278–94.

Brune K, Hinz B. 2004. The discovery and development of antiinflammatory drugs. *Arthritis Rheum* 50: 2391–99.

Bucay N, Sarosi I, Dunstan CR, Morony S, Tarpley J, Capparelli C, Scully S, Tan HL, Xu W, Lacey DL, Boyle WJ, Simonet WS. 1998. Osteoprotegerin-deficient mice develop early onset osteoporosis and arterial calcification. *Genes Dev* 12: 1260–68.

Bügel S. 2003. Vitamin K and bone health. *Proc Nutr Soc* 62: 839–43.

Burr DB. 2004. The importance of subchondral bone in the progression of osteoarthritis. *J Rheumatol Suppl.* 70: 77–80.

Butler RN. 2004. Commentary. *J Gerontol A: Biol Sci Med Sci* 59: 1158–59.

Butler RN, Sprott R, Warner H, Bland J, Feuers R, Forster M, Fillit H, Harman SM, Hewitt M, Hyman M, Johnson K, Kligman E, McClearn G, Nelson J, Richardson A, Sonntag W, Weindruch R, Wolf N. 2004. Biomarkers of aging: From primitive organisms to humans. *J Gerontol A Biol Sci Med Sci* 59: B560–67.

Caddel JL. 2000. Geriatric cachexia: a role for magnesium deficiency as well as for cytokines? *Am J Clin Nutr* 71: 851–53.

Campion GV. 1994. The prospect for cytokine based therapeutic strategies in rheumatoid arthritis. *Ann Rheum Dis* 53: 485–87.

Campisi J. 1997. Aging and cancer: the double-edged sword of replicative senescence. *J Am Geriatr Soc* 45: 482–88.

Campisi J. 2003. Cancer and ageing: rival demons? *Nat Rev Cancer* 3: 339–49.

Campisi J. 2004. Proliferative senescence and cancer. In: Balducci L, Lyman GH, Ershler WB, Extermann M (eds.), *Comprehensive geriatric oncology,* 2nd ed. UK: Taylor & Francis Group, pp. 127–37.

Campisi J. 2005. Senescent cells, tumor suppression, and organismal aging: good citizens, bad neighbors. *Cell* 120: 513–22.

Campos P, Saguy A, Ernsberger P, Oliver E, Gaesser G. 2006. The epidemiology of overweight and obesity: public health crisis or moral panic? *Int J Epidemiol* 35: 55–60.

Caplan AL. 2005. Death as an unnatural process. Why is it wrong to seek a cure for aging? *EMBO Rep* 6: S72–75.

Cappola AR, Xue QL, Ferrucci L, Guralnik JM, Volpato S, Fried LP. 2003. Insulin-like growth factor I and interleukin-6 contribute synergistically to disability and mortality in older women. *J Clin Endocrinol Metab* 88: 2019–25.

Cave AC, Brewer AC, Narayanapanicker A, Ray R, Grieve DJ, Walker S, Shah AM. 2006. NADPH oxidases in cardiovascular health and disease. *Antioxid Redox Signal* 8(5–6): 691–728.

Cenci S, Weitzmann MN, Roggia C, Namba N, Novack D, Woodring J, Pacifici R. 2000. Estrogen deficiency induces bone loss by enhancing T-cell production of TNF-α. *J Clin Invest* 106: 1229–37.

Cesari M, Kritchevsky SB, Baumgartner RN, Atkinson HH, Penninx BW, Lenchik L, Palla SL, Ambrosium WT, Tracy RP, Pahor M. 2005. Sarcopenia, obesity, and inflammation: results from the Trial of angiogensin converting enzyme inhibition and novel cardiovascular risk factors study. *Am J Clin Nutr* 82: 428–34.

Chakravarthy MV, Joyner MJ, Booth FW. 2002. An obligation for primary care physicians to prescribe physical activity to sedentary patients to reduce the risk of chronic health conditions. *Mayo Clin Proc* 77: 165–75.

Chan DC. 2006. Mitochondria: Dynamic organelles in disease, aging, and development. *Cell* 125: 1241–52.

Chandran M, Phillips SA, Ciaraldi T, Henry RR. 2003. Adiponectin: More than just another fat cell hormone? *Diabetes Care* 26: 2442–50.

Charmandari E, Tsigos C, Chrousos G. 2005. Endocrinology of the stress response. *Annu Rev Physiol* 67: 259–84.

Charo IF, Taubman MB. 2004. Chemokines in the pathogenesis of vascular disease. *Circ Res* 95: 858–66.

Cher ML, Towler DA, Rafii S, Rowley D, Donahue HJ, Keller E, Herlyn M, Cho EA, Chung LWK. 2006. Cancer interaction with the bone microenvironment. A workshop of the National Institutes of Tumor Microenvironment Study Section. *Am J Pathol* 168: 1405–12.

Chien KR, Karsenty G. 2005. Longevity and lineages: Toward the integrative biology of degenerative diseases in the heart, muscle, and bone. *Cell* 120: 533–44.

Childs B. 1999. *Genetic medicine: a logic of disease.* Baltimore: Johns Hopkins University Press.

Chilton R, Chiquette E. 2005. Thiazolidinediones and cardiovascular disease. *Curr Atheroscler Rep* 7: 115–20.

Chinetti G, Fruchart JC, Staels B. 2003. Peroxisome proliferator-activated receptors and inflammation: from basic science to clinical applications. *Int J Obes Relat Metab Disord* 27 (Suppl. 3): S41–45.

Chitu V, Stanley ER. 2006. Colony-stimulating factor-1 in immunity and inflammation. *Curr Opin Immunol* 18: 39–48.

Choi J, Enis DR, Koh KP, Shiao SL, Pober JS. 2004. T lymphocyte-endothelial cell interactions. *Annu Rev Immunol* 22: 683–709.

Chung HY, Jung KJ, Yu BP. 2005. Molecular inflammation as an underlying mechanism of aging: the anti-inflammatory action of caloric restriction. In: Surh Y-J, Packer L (eds.), *Oxidative stress, inflammation, and health.* New York: Taylor & Francis, pp. 389–421.

Citron M. 2001. Human beta-secretase and Alzheimer's disease. *Expert Opin Ther Targets* 5(3): 341–48.

Clark LP, Dion DM, Barker WH. 1990. Taking to bed: Rapid functional decline in an independently mobile older population living in an intermediate-care facility. *J Am Geriatr Soc* 38: 967–72.

Clemmons DR. 2004. The relative roles of growth hormone and IGF-1 in controlling insulin sensitivity. *J Clin Invest* 113: 25–27.

Clevers H. 2004. At the crossroads of inflammation and cancer. *Cell* 118: 671–74.

Cock TA, Auwerx J. 2003. Leptin: cutting the fat off the bone. *Lancet* 362: 1572–74.

Cockayne S, Adamson J, Lanham-New S, Shearer MJ, Gilbody S, Torgerson DJ. 2006. Vitamin K and the prevention of fractures. *Arch Intern Med* 166: 1256–61.

Cohen HJ. 2000. Editorial: In search of the underlying mechanisms of frailty. *J Gerontol Med Sci* 55A: M706–8.

Cohen HJ, Peiper CF, Harris T, Rao KMK, Currie MS. 1997. The association of plasma IL-6 levels with functional disability in community-dwelling elderly. *J Gerontol A Biol Sci Med Sci* 52: M201–8.

Cohen JC, Boerwinkle E, Mosley TH Jr, Hobbs HH. 2006. Sequence variations in PCSK9, low LDL, and protection against coronary heart disease. *N Engl J Med* 354: 1264–72.

Cohen MM Jr. 2006. The new bone biology: Pathologic, molecular, and clinical correlates. Am J Med Genet Part A 140A: 2646–2706.

Cohen P. 2006. Overview of the IGF-1 system. *Horm Res* 65 (Suppl. 1): 3–8.

Coles LS. 2004. Demography of human super centenarians. *J Gerontol A Biol Sci Med Sci* 59: B579–84.

Conaghan PG, Vanharanta H, Dieppe PA. 2005. Is progressive osteoarthritis an atheromatous vascular disease? *Ann Rheum Dis* 64: 1539–41.

Condeelis J, Pollard JW. 2006. Macrophages: obligate partners for tumor cell migration, invasion, and metastases. *Cell* 124: 263–66.

Coons JC. 2002. Hydroxymethylglutaryl-coenzyme A reductase inhibitors in osteoporosis management. *Ann Pharmacother* 36: 326–30.

Coresh J, Astor B. 2006. Decreased kidney function in the elderly: clinical and preclinical, neither benign. *Ann Intern Med* 145: 299–301.

Corton JC, Brown-Borg HM. 2005. Peroxisome proliferator-activated receptor _ coactivator 1 in caloric restriction and other models of longevity. *J Gerontol A Biol Sci Med Sci* 60: 1494–1509.

Costenbader KH, Coblyn JS. 2005. Statin therapy in rheumatoid arthritis. *South Med J* 98: 534–40.

Cramer JA, Silverman S. 2006. Persistence with bisphosphonate treatment for osteoporosis: finding the root of the problem. *Am J Med* 119 (4 Suppl. 1): S12–17.

Croft P, Jordan K, Jinks C. 2005. "Pain elsewhere" and the impact of knee pain in older people. *Arthritis Rheum* 52: 2350–54.

Crow MT. 2004. Surviving cardiac stress: cardioprotection mediated by a longevity gene. *Circ Res* 95: 953–56.

Crum NF, Lederman ER, Wallace MR. 2005. Infections associated with tumor necrosis factor-alpha antagonists. *Medicine* 84: 291–302.

Cuchel M, Rader DJ. 2006. Macrophage reverse cholesterol transport: key to the regression of atherosclerosis? *Circulation* 113: 2548–55.

Cummings DE, Merriam GR. 2003. Growth hormone therapy in adults. *Annu Rev Med* 54: 513–33.

Cundy T, Davidson J, Rutland MD, Steward C, DePaoli AM. 2005. Recombinant osteo-protegerin for juvenile Paget's disease. *N Engl J Med* 353: 918–23.

Cushenberry LM, Rodriguez de Bittner M. 2002. Potential use of HMG-CoA reductase inhibitors for osteoporosis. *Ann Pharmacother* 36: 671–78.

Cutler DM. 2001. The reduction in disability among the elderly. *Proc Natl Acad Sci USA* 98: 6546–47.

Dale KM, Coleman CI, Henyan NN, Kluger J, White CM. 2006. Statins and cancer risk: a meta-analysis. *JAMA* 295: 74–80.

Danenberg HD. Golomb G, Groothuis A, Gao J, Epstein H, Swaminathan RV, Seifert P, Edelman ER. 2003. Liposomal alendronate inhibits systemic innate immunity and reduces in-stent neointimal hyperplasia in rabbits. *Circulation* 108: 2798–2804.

Danielpour D, Song K. 2006. Cross-talk between IGF-1 and TGF-β signaling pathways. *Cytokine Growth Factor Rev* 17: 59–74.

David A, Metherell La, Clark AJ, Camacho-Hubner C, Savage MO. 2005. Diagnostic and therapeutic advances in growth hormone insensitivity. *Endocrinol Metab Clin North Am* 34: 581–95.

Davies P. 1994. Neuronal abnormalities, not amyloid, are the cause of dementia in Alzheimer's disease. In: Katzman R, Terry RD, Bick, C (eds.), *Alzheimer's disease.* New York: Raven Press, pp. 327–33.

Daviglus ML, Liu K. 2004. We must focus on achieving favorable levels of all risk factors simultaneously. *Arch Intern Med* 164: 2086–87.

Davis RM. 2006. Measuring the health impact of smoking and health care providers' performance in addressing the problem. *Ann Intern Med* 144: 444–46.

Day TF, Guo X, Garrett-Beal L, Yang Y. 2005. Wnt/beta-catenin signaling in mesenchymal progenitors controls osteoblast and chrondrocyte differentiation during vertebrate skeletogenesis. *Dev Cell* 8: 739–50.

DeBuyzere ML, Rietzschel E-R. 2006. C-reactive protein's place on the cardiovascular stage: prima ballerina or chorus girl? *J Hypertens* 24: 627–32.

de Crombrugghe B. 2005. Osteoblasts clock in for their day job. *Cell* 122: 651–53.

DeFronzo RA. 2006. Is insulin resistance atherogenic? Possible mechanisms. *Atheroscler Suppl.* 7: 11–15.

de Lange T. 2006. Lasker laurels for telomerase. *Cell* 126: 1017–20.

Dempster DW, Shane E, Horbert W, Lindsay R. 1986. A simple method for correlative light and scanning electron microscopy of human iliac crest bone biopsies: qualitative observations in normal and osteoporotic subjects. *J Bone Miner Res* 1: 15–21.

Denke MA. 2005. Diet, lifestyle, and nonstatin trials: review of time to benefit. *Am J Cardiol* 96: 3–10F.

Dentino AN, Pieper CF, Rao KMK, Currie MS, Harris T, Blaze DG, Cohen HJ. 1999. Association of interleukin-6 and other biologic variables with depression in older people living in the community. *J Am Geriatr Soc* 47: 6–11.

DePinho RA. 2000. The age of cancer. *Nature* 408: 248–54.

Dequeker J. 1985. The relationship between osteoporosis and osteoarthritis. *Clin Rheum Dis* 11: 271–96.

Deroo BJ, Korach KS. 2006. Estrogen receptors and human disease. *J Clin Invest* 116: 561–70.

Deshpande A, Hinds PW. 2006. The retinoblastoma protein in osteoblast differentiation and osteosarcoma. *Curr Mol Med* 6: 809–17.

Després JP. 2005. Our passive lifestyle, our toxic diet, and the atherogenic/diabetogenic metabolic syndrome: can we afford to be sedentary and unfit? *Circulation* 112: 453–55.

Després JP. 2006. Is visceral obesity the cause of the metabolic syndrome? *Ann Med* 38: 52–63.

Després JP, Golay A, Sjöström L. 2005. Effects of rimonabant on metabolic risk factors in overweight patients with dyslipidemia. *N Engl J Med* 353: 2121–34.

Després JP, Lemieux I, Almeras N. 2006. Contribution of CB1 blockage to the management of high-risk abdominal obesity. *Int J Obes* 30: 544–52.

Dewachter I, Van Leuven F. 2002. Secretases as targets for the treatment of Alzheimer's disease: the prospects. *Lancet Neurol* 1(7): 409–16.

Dickinson HO, Mason JM, Nicolson DJ, Campbell F, Boyer FR, Cook JV, Williams B, Ford GA. 2006. Lifestyle interventions to reduce raised blood pressure: a systematic review of randomized controlled trials. *J Hypertens* 24: 215–33.

Dickson DW, Crystal HA, Bevona C, Honer W, Vincent I, Davies P. 1995. Correlations of synaptic and pathological markers with cognition of the elderly. *Neurobiol Aging* 16(3): 285–98; discussion 298–304.

Dieppe P. 1984. Osteoarthritis: Are we asking the wrong questions? *Br J Rheumatol* 23: 161–65.

Dieppe P. 2005. Disease modification in osteoarthritis: are drugs the answer? *Arthritis Rheum* 52: 1956–59.

Dieppe P, Cushnagahan J, Young P, Kirwan J. 1993. Prediction of the progression of joint space narrowing in osteoarthritis of the knee by bone scintigraphy. *Ann Rheum Dis* 52: 557–63.

Dieppe PA. 2004. Relationship between symptoms and structural changes in osteoarthritis: what are the important targets for therapy? *J Rheumatol Suppl.* 70: 50–53.

Diez JJ, Iglesias P. 2003. The role of the novel adipocyte-derived adiponectin in human disease. *Eur J Endocrinol* 148: 293–300.

Dimri GP. 2004. The search for biomarkers of aging: Next stop INK4a/ARF locus. *Sci Aging Know Environ* 44: 40.

Dinarello CA. 2005. Blocking IL-1 in systemic inflammation. *J Exp Med* 201: 1355–59.

Dinarello CA. 2006. Interleukin-1 and interleukin-18 as mediators of inflammation and the aging process. *Am J Clin Nutr* 83: 447–55S.

Dirks AJ, Leeuwenburgh C. 2005. Caloric restriction in humans: potential pitfalls and health concerns. *Mech Ageing Dev* 127: 1–7.

Dirks AJ, Leeuwenburgh C. 2006. Tumor necrosis factor α signaling in skeletal muscle: effects of age and caloric restriction. *J Nutr Biochem* 17: 501–8.

Douchi T, Yamamoto S, Oki T, Maruta K, Kuwahata R, Yamasaki H, Nagata Y. 2000. Difference in the effect of adiposity on bone density between pre- and postmenopausal women. *Maturitas* 34: 261–66.

Drissi H, Zuscik M, Rosier R, O'Keefe R. 2005. Transcriptional regulation of chondrocyte maturation: potential involvement of transcription factors in OA pathogenesis. *Mol Aspects Med* 26: 169–79.

Du W, Pogoriler J. 2006. Retinoblastoma family genes. *Oncogene* 25: 5190–5200.

DuBois RN, Abramson SR, Crofford L, Gupta RA, Simon LS, van de Putte LBA, Lipsky PE. 1998. Cyclooxygenase in biology and disease. *FASEB J* 12: 1063–73.

Ducy P, Amling M, Takeda S, Priemel M, Schilling AF, Beil FT, Shen J, Vinson C, Rueger JM, Karsenty G. 2000. Leptin inhibits bone formation through a hypothalamic relay: a central control of bone mass. *Cell* 100: 197–207.

Duff GW. 2006. Evidence for genetic variation as a factor in maintaining health. *Am J Clin Nutr* 83: 431–35S.

Duff GW, Livvy P, Ordovas JM, Reilly PR. 2006. The future of living well to 100. *Am J Clin Nutr* 83: 488–90S.

Duff K. 2001. Transgenic mouse models of Alzheimer's disease: phenotype and mechanisms of pathogenesis. *Biochem Soc Symp* 67: 195–202.

Dunford JE, Rogers MJ, Ebetino FH, Phipps RJ, Coxon FP. 2006. Inhibition of protein prenylation by bisphosphonates causes sustained activation of Rac, Cdc42, and RhoGTPases. *J Bone Miner Res* 21: 684–94.

Durrington P. 2004. Statin policy and target populations: evidence-based or evidence-biased. *Curr Opin Lipidol* 15: 381–86.

Duthie EH Jr. 2004. Physiology of aging: relevance to symptoms, perceptions, and treatment tolerance. In: Balducci L, Lyman GH, Ershler WB, Extermann M (eds.), *Comprehensive geriatric oncology,* 2nd ed. UK: Taylor & Francis Group, pp. 207–22.

Dwyer J. 2006. Starting down the right path: nutrition connections with chronic diseases of later life. *Am J Clin Nutr* 83: 415–20S.

Dzau VJ. 2004. Markers of malign across the cardiovascular continuum. Interpretation and application. *Circulation* 109 (Suppl. IV): IV-1-2.

Edwards CJ, Russell RGG, Spector DT. 2001. Statins and bone: myth or reality. *Calcif Tissue Int* 69: 63–66.

Einhorn D, Aroda VR, Henry RR. 2004. Glitazones and the management of insulin resistance: what they do and how might they be used. *Endocrinol Metab Clin North Am* 33: 595–616.

Elefteriou F, Ahn JF, Takeda S, Starbuck M, Yang X, Liu X, Kondo H, Richards WG, Bannon TW, Noda M, Clement K, Vaisse C, Karsenty G. 2005. Leptin regulation of bone resorption by the sympathetic nervous system and CART. *Nature* 434: 514–20.

Emkey RD, Ettinger M. 2006. Improving compliance and persistence with bisphosphonate therapy for osteoporosis. *Am J Med* 119 (4 Suppl. 1): S18–24.

Engelman JA, Luo J, Cantley LC. 2006. The evolution of phosphatidylinositol 3-kinases as regulators of growth and metabolism. *Nat Rev Genet* 7: 606–19.

Epel ES, Blackburn EH, Lin J, Dhabhar FS, Adler NE, Morrow JD, Cawthon RM. 2004. Accelerated telomere shortening in response to life stress. *Proc Natl Acad Sci USA* 101: 17312–15.

Erdemli B, Serin-Kilicoglu S, Erdemli E. 2005. A new approach to the treatment of osteoporosis. *Orthopedics* 28: 59–62.

Ershler WB. 1993. Interleukin-6: a cytokine for gerontologists. *J Am Geriatr Soc* 41: 176–81.

Ershler WB. 2004. Biology of aging and cancer. In: Balducci L, Lyman GH, Ershler WB, Extermann M (eds.), *Comprehensive geriatric oncology,* 2nd ed. UK: Taylor & Francis Group, pp. 67–74.

Evans JF, Kargman SL. 2004. Cancer and cyclooxygenase-2 (COX-2) inhibition. *Curr Pharm Des* 10: 627–34.

Evans WJ. 1995. What is sarcopenia? *J Gerontol A Biol Sci Med Sci* 50 Spec: 5–8.

Evans WJ. 2004. Protein nutrition, exercise and aging. *J Am Coll Nutr* 23 (6 Suppl.): 601–9S.

Ezzati M, Henley SJ, Thun MJ, Lopez AD. 2005. Role of smoking in global and regional cardiovascular mortality. *Circulation* 112: 489–97.

Fabrizio P, Gattazzo C. Battistella L, Wei M, Cheng C, McGrew K, Longo VD. 2005. Sir2 blocks extreme life-span extension. *Cell* 123: 655–67.

Fagerberg B, Bondjers L, Nilsson P. 2004. Low birth weight in combination with catch-up growth predicts the occurrence of the metabolic syndrome in men at late middle age: the Atherosclerosis and Insulin Resistance Study. *J Intern Med* 256: 254–59.

Fassbender K, Simons M, Bergmann C, Stroick M, Lütjohann D, Keller P, et al. 2001. Simvastatin strongly reduces levels of Alzheimer's disease β-amyloid peptides Aβ42 and Aβ40 in vitro and in vivo. *Proc Natl Acad Sci USA* 98: 5856–61.

Fata JE, Kong YY, Li J, Sasaki T, Irie-Sasaki J, Moorehead RA, et al. 2000. The osteoclast differentiation factor osteoprotegerin-ligand is essential for mammary gland development. *Cell* 103: 41–50.

Felson DT. 2004. Obesity and vocational and avocational overload of the joint as risk factors for osteoarthritis. *J Rheumatol Suppl.* 70: 2–5.

Felson DT. 2006. Osteoarthritis of the knee. *N Engl J Med* 354: 841–48.

Felson DT, Chaisson CE, Hill CL, Totterman SMS, Gale E, Skinner KM, Kazis L, Gale DR. 2001. The association of bone marrow lesions with pain in knee osteoarthritis. *Ann Intern Med* 134: 541–49.

Felson DT, Neogi T. 2004. Osteoarthritis: Is it a disease of cartilage or of bone? *Arthritis Rheum* 50: 341–44.

Ferrari SL, Deutsch S, Antonarakis SE. 2005. Pathogenic mutations and polymorphisms in the lipoprotein receptor-related protein 5 reveal a new biological pathway for the control of bone mass. *Curr Opin Lipidol* 16: 207–13.

Ferrari SL, Rizzoli R. 2005. Gene variants for osteoporosis and their pleiotropic effects in aging. *Mol Aspects Med* 26: 145–67.

Ferrario CM, Strawn WB. 2006. Role of the renin-angiotensin-aldosterone system and proinflammatory mediators in cardiovascular disease. *Am J Cardiol* 98: 121–28.

Ferreira A, Lu Q, Orecchio L, Kosik KS. 1997. Selective phosphorylation of adult tau isoforms in mature hippocampal neurons exposed to fibrillar A beta. *Mol Cell Neurosci* 9(3): 220–34.

Ferri N, Paoletti R, Corsini A. 2006. Biomarkers for atherosclerosis: pathophysiological role and pharmacological modulation. *Curr Opin Lipidol* 17: 495–501.

Ferrin M. 1999. Stress and the reproductive cycle. J Clin Endocrinol Metab 84: 1768–74.

Ferrucci L, Ble A, Bandinelli S, Windham BG, Simonsick EM. 2005. Inflammation: the fire of frailty? In: Carey JR, Robine J-M, Michel J-P, Christen Y (eds.), *Longevity and frailty.* Berlin-Heidelberg: Springer-Verlag, pp. 91–98.

Ferrucci L, Corsi A, Lauretani F, Bandinelli S, Bartali B, Taub DD, Guralnik JM, Longo DL. 2005. The origins of age-related proinflammatory state. *Blood* 105: 2294–99.

Ferrucci L, Guranik JM, Studenski S Fried LP, Cutler GB Jr, Walston JD, Interventions on Frailty Working Group. 2004. Designing randomized, controlled trials aimed at preventing or delaying functional decline and disability in frail, older persons: a consensus report. *J Am Geriatr Soc* 52: 625–34.

Fiatarone MA, O'Neill EF, Ryan ND, Clements KM, Solares GR, Nelson ME, Roberts SB, Kehayias JJ, Lipsitz LA, Evans WJ. 1994. Exercise training and nutritional supplementation for physical frailty in very elderly people. *N Engl J Med* 330: 1769–75.

Finch CE, Crimmins EM. 2004. Inflammatory exposure and historical changes in human life-spans. *Science* 305: 1736–39.

Finch CE, Kirkwood TBL. 2000. Limits of determinism in aging. In: Finch CE, Kirkwood TBL (eds.), *Chance, development and aging.* New York: Oxford University Press, pp. 186–211.

Finkel T, Holbrook NJ. 2000. Oxidants, oxidative stress and the biology of ageing. *Nature* 408: 239–47.

Finucane TE, Loo TS. 1999. Textbook geriatrics: not yet of the essence. *Ann Intern Med* 130: 782–86.

Firestein GS. 2004. NF-kB: Holy grail for rheumatoid arthritis? *Arthritis Rheum* 50: 2381–86.

Fisher AL. 2004. Of worms and women: sarcopenia and its role in disability and mortality. *J Am Geriatr Soc* 52: 1185–90.

Fitzgerald GA. 2004. Coxibs and cardiovascular disease. *N Engl J Med* 351: 2875–78.

Fleisch H. 1998. Bisphosphonates: mechanisms of action. *Endocr Rev* 1998; 19: 80–100.

Flier JS. 1998. Clinical review 94: What's in a name? In search of leptin's physiological role. *J Clin Endocrinol Metab* 83: 1407–13.

Flier JS. 2004. Obesity wars: molecular progress confronts an expanding epidemic. *Cell* 116: 337–50.

Folkman J, Kalluri R. 2004. Concept cancer without disease. *Nature* 427: 787.

Fontana L. 2006. Excessive adiposity, calorie restriction, and aging. *JAMA* 295: 1577–78.

Foo RS-Y, Mani K, Kitsis RN. 2005. Death begets failure in the heart. *J Clin Invest* 115: 565–71.

Foreyt JP. 2005. Need for lifestyle intervention: how to begin. *Am J Cardiol* 96: 11–14.

Förstermann U, Münzel T. 2006. Endothelial nitric oxide synthase in vascular disease: from marvel to menace. *Circulation* 113: 1708–14.

Fossel MB. 2004. *Cells, aging and human disease.* New York: Oxford University Press.

Foster D. 2004. The demise of disease? I don't think so. *Am J Med* 116: 186–87.

Franceschi C, Bonafe M, Valensin S, Olivieri F, De Luca M, Ottaviani E, De Benedictis G. 2000. Inflamm-aging. An evolutionary perspective on immunosenescence. *Ann N Y Acad Sci* 908: 244–54.

Francis GA, Fayard E, Picard F, Auwerx J. 2003. Nuclear receptors and the control of metabolism. *Annu Rev Physiol* 65: 261–11.

Frayn KN, Karpe F, Fielding BA, Macdonald IA, Coppack SW. 2003. Integrative physiology of human adipose tissue. *Int J Obes Relat Metab Dis* 27: 875–88.

Freedman KB, Kaplan KS, Bilker WB, Strom BL, Lowe RA. 2000. Treatment of osteoporosis: are physicians missing an opportunity? *J Bone Joint Surg* 82A: 1063–70.

Fried LP, Ettinger WH, Lind B, Newman AB, Gardeu J. 1994. Physical disability in older adults: a physiological approach. *J Clin Epidemiol* 47: 747–60.

Fried LP, Ferrucci L, Darer J, Williamson JD, Anderson G. 2004. Untangling the concepts of disability, frailty, and comorbidity: Implications for improved targeting and care. *J Gerontol Med Sci* 59: 255–63.

Fried LP, Hadley EC, Walston JD, Newman A, Guralnik JM, Studenski S, Harris TB, Ershler WB, Ferrucci L. 2005. From bedside to bench: research agenda for frailty. *Sci Aging Knowledge Environ* pe24.

Fried LP, Tangen CM, Walston J, Newman AB, Hirsch C, Gottdiener J, Seeman T, Tracy R, Kop WJ, Burke G, McBurnie MA, Cardiovascular Health Study Collaborative Research Group. 2001. Frailty in older adults: Evidence for a phenotype. *J Gerontol Med Sci* 56A: M146–56.

Fries JF. 1980. Aging, natural death, and the compression of morbidity. *N Engl J Med* 303: 130–35.

Fries JF. 1988. Aging, illness, and health policy: Implications of the compression of morbidity. *Perspect Biol Med* 31: 407–28.

Fries JF. 1997. Editorial: can preventive gerontology be on the way? *Am J Public Health* 87: 1591–93.

Frisard M, Ravussin E. 2006. Energy metabolism and oxidative stress. *Endocrine* 29: 27–32.

Fritz G. 2005. HMG-CoA reductase inhibitors (statins) as anticancer drugs. *Int J Oncol* 27: 1401–9.

Frystyk J. 2005. Aging somatotropic axis: mechanisms and implications of insulin-like growth factor-related binding protein adaptation. *Endocrinol Metab Clin North Am* 34: 865–76.

Fu L, Patel MS, Bradley A, Wagner EF, Karsenty G. 2005. The molecular clock mediates leptin-regulated bone formation. *Cell* 122: 803–15.

Fukuhara A, Matsuda M, Nishizawa M, Segawa K, Tanaka M, Kishimoto K, Matsuki Y, Murakami M, Ichisaka T, Murakami H, Watanabe E, Takagi T, Akiyoshi M, Ohtsubo T, Kihara S, Yamashita S, Makishima M, Funahashi T, Yamanaka S, Hiramatsu R, Mtsuzawa Y, Shimomura I. 2005. Visfatin: A protein secreted by visceral fat that mimics the effects of insulin. *Science* 307: 426–30.

Fukui N, Zhu Y, Maloney WJ, Clohisy J, Sandell LJ. 2003. Stimulation of BMP-2 expression by pro-inflammatory cytokines IL-1 and TNF-α in normal and osteoarthritic chondrocytes. *J Bone Joint Surg* 85-A: 59–66.

Gable DR, Hurel SJ, Humphries SE. 2006. Adiponectin and its gene variants as risk factors for insulin resistance, the metabolic syndrome and cardiovascular disease. *Atherosclerosis* 188: 231–44.

Gabriely I, Barzilai N. 2003. Surgical removal of visceral adipose tissue: effects on insulin action. *Curr Diab Rep* 3: 201–6.

Gale EAM. 2005. The myth of the metabolic syndrome. *Diabetologia* 48: 1679–83.

Gallou-Kabani C, Junien C. 2005. Nutritional epigenomics of metabolic syndrome: new perspective against the epidemic. *Diabetes* 54: 1899–1906.

Games D, Buttini M, Kobayshi D, Schenk D, Seubert P. 2006. Mice as models: transgenic approaches and Alzheimer's disease. *J Alzheimer's Dis* 9: 133–49.

Gans ROB. 2006. The metabolic syndrome, depression, and cardiovascular disease: Interrelated conditions that share pathophysiologic mechanisms. *Med Clin North Am* 90: 573–91.

Garrett IK, Gutierrez G, Mundy GR. 2001. Statins and bone formation. *Curr Pharm Des* 7: 715–36.

Gasparini L, Ongini E, Wenk G. 2004. Non-steroidal anti-inflammatory drugs (NSAIDs) in Alzheimer's disease: old and new mechanisms of action. *J Neurochem* 91: 521–36.

Gass M, Dawson-Hughes B. 2006. Preventing osteoporosis-related fractures: an overview. *Am J Med* 119 (4 Suppl. 1): S3–11.

Gehlbach SH, Bigelow C, Heimisdottir M, May S, Walker M, Kirkwood JR. 2000. Recognition of vertebral fracture in a clinical setting. *Osteoporos Int* 11: 577–82.

Gennari L, Nuti R, Bilezikian JP. 2004. Aromatase activity and bone homeostasis in men. *J Clin Endocrinol Metab* 89: 5898–5907.

Genovese C, Trani D, Caputi M, Claudio PP. 2006. Cell cycle control and beyond: emerging roles for the retinoblastoma gene family. *Oncogene* 25: 5210–9.

Gerstenfeld LC, Einhorn TA. 2004. COX inhibitors and their effects on bone healing. *Expert Opin Drug Saf* 2: 131–36.

Gestwicki JE, Crabtree GR, Graef IA. 2004. Harnessing chaperones to generate small-molecule inhibitors of amyloid beta aggregation. *Science* 306(5697): 865–69.

Giacinti C, Giordano A. 2006. RB and cell cycle progression. *Oncogene* 25: 5220–27.

Gilley D, Tanaka H, Herbert BS. 2005. Telomere dysfunction in aging and cancer. *Int J Biochem Cell Biol* 37: 1000–1013.

Gillick M. 2001. Guest editorial: pinning down frailty. *J Gerontol Med Sci* 56A: M134–35.

Gimble JM, Zvonic S, Floyd S, Floyd E, Kassem M, Nuttall ME. 2006. Playing with bone and fat. *J Cell Biochem* 98: 251–66.

Giono LE, Manfredi JJ. 2006. The p53 tumor suppressor participates in multiple cell cycle checkpoints. *J Cell Physiol* 209: 13–20.

Giordano R, Lanfranco F, Bo M, Pellegrino M, Picu A, Baldi M, Balbo M, Bonelli L, Grottoli S, Ghigo E, Arvat E. 2005. Somatopause reflects age-related changes in the neural control of GH/IGF-1 axis. *J Endocrinol Invest* 28: 94–98.

Giresi PG, Stevenson EJ, Theilhaber J, Koncarevic A, Parkington J, Fielding RA, Kandarian SC. 2005. Identification of a molecular signature of sarcopenia. *Physiol Genomics* 21: 253–63.

Gizard R, Amant C, Barbier O, Belloosta S, Robillard R, Percevault F, Sevestre H, Krimpenfort P, Corsini A, Rochette J, Glineur C, Fruchart JC, Torpier G, Staels B. 2005. PPAR alpha inhibits vascular smooth muscle cell proliferation underlying intimal hyperplasia by inducing the tumor suppressor p16INK4a. *J Clin Invest* 115: 3228–38.

Glaser R, Keicolt-Glaser JK. 2005. Stress-induced immune dysfunction: implications for health. *Nat Immunol* 5: 243–51.

Glass CK. 2006. Going nuclear in metabolic and cardiovascular disease. *J Clin Invest* 116: 556–60.

Glass DA 2nd, Bialek P, Ahn JD, Starbuck M, Patel MS, Clevers HS, Taketo MM, Long F, McMahon AP, Lang RA, Karsenty G. 2005. Canonical Wnt signaling in differentiated osteoblasts controls osteoclast differentiation. *Dev Cell* 8: 751–64.

Gluckman PD, Hanson MA. 2004. Living with the past: evolution, development, and patterns of disease. *Science* 305: 1733–36.

Golay A, Ybarra J. 2005. Link between obesity and type 2 diabetes. *Best Pract Res Clin Endocrinol Metab* 19: 649–63.

Gold DT, McClung B. 2006. Approaches to patient education: emphasizing the long-term value of compliance and persistence. *Am J Med* 119 (4 Suppl. 1): S32–37.

Golden LH, Insogna KL. 2004. The expanding role of p13-kinase in bone. *Bone* 34: 3–12.

Goldring MB. 2006. Are bone morphogenetic proteins effective inducers of cartilage repair? *Arthritis Rheum* 54: 387–89.

Goldring MB, Tsuchimochi K, Ijiri K. 2006. The control of chondrogenesis. *J Cell Biochem* 97: 33–44.

Gong Y, Slee RB, Fukai N, Rawadi G, Roman-Roman S, Reginato AM, Wang H, Cundy T, Glorieux FH, Lev D, Zacharin M, Oexle K, Marcelino J, Suwairi W, Heeger S, Sabatakos G, Apte S, Adkins WN, Allgrove J, Arslan-Kirchner M, Batch JA, Beighton P, Black GC, Boles RG, Boon LM, Borrone C, Brunner HG, Carle GF, Dallapiccola B, De Paepe A, Floege B, Halfhide ML, Hall B, Hennekam RC, Hirose T, Jans A, Juppner H, Kim CA, Keppler-Noreuil K, Kohlschuetter A, LaCombe D, Lambert M, Lemyre e, Letteboer T, Peltonen L, Ramesar RS, Romanengo M, Somer H, Steichen-Gersdorf E, Steinmann B, Sullivan B, Superti-Furga A, Swoboda W, van den Boogaard MJ, Van Hul W, Vikkula M, Votruba M, Zabel B, Garcia T, Baron R, Olsen BR, Warman JL, Osteoporosis-Pseudoglioma Syndrome Collaborative Group. 2001. LDL receptor-related protein 5 (LRP5) affects bone accrual and eye development. *Cell* 107: 513–23.

Gonyeau MJ. 2005. Statins and osteoporosis: a clinical review. *Pharmacotherapy* 25: 228–43.

Gonzalez-Gay MA, Gonzalez-Juanetey C, Martin J. 2005. Rheumatoid arthritis: a disease associated with accelerated atherogenesis. *Semin Arthritis Rheum* 35: 8–17.

Goodpaster BH, Krishnaswami S, Harris TB, Katsiaras A, Kritchevsky SB, Simonsick EM, Nevitt M, Holvoet P, Newman AB. 2005. Obesity, regional body fat distribution, and the metabolic syndrome in older men and women. *Arch Intern Med* 165: 777–83.

Goodwin JS. 1999. Geriatrics and the limits of modern medicine. *N Engl J Med* 340: 1283–85.

Gorenne I, Kavurma M, Scott S, Bennett M. 2006. Vascular smooth muscle cell senescence in atherosclerosis. *Cardiovasc Res* 72: 9–17.

Graham DJ, Campen D, Hui R, Spence M, Cheetham C, Levy G, Shoor S, Ray WA. 2005. Risk of acute myocardial infarction and sudden cardiac death in patients treated with cyclo-oxygenase 2 selective and non-selective non-steroidal anti-inflammatory drugs: nested case-control study. *Lancet* 365: 475–81.

Greenberg AS, Obin MS. 2006. Obesity and the role of adipose tissue inflammation and metabolism. *Am J Clin Nutr* 83: 461–65S.

Greer EL, Brunet A. 2005. FOXO transcription factors at the interface between longevity and tumor suppression. *Oncogene* 24: 7410–25.

Greten FR, Eckmann L, Greten TF, Park JM, Li ZW, Egan WJ , et al. 2004. IKKβ links inflammation and tumorigenesis in a mouse model of colitis-associated cancer. *Cell* 118: 285–96.

Grey A, Cundy T. 2006. Bisphosphonates and osteonecrosis of the jaw. Ann Intern Med 145: 791.

Griendling KK, Minieri CA, Ollerenshaw JD, Alexander RW. 1994. Angiotensin II stimulates NADH and NADPH oxidase activity in cultured vascular smooth muscle cells. *Circ Res* 74: 1141–48.

Grimley Evans J. 2005. Scientific aspects of ageing: a lordly report. *J R Soc Med* 98: 482–83.

Groopman J. 2002. Hormones for men. *New Yorker,* 7/29/02.

Grosser T, Fries S, FitzGerald GA. 2006. Biological basis for the cardiovascular consequences of COX-2 inhibition: therapeutic challenges and opportunities. *J Clin Invest* 116: 4–5.

Grundy SM. 2003. Inflammation, hypertension, and the metabolic syndrome. *JAMA* 290: 3000–3002.

Grundy SM. 2004. Metabolic Syndrome: Part I. *Endocrinol Metab Clin North Am* 33: ix–xi.

Grundy SM. 2006a. Diabetes and coronary risk equivalency. What does it mean? *Diabetes Care* 29: 457–60.

Grundy SM. 2006b. Does a diagnosis of metabolic syndrome have value in clinical practice? *Am J Clin Nutr* 83: 1248–51.

Grunfeld C, Feingold KR. 1992. Metabolic disturbances and wasting in the acquired immunodeficiency syndrome. *N Engl J Med* 327: 329–36.

Guarente L. 2005. Caloric restriction and SIR2 genes: Towards a mechanism. *Mech Ageing Dev* 126: 923–28.

Guarente L, Kenyon C. 2000. Genetic pathways that regulate ageing in model organisms. *Nature* 408: 255–62.

Guarente L, Picard L. 2005. Calorie restriction: the SIR2 connection. *Cell* 120: 473–82.

Gullestad L, Aukrust P. 2005. Review of trials in chronic heart failure showing broad-spectrum anti-inflammatory approaches. *Am J Cardiol* 95: 17–23C.

Guralnik JM. 2005. The evolution of research on disability in old age. *Aging Clin Exp Res* 17: 165–67.

Gurnell M. 2005. Peroxisome proliferator-activated receptor gamma and the regulation of adipocyte function: lessons from human genetic studies. *Best Pract Res Clin Endocrinol Metab* 19: 501–23.

Gutterman DD. 2005. Mitochondria and reactive oxygen species: an evolution in function. *Circ Res* 97: 302–4.

Guttmacher AE, Collins FS, Drazen JM. 2004. *Genomic medicine: articles from the New England Journal of Medicine.* Baltimore: Johns Hopkins University Press.

Haass C, De Strooper B. 1999. The presenilins in Alzheimer's disease: proteolysis holds the key. *Science* 286(5441): 916–19.

Hadler NM. 1985. Osteoarthritis as a public health problem. *Clin Rheum Dis* 11: 175–85.

Hadler NM. 1992. Knee pain is the malady: not osteoarthritis. *Ann Rheum Dis* 53: 143–46.

Hadley EC, Lakatta EG, Morrison-Bogorad M, Warner HR, Hodes RJ. 2005. The future of aging therapies. *Cell* 120: 557–67.

Haffner SM. 2006a. Risk constellations in patients with the metabolic syndrome: epidemiology, diagnosis, and treatment patterns. *Am J Med* 119: 3–9S.

Haffner SM. 2006b. The metabolic syndrome: Inflammation, diabetes mellitus, and cardiovascular disease. *Am J Cardiol* 97: 3–11.

Haffner SM, Ruilope L, Dahlöf B, Adabie E, Kupfer S, Zannad F. 2006. Metabolic syndrome, new onset diabetes, and new end points in cardiovascular trials. *J Cardiovasc Pharmacol* 47: 469–75.

Haigis MC, Guarente LP. 2006. Mammalian sirtuins—emerging roles in physiology, aging, and calorie restriction. Genes Dev 20: 2913–21.

Hajjar I, Schumpert J, Hirth V, Wieland D, Eleazer GP. 2002. The impact of the use of statins on the prevalence of dementia and the progression of cognitive impairment. *J Gerontol Med Sci* 57A: M414–18.

Hall A. 1998. Rho GTPases and the actin cytoskeleton. *Science* 279: 509–13.

Haluzík M, Pa ízková J, Haluzík MM. 2004. Adiponectin and its role in the obesity-induced insulin resistance and related complications. *Physiol Rev* 53: 123–29.

Hamerman D. 1989. Invited review: mechanisms of disease: the biology of osteoarthritis. *N Engl J Med* 320: 1322.

Hamerman D. 1993. Invited Review for Geriatrics Bioscience. Aging and osteoarthritis: basic mechanisms. *J Am Geriatr Soc* 41: 760–77.

Hamerman D. 1995 Clinical implications of osteoarthritis and aging. *Ann Rheum Dis* 54: 82–85.

Hamerman D. 1997. The aging skeleton. Osteoarthritis and osteoporosis. In: Hamerman D (ed.), *Osteoarthritis: public health implications for an aging population.* Baltimore: Johns Hopkins University Press, pp. 99–121.

Hamerman D. 1999. Toward an understanding of frailty. *Ann Intern Med* 130: 945–50.

Hamerman D. 2002a. Geriatric practice revisited: toward collaboration with primary care medicine. *J Am Geriatr Soc* 50: 971–72.

Hamerman D. 2002b. Molecular-based therapeutic approaches in treatment of anorexia of aging and cancer cachexia. *J Gerontol Med Sci* 57A: M5111–18.

Hamerman D. 2004. Frailty, cancer cachexia and near death. In: Balducci L, Lyman GH, Ershler WB, Extermann M (eds.), *Comprehensive geriatric oncology,* 2nd ed. UK: Taylor and Francis Group, pp. 236–94.

Hamerman D. 2005a. Bone health across the generations: a primer for health providers concerned with osteoporosis prevention. *Maturitas* 50: 1–7.

Hamerman D. 2005b. Osteoporosis and atherosclerosis: biological linkages and the emergence of dual-purpose therapies. *Q J Med* 98: 467–84.

Hamerman D, Dubler NN, Kennedy GJ, Masdeu J. 1986. Geriatric grand rounds. Decision making in response to an elderly woman with dementia who refused surgical care of her fractured hip. *J Am Geriatr Soc* 34: 234–39.

Hamerman D, Fox A. 1992. Responses of the health professions to the demographic revolution: a multidisciplinary approach. *Perspect Biol Med* 35: 583–93.

Hamerman D, Freeman K, Stanley ER. 1998. Colony stimulating factor-1 in synovial fluids from osteoarthritis and injured knees. *Ann Rheum Dis* 57: 260–61.

Hamerman D, Maklan CW. 1987. Editorial. Geriatric practice: taking up where primary care leaves off. *Am J Med* 82: 525–28.

Hamerman D, Sherlock L, Damus K, Habermann ET. 1988. Research in the Teaching Nursing Home. An approach to assessing osteoarthritis of the hips and knees in residents of a health related facility. *Einstein Q J Biol Med* 6: 110–19.

Hamerman D, Stanley ER. 1996. Implications of increased bone density in osteoarthritis. *J Bone Miner Res* 11: 1205–8.

Hamerman D, Zeleznik J. 2001. Translating basic research into geriatric health care. *Exp Gerontol* 36: 193–203.

Hamrick MW, Della-Fera MA, Choi YH, Pennington C, Hartzell D, Baile CA. 2005. Leptin treatment induces loss of bone marrow adipocytes and increases bone formation in leptin-deficient ob/ob mice. *J Bone Miner Res* 20: 994–1001.

Hanahan D, Weinberg RA. 2000. The hallmarks of cancer. *Cell* 100: 57–70.

Hansson GK. 2005. Inflammation, atherosclerosis, and coronary heart disease. *N Engl J Med* 352: 1685–95.

Hansson GK, Libby P, Schonbeck U, Yan ZQ. 2002. Innate and adaptive immunity in the pathogenesis of atherosclerosis. *Circ Res* 91: 281–91.

Harada SI, Rodan GA. 2003. Control of osteoblast function and regulation of bone mass. *Nature* 423: 349–55.

Hardy J. 1997. Amyloid, the presenilins and Alzheimer's disease. *Trends Neurosci* 20(4): 154–59.

Hardy J. 2004. Toward Alzheimer therapies based on genetic knowledge. *Annu Rev Med* 55: 15–25.

Hardy J, Selkoe DJ. 2002. The amyloid hypothesis of Alzheimer's disease: progress and problems on the road to therapeutics. *Science* 297(5580): 353–56.

Hardy JA, Higgins GA. 1992. Alzheimer's disease: the amyloid cascade hypothesis. *Science* 256(5054):184–85.

Harman D. 1956. Aging: a theory based on free radical and radiation chemistry. *J Gerontol* 11: 298–300.

Harper JM, Salmon AB, Chang Y, Bonkowski M, Bartke A, Miller RA. 2006. Stress resistance and aging: Influence of genes and nutrition. *Mech Ageing Dev* 17: 687–94.

Harris D, Haboubi N. 2005. Malnutrition screening in the elderly population. *J R Soc Med* 98: 411–14.

Harris RBS. 2000. Leptin-much more than a satiety signal. *Annu Rev Nutr* 20: 45–75.

Harris RE, Beebe-Donk J, Doss H, Burr Doss D. 2005. Aspirin, ibuprofen, and other non-steroidal anti-inflammatory drugs in cancer prevention: a critical review of non-selective COX-2 blockage (review). *Oncol Rep* 13: 559–83.

Hart P, Cronin C, Daniels M, Worthy T, Doyle DV, Spector TD. 2002. The relationship of bone density and fracture incidence to incident and progressive radiographic osteoarthritis of the knee. The Chingford Study. *Arthritis Rheum* 46: 92–99.

Haslam DW, James WPT. 2005. Obesity. *Lancet* 366: 1197–1209.

Hatzigeorgiou C, Jackson JL. 2005. Hydroxymethylglutaryl-coenzyme A reductase inhibitors and osteoporosis: a meta-analysis. *Osteoporos Int* 16: 990–98.

Hawkins M, Rossetti L. 2005. Insulin resistance and its role in the pathogenesis of type 2 diabetes. In: Kahn R et al. (eds.), *Joslin's diabetes mellitus,* 14th ed. Boston: Lippincott, Williams and Wilkins, pp. 425–48.

Hawkins SA, Wiswell RA, Marcell TJ. 2003. Exercise and the master athlete: a model of successful aging? *J Gerontol A Biol Sci Med Sci* 58: 1009–11.

Hayflick L. 1965. The limited in vitro lifetime of human diploid cell strains. *Exp Cell Res* 37: 614–36.

Hayflick L. 2000. The future of ageing. *Nature* 408: 267–69.

Hayflick L. 2004a. Response to Dr. Holliday. *J Gerontol A Biol Sci Med Sci* 59: 552–53.

Hayflick L. 2004b. The not-so-odd close relationship between biological aging and age-associated pathologies in humans. *J Gerontol A Biol Sci Med Sci* 59: B547–50.

Hazzard WR. 1983. Preventive gerontology. Strategies for health aging. *Postgrad Med* 74: 279–87.

Hazzard WR, Woolard N, Regenstreif DI. 1997. Integrating geriatrics into the subspecialties in internal medicine. The Hartford Foundation/American Geriatrics Society/Wake Forest University Bowman Gray School of Medicine Initiative. *J Am Geriatr Soc* 45: 638–40.

Hegener HH, Lee IM, Cook NR, Ridker PM, Zee RYL. 2006. Association of adiponectin gene variations with risk of incident myocardial infarction and ischemic stroke: A nested case-control study. *Clin Chem* 52: 2021–27.

Heilbronn LK, de Jonge L, Frisard MI. De Lany JP, Larson-Meyer DE, Rood J, Nguyen T, Martin CK, Volaufova J, Most MM, Greenway FL, Smith SR, Deutsch WA, Williamson DA, Ravussin E. 2006. Effect of 6-month calorie restriction on biomarkers of longevity, metabolic adaptation, and oxidative stress in overweight individuals. *JAMA* 295: 1529–48.

Heistad DD. 2006. Oxidative stress and vascular disease. 2005 Duff Lecture. *Arterioscler Thromb Vasc Biol* 26: 689–95.

Helmbold H, Deppert W, Bohn W. 2006. Regulation of cellular senescence by Rb2/p130. *Oncogene* 25: 5257–62.

Herman MA, Kahn BB. 2006. Glucose transport and sensing in the maintenance of glucose homeostasis and metabolic harmony. *J Clin Invest* 116: 1767–75.

Herrup K, Arendt T. 2002. Re-expression of cell cycle proteins induces neuronal cell death during Alzheimer's disease. *J Alzheimer's Dis* 4(3): 243–47.

Higashi T, Shekelle PG, Adams JL, Kamberg CJ, Roth CP, Solomon DH, Reuben DB, Chiang L, MacLean CH, Chang JT, Young RT, Saliba DM, Wenger NS. 2005. Quality of care is associated with survival in vulnerable older patients. *Ann Intern Med* 143: 274–81.

Hijazi RA, Cunningham GR. 2005. Andropause: is androgen replacement therapy indicated for the aging male? *Annu Rev Med* 56: 117–37.

Hochberg MC, Lebwohl MG, Plevy SE, Hobbs KF, Yocum DE. 2005. The benefit/risk profile of TNF-blocking agents: findings of a consensus panel. *Semin Arthritis Rheum* 34: 819–36.

Hoffman AR. 2005. Treatment of the adult growth hormone deficiency syndrome: directions for future research. *Growth Horm IGF Res* 15: S48–52.

Hogan DB, MacKnight C, Bergman H, Steering Committee, Canadian Initiative on Frailty and Aging. 2003. Models, definitions, and criteria of frailty. *Aging Clin Exp Res* 15: 1–29.

Holen I, Shipman CM. 2006. Role of osteoprotegerin in cancer. *Clin Sci* 110: 279–91.

Holliday R. 2004. Response to Dr. Hayflick. *J Gerontol A Biol Sci Med Sci* 59: 551.

Holzenberger M, Dupont J, Ducos B, Leneuve P, Géloën, Even PC, Cervera P, Le Bouc Y. 2003. IGF-1 receptor regulates lifespan and resistance to oxidative stress in mice. *Nature* 421: 182–86.

Honore P, Luger NM, Sabina MC, Schwei MJ, Rogers SD, Mach DB, O'Keefe PF, Ramnaraine ML, Clohisy DR, Mantyh PW. 2000. Osteoprotegerin blocks bone cancer-induced skeletal destruction, skeletal pain and pain-related neurochemical reorganization of the spinal cord. *Nat Med* 6: 521–28.

Horowitz M. 2003. Matrix proteins versus cytokines in the regulation of osteoblast function and bone formation. *Calcif Tissue Int* 72: 5–7.

Hotamisligil GS. 2003. Inflammatory pathways and insulin action. *Int J Obes Relat Metab Disord* 27 (Suppl. 3): S53–55.

Hu H, Hilton MJ, Tu X, Yu K, Ornitz DM, Long F. 2005. Sequential roles of hedgehog and Wnt signaling in osteoblast development. *Development* 132: 49–60.

Huls A. 2001. Assessment of nutritional status in the older adult. In: Watson RR (ed.), *Handbook of nutrition in the aged,* 3rd ed. Boca Raton: CRC Press, pp. 107–14.

Hunter GR, McCarthy JP, Bamman MM. 2004. Effects of resistance training on older adults. *Sports Med* 34: 329–48.

Hursting SD, Lavigne JA, Berrigan D, Perkins SN, Barrett JC. 2003. Caloric restriction, aging, and cancer prevention: Mechanisms of action and applicability to humans. *Annu Rev Med* 54: 131–52.

Hutton M, Lendon CL, Rizzu P, Baker M, Froelich S, Houlden H, Pickering-Brown S, Chakraverty S, Isaacs A, Grover A, Hackett J, Adamson J, Lincoln S, Dickson D, Davies P, Petersen RC, Stevens M, de Graaff E, Wauters E, van Baren J, Hillebrand M, Joosse M, Kwon JM, Nowotny P, Heutink P, et al. 1998. Association of missense and 5'-splice-site mutations in tau with the inherited dementia FTDP-17. *Nature* 393(6686): 702–5.

Huyse FJ, Stiefel FC. 2006. Integrated care for the complex mentally ill. *Med Clin North Am* 90.

Iannuzzi-Sucich M, Prestwood KM, Kenny AM. 2002. Prevalence of sarcopenia and predictors of skeletal muscle mass in healthy, older men and women. *J Gerontol A Biol Sci Med Sci* 57: M772–77.

Idris AI, van't Hof RJ, Greig IR, Ridge SA, Baker D, Ross RA, Ralston SH. 2005. Regulation of bone mass, bone loss and osteoclast activity by cannabinoid receptors. *Nat Med* 11: 774–79.

Ikeda F, Nishimura R, Matsubara T, Tanaka S, Inoue Ji, Reddy SV, Hata K, Yamashita K, Hiraga R, Watanabe T, Kukito T, Yoshioka K, Rao A, Yoneda T. 2004. Critical roles of c-Jun signaling in regulation of NFAT family and RANKL-regulated osteoclast differentiation. *J Clin Invest* 114: 475–84.

Ingelsson E, Arnlov J, Sundstrom J, Lind L. 2005. Inflammation, as measured by the erythrocyte sedimentation rate, is an independent predictor for the development of heart failure. *J Am Coll Cardiol* 45: 1802–6.

Irminger-Finger I. 2005. Cancer and aging at the crossroads. *Int J Biochem Cell Biol* 37: 912.

Isacson O, Seo H, Lin L, Albeck D, Granholm AC. 2002. Alzheimer's disease and Down's syndrome: roles of APP, trophic factors and ACh. *Trends Neurosci* 25(2): 79–84.

Isidori AM, Lenzi A. 2005. Risk factors for androgen decline in older males: lifestyle, chronic diseases and drugs. *J Endocrinol Invest* 28: 14–22.

Isley WL. 2002. Growth hormone therapy for adults: not ready for prime time? *Ann Intern Med* 137: 190–96.

Ito K, Yamamura S, Essilfie-Quaye S, Cosio B, Ito M, Barnes PJ, Adock IM. 2006. Histone deacetylase 2-mediated deacetylation of the glucocorticoid receptor enables NFkB suppression. *J Exp Med* 203: 7–13.

Jacobson L. 2005. Hypothalamic-pituitary-adrenocortical axis regulation. *Endocrinol Metab Clin North Am* 34: 271–92.

Jaddoe VW, Witteman JC. 2006. Hypothesis on the fetal origins of adult diseases: contributions of epidemiological studies. *Eur J Epidemiol* 21: 91–102.

Janssen I, Heymsfield SB, Allison DB, Kotler DP, Ross R. 2002. Body mass index and waist circumference independently contribute to the prediction of nonabdominal, abdominal subcutaneous, and visceral fat. *Am J Clin Nutr* 75: 683–88.

Janssen I, Shepard DS, Katzmarzyk PT, Roubenoff R. 2004. The healthcare costs of sarcopenia in the United States. *J Am Geriatr Soc* 52: 80–85.

Janssens K, ten Dijke P, Janssens S, Van Hul W. 2005. Transforming growth factor-β1 to the bone. Endocr Rev 26: 743–74.

Janzen V, Forkert R, Fleming HE, Saito Y, Waring MT, Dombkowski DM, Cheng T, DePinho RA, Sharpless NE, Scadden DT. 2006. Stem-cell ageing modified by the cyclin-dependent kinase inhibitor p16[INK4a]. *Nature* 443: 421–26.

Jarrett PG, Rockwood K, Carver D, Stollee P, Cosway S. 1995. Illness presentation in elderly patients. *Arch Intern Med* 155: 1060–64.

Javaid MK, Godfrey KM, Taylor P, Robinson SM, Crozier SR, Dennison EM, Robinson JS, Breier BR, Arden NK, Cooper C. 2005. Umbilical cord leptin predicts neonatal bone mass. *Calcif Tissue Int* 76: 341–47.

Jen KL, Buison A, Darga L, Nelson D. 2005. The relationship between blood leptin level and bone density is specific to ethnicity and menopausal status. *J Lab Clin Med* 146: 18–24.

Jobst EE, Enriori PJ, Sinnayah P, Cowley MA. 2006. Hypothalamic regulatory pathways and potential obesity treatment targets. *Endocrine* 29: 33–48.

Jockenhövel F, Blum WF, Vogel E, Englaro P, Muller-Wieland D, Reinwein D, Rascher W, Crone W. 1997. Testosterone substitution normalizes elevated serum leptin levels in hypogonadal men. *J Clin Endocrinol Metab* 82: 2510–13.

Johnson ML, Harnish K, Nusse R, Van Hul W. 2004. LRP5 and Wnt signaling: a union made for bone. *J Bone Miner Res* 19: 1749–57.

Johnson TE. 2005. Genes, phenes, and dreams of immortality: the 2003 Kleemeier Award lecture. *J Gerontol A Biol Sci Med Sci* 60: 680–87.

Jones G, Ding C, Scott F, Cicuttini F. 2004. Genetic mechanisms of knee osteoarthritis: a population based case-control study. *Ann Rheum Dis* 63: 1255–59.

Jørgensen JOL. 2006. Editorial: A simple twist of science: the convoluted tale of ghrelin continues. *J Clin Endocrinol Metab* 91: 3279–80.

Joseph C, Kenney AM, Taxel P, Lorenzo JA, Duque G, Kuchel GA. 2005. Role of endocrine-immune dysregulation in osteoporosis, sarcopenia, frailty and fracture risk. *Mol Aspects Med* 26: 181–201.

Judge JO. 1997. The principles and prescription of exercise in frail older persons. In: Hamerman D (ed.), *Osteoarthritis: public health implications for an aging population*. Baltimore: Johns Hopkins University Press, pp. 127–49.

Juge-Aubrey CE, Henrichot E, Meier CA. 2005. Adipose tissue: a regulator of inflammation. *Best Pract Res Clin Endocrinol* 19: 547–66.

Jurimae J, Rembel K, Jurimae T, Rehand M. 2005. Adiponectin is associated with bone mineral density in perimenopausal women. *Horm Metab Res* 37: 297–302.

Kadowaki T, Yamauchi T, Kubota N, Hara K, Ueki K, Tobe K. 2006. Adiponectin and adiponectin receptors in insulin resistance, diabetes, and the metabolic syndrome. *J Clin Invest* 116: 1784–92.

Kahlem P, Dorken b, Schmitt CA. 2004. Cellular senescence in cancer treatment: friend or foe? *J Clin Invest* 113: 169–74.

Kahn BB, Flier JS. 2000. Obesity and insulin resistance. *J Clin Invest* 106: 473–81.

Kahn R, Buse J, Ferrannini E, Stern M. 2005. The metabolic syndrome: time for a critical appraisal. *Diabetes Care* 28: 2289–2304.

Kajinami K, Okabayashi M, Sata R, Polisecki E, Schaefer EJ. 2005. Statin pharmacogenomics: what have we learned, and what remains unanswered? *Curr Opin Lipidol* 16: 606–13.

Kajstura J, Belli R, Sonnenblick EH, Anversa P, Leri A. 2006. Cause of death: suicide. *J Mol Cell Cardiol* 40: 425–37.

Kakuda N, Funamoto S, Yagishita S, Takami M, Osawa S, Dohmae N, Ihara Y. 2006. Equimolar production of amyloid beta-protein and amyloid precursor protein intracellular domain from beta-carboxyl-terminal fragment by gamma-secretase. *J Biol Chem* 281(21): 14776–86.

Kaliora AC, Dedoussis GVZ, Schmidt H. 2006. Dietary antioxidants in preventing atherogenesis. *Atherosclerosis* 187: 1–17.

Kamel HK. 2003. Sarcopenia and aging. *Nutr Rev* 61: 157–67.

Kanaya AM, Fyr CW, Vittinghoff E, Harris TB, Park SW, Goodpaster BH, Tylavsky F, Cumming SR. 2006. Adipocytokines and incident diabetes mellitus in older adults: The independent effect of plasminogen activator inhibitor 1. *Arch Intern Med* 166: 350–56.

Kane RL. 2002. The future history of geriatrics: geriatrics at the crossroads. *J Gerontol Med Sci* 57A: M803–5.

Kaneto H, Nakatani Y, Kawamori D, Miyatsuka T, Matsuoka T-a, Matsuhisa M, Yamasaki Y. 2006. Role of oxidative stress, endoplasmic reticulum stress, and c-Jun N-terminal kinase in pancreatic β-cell dysfunction and insulin resistance. *Int J Biochem Cell Biol* 38: 782–93.

Kanzaki M. 2006. Insulin receptor signals regulating GLUT4 translocation and actin dynamics. *Endocr J* 53: 267–93.

Karakelides H, Sreekumaran NK. 2005. Sarcopenia of aging and its metabolic impact. *Curr Top Dev Biol* 68: 123–48.

Karin M. 2006. Role for IKK2 in muscle: waste not, want not. *J Clin Invest* 116: 2866–68.

Karin M, Greten FR. 2005. NF-kappaB: linking inflammation and immunity to cancer development and progression. *Nat Rev Immunol* 5: 749–59.

Karsenty G. 1999. The genetic transformation of bone biology. *Genes Dev* 13: 3037–51.

Karsenty G. 2001. Editorial: The not-so-odd couple: the clinician and the experimentalist. *J Clin Endocrinol Metab* 86: 1882–83.

Karsenty G. 2003. The complexities of skeletal biology. *Nature* 423: 316–18.

Karsenty G. 2005. An aggrecanase and osteoarthritis. *N Engl J Med* 353: 522–23.

Karsenty G. 2006. Convergence between bone and energy homeostases. Leptin regulation of bone mass. Cell Metab 4: 341–8.

Karsenty G, Wagner EF. 2002. Reaching a genetic and molecular understanding of skeletal development. *Dev Cell* 2: 389–406.

Kasuga M. 2006. Insulin resistance and pancreatic β cell failure. *J Clin Invest* 116: 1756–60.

Katic M, Kahn CR. 2005. The role of insulin and IGF-1 signaling in longevity. *Cell Mol Life Sci* 62: 320–43.

Katzman R. 1988. Discussion. In: Evered D, Whelen J (eds.), *Research and the ageing population*. Ciba Foundation Symposium 134. New York: John Wiley, p. 47.

Kaufman JM, Vermeulen A. 2005. The decline of androgen levels in elderly men and its clinical and therapeutic implications. *Endocr Rev* 26: 833–76.

Kaufman SR. 1994. The social construction of frailty, an anthropological perspective. *J Aging Studies* 8: 45–58.

Kavanaugh A. 2005. Health economics: implications for novel antirheumatic therapies. *Ann Rheum Dis* 64 (Suppl. 4): iv65–69.

Keller L, Jemielity S. 2006. Social insects as a model to study the molecular basis of ageing. *Exp Gerontol* 41: 553–56.

Kennedy D. 2004. Longevity, quality, and the one-hoss shay. *Science* 305: 1369.

Kennedy ET. 2006. Evidence for nutritional benefits in prolonging wellness. *Am J Clin Nutr* 83: 410–14S.

Kenyon C. 2005. The plasticity of aging: insights from long-lived mutants. *Cell* 120: 449–60.

Kenyon C. 2006. Surviving longer. 21st Annual Darwin College Lecture Series, in press.

Kershaw EE, Flier JS. 2004. Adipose tissue as an endocrine organ. *J Clin Endocrinol Metab* 89: 2548–56.

Khanna D, McMahon M, Furst DE. 2004. Anti-tumor necrosis factor α therapy and heart failure. What have we learned and where do we go from here? *Arthritis Rheum* 50: 1040–50.

Khosla S. 2002. Editorial: Leptin: Central or peripheral to the regulation of bone metabolism? *Endocrinology* 143: 4161–64.

Khosla S. 2004. Role of hormonal changes in the pathogenesis of osteoporosis in men. *Calcif Tissue Int* 75: 110–13.

Kiecolt-Glaser JK, Preacher KJ, MacCallum RC, Atkinson C, Malarkey WB, Glaser R. 2003. Chronic stress and age-related increases in the proinflammatory cytokine IL-6. *Proc Natl Acad Sci* USA 100: 9090–95.

Kim J-a, Montagnani M, Koh KK, Quon MJ. 2006. Reciprocal relationships between insulin resistance and endothelial dysfunction. Molecular and pathophysiological mechanisms. *Circulation* 113: 1888–1904.

Kim S, Popkin BM. 2006. Commentary: Understanding the epidemiology of overweight and obesity: a real global public health concern. *Int J Epidemiol* 35: 60–67.

Kim Y, Sharpless NE. 2006. The regulation of INK4/ARF in cancer and aging. Cell 127: 265–75.

Kinney JM. 2004. Nutritional frailty, sarcopenia and falls in the elderly. *Curr Opin Clin Nutr Metab Care* 7: 15–20.

Kirkwood TB. 2005a. Understanding the odd science of aging. *Cell* 120: 437–47.

Kirkwood TB. 2005b. Time of our lives. What controls the length of life? *EMBO Rep* 6 Spec No: S4–8.

Kirkwood TB, Shanley DP. 2005. Food restriction, evolution and ageing. *Mech Ageing Dev* 126: 1011–16.

Kishimoto T. 2005. Interleukin-6: from basic science to medicine-40 years in immunology. *Annu Rev Immunol* 23: 1–21.

Kitamura T, Feng Y, Kitamura YI, Chua SC, Jr, Xu AW, Barsh GS, Rossetti L, Accili D. 2006. Forkhead protein FoxO1 mediates Agrp-dependent effects of leptin on food intake. *Nat Med* 12: 534–40.

Klegeris A, McGeer PL. 2005. Non-steroidal anti-inflammatory drugs (NSAIDs) and other anti-inflammatory agents in the treatment of neurodegenerative disease. *Curr Alzheimer Res* 2: 355-65.

Klein BEK, Klein RK, Knudtson MD, Lee KE. 2005. Frailty, morbidity and survival. *Arch Gerontol Geriatr* 41: 141–49.

Klurfeld DM. 2005. Response: Snack foods, obesity and realistic recommendations. *J Am Coll Nutr* 24: 156–57.

Knight JA. 2000. The biochemistry of aging. *Adv Clin Chem* 35: 1-62.

Knouff C, Auwerx J. 2004. Peroxisome proliferator-activated receptor-_ calls for activation in moderation: Lessons from genetics and pharmacology. *Endocr Rev* 25: 899–918.

Knudsen ES. 2006. Overview. *Curr Mol Med* 6: 703.

Knudsen ES, Knudsen KE. 2006. Retinoblastoma tumor suppressor: where cancer meets the cell cycle. *Exp Biol Med* 231: 1271–81.

Koay MA, Woon PY, Zhang Y, Miles LJ, Duncan EL, Ralston SH, Compston JE, Cooper

C, Keen R, Langdahl BL, MacLelland A, O'Riordan J, Pols HA, Reid DM, Uitterlinden AG, Wass JA, Brown MA. 2004. Influence of LRP5 polymorphisms on normal variation in BMD. *J Bone Miner Res* 19: 1619–27.

Kobayashi K. 2005. Adipokines: therapeutic targets for metabolic syndrom. *Curr Drug Targets* 6: 525–29.

Koch AE. 2005. Chemokines and their receptors in rheumatoid arthritis. Future targets? *Arthritis Rheum* 52: 710–21.

Koehler CM, Beverly KN, Leverich EP. 2006. Redox pathways of the mitochondrion. *Antioxid Redox Signal* 8: 813–22.

Koerner A, Kratzsch J, Kiess W. 2005. Adipocytokines: leptin-the classical, resistin-the controversial, adiponectin-the promising, and more to come. *Best Pract Res Clin Endocrinol Metab* 19: 525–46.

Koh KK, Han SH, Quon MJ. 2005. Inflammatory markers and the metabolic syndrome: insights from therapeutic interventions. *J Am Coll Cardiol* 46: 1978–85.

Kojima M, Kangawa K. 2005. Ghrelin: structure and function. *Physiol Rev* 85: 495–522.

Komori T. 2005. Regulation of skeletal development by the Runx family of transcription factors. *J Cell Biochem* 95: 445–53.

Kornman KS. 2006. Interleukin-1 genetics, inflammatory mechanisms, and nutrigenic opportunities to modulate disease of aging. *Am J Clin Nutr* 83: 475–83S.

Kostenuik PJ. 2005. Osteoprotegerin and RANK L regulate bone resorption, density, geometry and strength. *Curr Opin Pharmacol* 5: 618–25.

Koubova J, Guarente L. 2003. How does calorie restriction work? *Genes Dev* 17: 313–21.

Kovacs P, Stumvoll M. 2005. Fatty acids and insulin resistance in muscle and liver. *Best Pract Res Clin Endocrinol Metab* 19: 625–35.

Krabbe KS, Pedersen M, Bruunsgaard H. 2004. Inflammatory mediators in the elderly. *Exp Gerontol* 39: 687–99.

Krishnan V, Bryant HU, Mac Dougald OA. 2006. Regulation of bone mass by Wnt signaling. *J Clin Invest* 116: 1202–9.

Kuan CY, Schloemer AJ, Lu A, Burns KA, Weng WL, Williams MT, Strauss KI, Vorhees CV, Flavell RA, Davis RJ, Sharp FR, Rakic P. 2004. Hypoxia-ischemia induces DNA synthesis without cell proliferation in dying neurons in adult rodent brain. *J Neurosci* 24(47): 10763–72.

Kumada M, Kihara S, Ouchi N, Kobayashi H, Okamoto Y, Ohashi K, Maeda K, Nagaretani H, Kishida K, Maeda N, Nagasawa A, Funahashi t, Matsuzawa Y. 2004. Adiponectin specifically increased tissue inhibitor of metalloproteinase-1 through interleukin-10 expression in human macrophages. *Circulation* 109: 2046–49.

Kumada M, Kihara S, Sumitsuji S, Kawamoto T, Matsumoto S, Ouchi N, Arita Y, Okamoto Y, Shimomura I, Hiraoka H, Nakamura T, Funahashi T, Matsuzawa Y, Osaka CAD Study Group. Coronary artery disease. 2003. Association of hypoadiponectinemia with coronary artery disease. *Arterioscler Thromb Vasc Biol* 23: 85–89.

Kurosu H, Yamamoto M, Clark JD, Pastor JV, Nandi A, Gurnani P, McGuinness OP, Chikuda H, Yamaguchi M, Kawaguchi H, Shimomura I, Takayama Y, Herz J, Kahn CR, Rosenblatt KP, Kuro-o M. 2005. Suppression of aging in mice by the hormone Klotho. *Science* 309: 1829–33.

Kurt MA, Davies DC, Kidd M, Duff K, Rolph SC, Jennings KH, Howlett DR. 2001. Neuro-degenerative changes associated with beta-amyloid deposition in the brains of mice carrying mutant amyloid precursor protein and mutant presenilin-1 transgenes. *Exp Neurol* 171(1): 59–71.

Kurtz TW. 2006. New treatment strategies for patients with hypertension and insulin resistance. *Am J Med* 119(5A): 24–30S.

Kusminski CM, McTernan PG, Kuman S. 2005. Role of resistin in obesity, insulin resistance and type II diabetes. *Clin Sci* 109: 243–56.

Laakso M, Kovanen PT. 2006. Metabolic syndrome: to be or not to be? *Ann Med* 38: 32–33.

Labrie F, Luu-The V, Belanger A, Lin SX, Simard J, Pelletier G, Labrie C. 2005. Is dehydroepiandrosterone a hormone? *J Endocrinol* 187: 169–96.

Lai LP, Mitchell J. 2006. Indian hedgehog: its roles and regulation in endochondral bone development. *J Cell Biochem* 96: 1162–73.

Lam L, Cao Y. 2005. Statins and their roles in cancer. *Drugs Today* (Barc) 41: 329–34.

Lamberts SWJ, van den Beld A, van der Lely AJ. 1997. The endocrinology of aging. *Science* 278: 419–24.

Landefeld CS, Callahan CM, Woolard N. 2003. General internal medicine and geriatrics: Building a foundation to improve the training of general internists in the care of older adults. *Ann Intern Med* 139: 609–14.

Landin-Wilhelmsen K, Nilsson A, Bosaeus I, Bengtsson B-A. 2003. Growth hormone increases bone mineral content in postmenopausal osteoporosis: a randomized, placebo-controlled trial. *J Bone Miner Res* 18: 393–405.

Laroche M, Delmotte A. 2005. Increased arterial calcification in Paget's disease of bone. *Calcif Tissue Int* 77: 129–33.

Larson EB, Wang LI, Bowen JD, McCormick WC, Teri L, Crane P, Kukull W. 2006. Exercise is associated with reduced risk for incident dementia among persons 65 years of age and older. *Ann Intern Med* 144: 73–81.

Lau DCW, Dhillon B, Yan H, Szmitko PE, Verma S. 2005. Adipokines: molecular links between obesity and atherosclerosis. *Am J Physiol Heart Circ Physiol* 288: H2031–41.

Lawlor DH, Chaturvedi N. 2006. Treatment and prevention of obesity: are there critical periods for intervention? *Int J Epidemiol* 35: 3–9.

Lawrence DM. 2005. Chronic disease care: rearranging the deck chairs. *Ann Intern Med* 143: 458–59.

Lazar MA. 2005. How obesity causes diabetes: not a tall tale. *Science* 307: 373–75.

Lee Y, Rodriguez C, Dionne RA. 2005. The role of COX-2 in acute pain and the use of selective COX-2 inhibitors for acute pain relief. *Curr Pharm Des* 11: 1737–55.

Lefebvre P, Chinetti G, Fruchart J-C, Staels B. 2006. Sorting out the roles of PPARα in energy metabolism and vascular homeostasis. *J Clin Invest* 116: 571–80.

Leishman AJ, Bundick RV. 2005. Immunology: turning basic research into a therapy. *Immunol Lett* 99: 1–7.

Lenfant C. 2003. Clinical research to clinical practice: lost in translation? *N Engl J Med* 349: 868–74.

Leslie BR. 2005. Metabolic syndrome: historical perspectives. *Am J Med Sci* 330: 264–68.

Leveille SG. 2004. Musculoskeletal aging. *Curr Opin Rheumatol* 16: 114–18.

Leveille SG, Fried LP, McMullen W, Guralnik JM. 2004. Advancing the taxonomy of disability in older adults. *J Gerontol A Biol Sci Med Sci* 59: M86–93.

Levey AS, Coresh J, Greene T, Stevens LA, Zhang YL, Hendriksen S, Kusek JW, Van Lente F, for the Chronic Kidney Disease Epidemiology Collaboration. 2006. Using standardized serum creatinine values in the modification of diet in renal disease study equation for estimating glomerular filtration rate. *Ann Intern Med* 145: 247–54.

Levine AJ. 1997. p53, the cellular gatekeeper for growth and division. *Cell* 88: 323–31.

Lewis J, McGowan E, Rockwood J, Melrose H, Nacharaju P, Van Slegtenhorst M, Gwinn-Hardy K, Paul Murphy M, Baker M, Yu X, Duff K, Hardy J, Corral A, Lin WL, Yen SH, Dickson DW, Davies P, Hutton M. 2000. Neurofibrillary tangles, amyotrophy and progressive motor disturbance in mice expressing mutant (P301L) tau protein. *Nat Genet* 25(4):402–5.

Li AC, Palinski W. 2006. Peroxisome proliferator-activated receptors: how their effects on mcrophages can lead to the development of new drug therapy against atherosclerosis. *Annu Rev Pharmacol Toxicol* 46: 1–39.

Li L, Pettit AR, Gregory LS, Forwood MR. 2006. Regulation of bone biology by prostaglandin endoperoxide H synthases (PGHS): a rose by any other name . . . *Cytokine Growth Factor Rev* 17: 203–16.

Liao JK. 2002. Isoprenoids as mediators of the biological effects of statins. *J Clin Invest* 110: 285–88.

Liao JK. 2005. Clinical implications for statin pleiotropy. *Curr Opin Lipidol* 16: 624–29.

Liao JK, Laufs U. 2005. Pleiotropic effects of statins. *Annu Rev Pharmacol Toxicol* 45: 89–118.

Liberman UA. 2006. Long-term safety of bisphosphonate therapy for osteoporosis: a review of the evidence. *Drugs Aging* 23: 289–98.

Libby P. 2002. Inflammation in atherosclerosis. *Nature* 420: 868–74.

Libby P. 2006. Inflammation and cardiovascular disease mechanisms. *Am J Clin Nutr* 83: 456–60S.

Libby P, Aikawa M. 2003. Mechanisms of plaque stabilization with statins. *Am J Cardiol* 91 (Suppl.): 4–8B.

Lim W, Hong S, Nelesen R, Dimsdale JE. 2005. The association of obesity, cytokine levels, and depressive symptoms with diverse measures of fatigue in healthy subjects. *Arch Intern Med* 165: 910–15.

Lin EY, Nguyen AV, Russell RG, Pollard JW. 2001. Colony-stimulating factor-1 promotes progression of mammary tumors to malignancy. *J Exp Med* 198: 727–39.

Lindeman RD, Tobin J, Shock NW. 1985. Longitudinal studies on the rate of decline in renal function with age. *J Am Geriatr Soc* 33: 278–85.

Linton MF, Fazio S. 2003. Macrophages, inflammation, and atherosclerosis. *Int J Obes Relat Metab Disord* 27 (Suppl. 3): S35–40.

Lipsitz LA. 2002. Dynamics of stability: The physiologic basis of functional health and frailty. *J Gerontol Biol Sci* 57A: B115–25

Lipsitz LA, Goldberger AL. 1992. Loss of "complexity" and aging. Potential applications of fractals and chaos theory to senescence. *JAMA* 267: 1806–9.

Lipsky PE. 1999. Specific COX-2 inhibitors in arthritis, oncology, and beyond: where is the science headed? *J Rheumatol* 26 (Suppl. 56): 25–30.

Little RD, Carulli JP, Del Mastro RG, Dupuis J, Osborne M, Folz C, Manning SP, Swain PM, Zhao SC, Eustace B, Lappe MM, Spitzer L, Zweier S, Braunschweiger K, Benchekroun Y, Hu X, Adair R, Chee L, FitzGerald MG, Tulig C, Caruso A, Tzellas N, Bawa A, Franklin B, McGuire S, Nogues X, Gong G, Allen KM, Anisowicz A, Morales AJ, Lomedico PT, Recker SM, Van Eerdewegh P, Recker RR, Johnson ML. 2002. A mutation in the LDL receptor-related protein 5 gene results in the autosomal dominant high-bone-mass trait. *Am J Hum Genet* 70: 11–19.

Liu G, Chen X. 2006. Regulation of the p53 transcriptional activity. *J Cell Biochem* 97: 448–58.

Liu G, Rondinone CM. 2005. JNK: bridging the insulin signaling and inflammatory pathway. *Curr Opin Investig Drugs* 6: 979–87.

Liu H, Bravata DM, Olkin I, Nayak S, Roberts B, Garber AM, Hoffman AR. 2007. Systematic review: The safety and efficacy of growth hormone in the healthy elderly. Ann Intern Med 146: 104–115.

Liu PY, Iranmanesh A, Nehra AX, Keenan DM, Veldhuis JD. 2005. Mechanisms of hypoandrogenemia in healthy aging men. *Endocrinol Metab Clin N Am* 34: 935–55.

Lleo A, Greenberg SM, Growdon JH. 2006. Current pharmacotherapy for Alzheimer's disease. *Annu Rev Med* 57:513–33.

Loeser RF. 2006. Molecular mechanisms of cartilage destruction: mechanics, inflammatory mediators, and aging collide. *Arthritis Rheum* 54: 1357–60.

Lombard DB, Chua KF, Mostoslavsky, Franco S, Gostissa M, Alt FW. 2005. DNA repair, genome stability, and aging. *Cell* 120: 497–512.

Lombardi G, Di Somma C, Rota F, Colao A. 2005. Associated hormonal decline in aging: is there a role for GH therapy in aging men? *J Endocrinol Invest* 28: 99–108.

Long YC, Zierath JR. 2006. AMP-activated protein kinase signaling in metabolic regulation. *J Clin Invest* 116: 1776–83.

Longo VD, Finch CE. 2003. Evolutionary medicine: from dwarf model systems to healthy centenarians? *Science* 299: 1342–46.

Longo VD, Kennedy BK. 2006. Sirtuins in aging and age-related disease. *Cell* 126: 257–68.

Look AHEAD Research Group. 2006. The Look AHEAD Study: a description of the lifestyle intervention and the evidence supporting it. *Obesity* 14: 737–52.

Loughlin J. 2005. Polymorphism in signal transduction is a major route through which osteoarthritis susceptibility is acting. *Curr Opin Rheumatol* 17: 629–33.

Lowell BB, Shulman GI. 2005. Mitochondrial dysfunction and type 2 diabetes. *Science* 307: 384–87.

Luo JL, Kamata H, Karin M. 2005. IKK/NF-kappaB signaling: balancing life and death: a new approach to cancer therapy. *J Clin Invest* 115: 2615–32.

Lynn J. 2004. Frailty might be delayed, but not prevented. Annals online, Oct 6, 2004 (Rapid Response to: Wilson JF, Frailty-and its dangerous effects-might be preventable. *Ann Intern Med* 141: 489–92).

MacIntosh CB, Andrews JM, Jones KL, Wishart JM, Morris HA, Jansen JB, Morley JE, Horowitz M, Chapman IM. 1999. Effects of age on concentrations of plasma cholecystokinin glucagon-like peptide 1, and peptide YY and their relation to appetite and pyloric motility. *Am J Clin Nutr* 69: 999–1006.

Mackie K. 2006a. Cannabinoid receptors as therapeutic targets. *Annu Rev Pharmacol Toxicol* 46: 101–22.

Mackie K. 2006b. Mechanisms of CB1 receptor signaling: endocannabinoid modulationof synaptic strength. *Int J Obes* 30: S19–23.

Maggio M, Guralnik JM, Longo DL, Ferrucci L. 2006. Interleukin-6 in aging and chronic disease: a magnificent pathway. *J Gerontol A Biol Sci Med Sci* 61: 575–84.

Mahley RW, Huang Y, Weisgraber KW. 2006. Putting cholesterol in its place: apoE and reverse cholesterol transport. *J Clin Invest* 116: 1226–29.

Maini RN. 2005. The 2005 International Symposium on Advances in Targeted Therapies: What have we learned in the 2000s and where are we going? *Ann Rheum Dis* 64: iv106–8.

Maini RN, Taylor PC. 2000. Anti-cytokine therapy for rheumatoid arthritis. *Annu Rev Med* 51: 207–9.

Makowski L, Hotamisligil GS. 2004. Fatty acid binding proteins: the evolutionary crossroads of inflammatory and metabolic responses. *J Nutr* 134: 2464–68S.

Malnick SDH, Knobler H. 2006. The medical complications of obesity. *Q J Med* 99: 565–79.

Manini TM, Everhart JE, Patel KV, Schoeller DA, Colbert LH, Visser M, Tylavsky F, Bauer DC, Goodpaster BH, Harris TB. 2006. Daily activity energy expenditure and mortality among older adults. *JAMA* 296: 171–79.

Mankin HJ, Dorfman H, Lippiello L, Zarins A. 1971. Biochemical and metabolic abnormalities in articular cartilage from osteoarthritic hips II. Correlation of morphology and biochemical and metabolic data. *J Bone Joint Surg Am* 53: 523–37.

Mann DL. 2003. Stress-activated cytokines and the heart: From adaptation to maladaptation. *Annu Rev Physiol* 65: 81–101.

Manton KG, Gu XL. 2001. Changes in the prevalence of chronic disability in the United States black and nonblack population above 65 from 1982–1999. *Proc Natl Acad Sci USA* 98: 6354–59.

Marcell TJ. 2003. Sarcopenia: causes, consequences, and preventions. *J Gerontol Med Sci* 58A: 911–16.

Marcum JA. 2005. From the molecular genetics revolution to gene therapy: translating basic research into medicine. *J Lab Clin Med* 146: 312–16.

Marcus AJ, Broekman MJ, Pinsky DJ. 2002. COX inhibotors and thromboregulation. *N Engl J Med* 347: 1025–26.

Martin GM. 2005. Genetic modulation of senescent phenotypes in *homo sapiens*. *Cell* 120: 523–32.

Martin GM. 2006. Keynote lecture: an update on the what, why and how questions of ageing. *Exp Gerontol* 41: 460–63.

Martin GM, LaMarco K, Strauss E, Kelner KL. 2003. Research on aging: the end of the beginning. *Science* 299: 1339–41.

Martin I, Grotewiel MS. 2006. Oxidative damage and age-related functional declines. *Mech Ageing Dev* 127: 411–23.

Martin JA, Brown TD, Heiner AD, Buckwalter JA. 2004. Chondrocyte senescence, joint loading, and osteoarthritis. *Clin Orthop Relat Res* 427 (Suppl.): S96–103.

Martin TJ. 2005. Osteoblast-derived PTHrP is a physiological regulator of bone formation. *J Clin Invest* 115: 2322–24.

Masoro EJ. 2006. Caloric restriction and aging: controversial issues. *J Gerontol A Biol Sci Med Sci* 61: 14–19.

Masse PG, Tranchant CC, Dosy J, Donovan SM. 2005. Coexistence of osteoporosis and cardiovascular disease risk factors in apparently health, untreated postmenopausal women. *Int J Vitam Nutr Res* 75: 97–106.

Mauras N, Haymond MW. 2005. Are the metabolic effects of GH and IGF-1 separable? *Growth Horm IGF Res* 15: 19–27.

Mazieres B, Garnero P, Guéguen A, Abbal M, Berdah L, Lequesne M, Nguyen M, Salles J-P, Vignon E, Dougados M. 2006. Molecular markers of cartilage breakdown and synovitis at baseline as predictors of structural progression of hip osteoarthritis. The ECHODIAH Cohort. *Ann Rheum Dis* 65: 354–59.

Mazziotti G, Bianchi A, Bonadonna S, Nuzzo M, Cimino V, Fusco A, De Marinis L, Giustina A. 2006. Increased prevalence of radiological spinal deformities in adult patients with GH deficiency: Influence of GH replacement therapy. *J Bone Miner Res* 21: 520–28.

McClung MR. 2005. Osteopenia: to treat or not to treat? *Ann Intern Med* 142: 796–97.

McClung MR, Lewiecki EM, Cohen SB, Bolognese MA, Woodson GC, Moffett AH, et al. 2006. Denosumab in postmenopausal women with low bone mineral density. *N Engl J Med* 354: 821–31.

McGinnis JM. 1988. The Tithonus syndrome: health and aging in America. In: Chernoff R, Lipschitz DA (eds.), *Health promotion and disease prevention in the elderly*. New York: Raven Press, pp. 1–5.

McMillen IC, Robinson JS. 2005. Developmental origins of the metabolic syndrome: prediction, plasticity, and programming. *Physiol Rev* 85: 571–633.

McTernan PG, Kusminski CM, Kumar S. 2006. Resistin. *Curr Opin Lipidol* 17: 170–75.

Mellström D, Johnell O, Ljunnggren Ö, Eriksson AL, Lorentzon M, Mallmin H, Holmberg A, Redlund-Johnell I, Orwoll E, Ohlsson C. 2006. Free testosterone is an independent predictor of BMD and prevalent fractures in elderly men: MrOS Sweden. *J Bone Miner Res* 21: 529–35.

Melton LJ, III. 1995. How many women have osteoporosis now? *J Bone Miner Res* 10: 175–77.

Melton LJ, Khosla S, Crowson CS, O'Connor MK, O'Fallon WM, Riggs BL. 2000. Epidemiology of sarcopenia. *J Am Geriatr Soc* 48: 625–30.

Merriam GR, Carney C, Smith LC, Kletke M. 2004. Adult growth hormone deficiency: current trends in diagnosis and dosing. *J Pediatr Endocrinol Metab* 17 (Suppl. 4): 1307–20.

Messier SP, Gutekunst DJ, Davis C, DeVita P. 2005. Weight loss reduces knee-joint loads in overweight and obese older adults with knee osteoarthritis. *Arthritis Rheum* 52: 2026–32.

Meyer MC, Rastogi P, Beckett CS, McHowat J. 2005. Phospholipase A2 inhibitors as potential anti-inflammatory agents. *Curr Pharm Des* 11: 1301–12.

Miida T, Takahashi A, Tanabe N, Ikeuchi T. 2005. Can statin therapy really reduce the risk of Alzheimer's disease and slow its progression? *Curr Opin Lipidol* 16: 619–23.

Miller RA. 1997. When will the biology of aging become useful? Future landmarks in biomedical gerontology. *J Am Geriatr Soc* 45: 1258–67.

Miller RA, Chang Y, Galecki AT, Al-Regaiey K, Kopchick JJ, Bartke A. 2002. Gene expression patterns in calorically restricted mice: Partial overlap with long-lived mutant mice. *Mol Endocrinol* 16: 2657–66.

Min H, Morony S, Sarosi I, Dunstan CR, Cappareli C, Scully S, Van G, Kaufman S, Kostenuik PJ, Lacey DL, Boyle WJ, Simonet WS. 2000. Osteoprotegerin reverses osteoporosis by inhibiting endosteal osteoclasts and prevents vascular calcification by blocking a process resembling osteoclastogenesis. *J Exp Med* 192: 463–74.

Mobbs CV. 1996. Neuroendocrinology of aging. In: Schneider EL, Rowe JW (eds.), *Handbook of the biology of aging,* 4th ed. New York: Academic Press, pp. 234–82.

Molitch ME, Clemmons DR, Malozowski S, Merriam GR, Shalet SM, Vance ML, for the Endocrine Society's Guidelines Subcommittee. 2006. Evaluation and treatment of adult growth hormone deficiency: an Endocrine Society Clinical Practice Guideline. *J Clin Endocrinol Metab* 91: 1621–34.

Moller DE, Kaufman KD. 2005. Metabolic syndrome: a clinical and molecular perspective. *Annu Rev Med* 56: 45–62.

Mooi WJ, Peeper DS. 2006. Oncogene-induced cell senescence: halting on the road to cancer. *N Engl J Med* 355: 1037–46.

Morand EF. 2005. New therapeutic target in inflammatory disease: macrophage migration inhibitory factor. *Intern Med J* 35: 419–26.

Morisco C, Lembo G, Trimarco B. 2006. Insulin resistance and cardiovascular risk: new insights from molecular and cellular biology. *Trends Cardiovasc Med* 16: 183–88.

Morley JE. 1997. Anorexia of aging: physiologic and pathologic. *Am J Clin Nutr* 66: 760–73.

Morley JE. 1999. Growth hormone: fountain of youth or death hormone. *J Am Geriatr Soc* 47: 1475–76.

Morley JE. 2001. Decreased food intake with aging. *J Gerontol Biol Sci Med Sci* 56A: 81–88.

Morley JE. 2003a. Editorial: anorexia and weight loss in older persons. *J Gerontol Med Sci* 58A: 131–37.

Morley JE. 2003b. Editorial: Sarcopenia revisited. *J Gerontol Med Sci* 58A: 909–10.

Morley JE. 2004. A brief history of geriatrics. *J Gerontol A: Biol Sci Med Sci* 59: 1132–52.

Morley JE, Baumgartner RN. 2004. Cytokine-related aging process. *J Gerontol Med Sci* 59A: 924–29.

Morley JE, Haren MT, Kim MJ, Kevorkian R, Perry HM III. 2005. Testosterone, aging, and quality of life. *J Endocrinol Invest* 28: 76–80.

Morley JE, Kraenzle D. 1994. Causes of weight loss in a community nursing home. *J Am Geriatr Soc* 42: 583–85.

Morley JE, Perry HM III, Baumgartner RP, Garry PJ. 1999. Leptin, adipose tissue and aging: is there a role for testosterone? *J Gerontol: Biol Sci* 54A: B108–9.

Morley JE, Perry HM III, Miller DK. 2002. Editorial: Something about frailty. *J Gerontol Med Sci* 57A: M698–704.

Morton GJ, Cummings DE, Baskin DG, Barsh GS, Schwartz MW. 2006. Central nervous system control of food intake and body weight. *Nature* 443: 289–95.

Muhlberg W, Sieber C. 2004. Sarcopenia and frailty in geriatric patients: implications for training and prevention. *Z Gerontol Geriatr* 37: 2–8.

Mui LW, Haramati LB, Alterman DD, Haramati N, Zelefsky MN, Hamerman D. 2003. Evaluation of vertebral fractures on lateral chest radiographs of inner-city postmenopausal women. *Calcif Tissue Int* 73: 550–54.

Mundy GR. 2003. Endothelin-1 and osteoblastic metastasis. *Proc Natl Acad Sci USA* 100: 10588–89.

Mundy GR. 2006. Nutritional modulators of bone remodeling during aging. *Am J Clin Nutr* 83: 427–30S.

Murphy JM, Dixon K, Beck S, Fabian D, Feldman A, Barry F. 2002. Reduced chondrogenic and adipogenic activity of mesenchymal stem cells from patients with advanced osteoarthritis. *Arthritis Rheum* 46: 704–13.

Musi N, Goodyear LJ. 2006. Insulin resistance and improvements in signal transduction. *Endocrine* 29: 73–80.

Mutch NJ, Wilson HM, Booth NA. 2001. Plasminogen activator inhibitor-1 and haemostasis in obesity. *Proc Nutr Soc* 60: 341–47.

Nair KS. 2005. Aging muscle. *Am J Clin Nutr* 81: 953–63.

Najjar SS, Scuteri A, Lakatta EG. 2005. Arterial aging: is it an immutable cardiovascular risk factor? *Hypertension* 46: 454–62.

Nakashima K, Zhou X, Kunkel G, Zhang Z, Deng JM, Behringer RR, de Crombrugghe B. 2002. The novel zinc finger-containing transcription factor Osterix is required for osteoblast differentiation and bone formation. *Cell* 108: 17–29.

Naslund J, Haroutunian V, Mohs R, Davis KL, Davies P, Greengard P, Buxbaum JD. 2000. Correlation between elevated levels of amyloid beta-peptide in the brain and cognitive decline. *JAMA* 283(12): 1571–77.

Nass R, Thorner MO. 2004. Life extension versus improving quality of life. *Best Prac Res Clin Endocrinol Metab* 18: 381–91.

Neel JV. 1999. The "thrifty genotype" in 1998. *Nutr Rev* 57: S2–9.

Neels JG, Olefsky JM. 2006. Inflamed fat: what starts the fire? *J Clin Invest* 116: 33–35.

Nestorov I. 2005. Clinical pharmacokinetics of TNF antagonists: how do they differ? *Semin Arthritis Rheum* 34: 12–18.

Neve RL, McPhie DL, Chen Y. 2000. Alzheimer's disease: a dysfunction of the amyloid precursor protein(1). *Brain Res* 886(1–2): 54–66.

Newman AB, Arnold AM, Naydeck BL, Fried LP, Burke GL, Enright P, et al. 2003. "Successful Aging." Effect of subclinical cardiovascular disease. *Arch Intern Med* 163: 2315–22.

Newman AB, Kupelian V, Visser M, Simonsick EM, Goodpaster BH, Kritchevsky SB, Tylavsky FA, Rubin SM, Harris TB. 2006. Strength, but not muscle mass, is associated with mortality in the health aging and body composition study cohort. *J Gerontol A Biol Sci Med Sci* 61: 72–77.

Newman AB, Yanez D, Harris T, Duxbury A, Enright PJ, Fried LP. 2001. Weight change in old age and its association with mortality. *J Am Geriatr Soc* 49:1309–18.

Nigro J, Osman N, Dart AM, Little PJ. 2006. Insulin resistance and atherosclerosis. *Endocr Rev* 27: 242–59.

Nikolić M, Bajek S, Bobinac D, Vranić TŠ, Jerković R. 2005. Aging of human skeletal muscles. *Coll Antropol* 29: 67–70.

Nishizawa Y, Nakamura T, Ohta H, Kushida K, Gorai I, Shiraki M, Fukunaga M, Hosoi T, Mili T, Chaki O, Ichimura S, Nakatsuka K, Miura M, Committee on the Guidelines for the Use of Biochemical Markers of Bone Turnover in Osteoporosis Japan Osteoporosis Society. 2005. Guidelines for the use of biochemical markers of bone turnover in osteoporosis (2004). *J Bone Miner Metab* 23: 97–104.

Nissen SE, Nicholls SJ, Sipahi I, Libby P, Raichlen JS, Ballantyne CM, Davignon J, Erbel R, Fruchart JC, Tardif JC, Schoenhagen P, Crowe T, Cain V, Wolski K, Goormastic M, Tuzcu EM. 2006. Effect of very high-intensity statin therapy on regression of coronary atherosclerosis. The ASTEROID trial. *JAMA* 295: 1556–65.

Noble W, Olm V, Takata K, Casey E, Mary O, Meyerson J, Gaynor K, LaFrancois J, Wang L, Kondo T, Davies P, Burns M, Veeranna, Nixon R, Dickson D, Matsuoka Y, Ahlijanian M, Lau LF, Duff K. 2003. Cdk5 is a key factor in tau aggregation and tangle formation in vivo. *Neuron* 38(4): 555–65.

Odierna E, Zeleznik J. 2003. Pressure ulcer education: a pilot study of the knowledge and clinical confidence of geriatric fellows. *Adv Skin Wound Care* 16: 26–30.

Ogden CL, Carroll MD, Curtin LR, McDowell MA, Tabak CJ, Flegal KM. 2006. Prevalence of overweight and obesity in the United States, 1999–2004. *JAMA* 295: 1549–55.

Ogura Y, Sutterwala FS, Flavell RA. 2006. The inflammasome: first line of the immune response to cell stress. *Cell* 126: 659–62.

Okamoto Y, Kihara S, Funahashi T, Matsuzawa Y, Libby P. 2006. Adiponectin: a key adipocytokine in metabolic syndrome. *Clin Sci* 110: 267–78.

O'Keefe JH, Jr, Cordain L, Jones PG, Abuissa H. 2006. Coronary artery disease prognosis and C-reactive protein levels improve in proportion to percent lowering of low-density lipoprotein. *Am J Cardiol* 98: 135–39.

Olofsson H, Byberg L, Mohsen R, Melhus H, Lithell H, Michaelsson K. 2005. Smoking and the risk of fracture in older men. *J Bone Miner Res* 20: 1208–15.

Olshansky SJ, Cassel CK. 1997. Implications of the accrual of chronic non-fatal conditions in very elderly persons. In: Hamerman D (ed.), *Osteoarthritis. public health implications for an aging population*. Baltimore: Johns Hopkins University Press, pp. 15–29.

Olshansky SJ, Hayflick L, Perls TT. 2004. Anti-aging medicine: the hype and the reality, Part I. *J Gerontol Biol Sci* 59A: 513–14.

Olson RE. 2000. Editorial. Osteoporosis and vitamin K intake. *Am J Clin Nutr* 71: 1031–32.

O'Rahilly S, Barroso I, Wareham NJ. 2005. Genetic factors in type 2 diabetes: the end of the beginning? *Science* 307: 370–72.

O'Rahilly S, Savill J. 1997. Science, medicine, and the future. Non-insulin dependent diabetes mellitus: the gathering storm. *BMJ* 314: 955–59.

Ordovas JM, Mooser V. 2005. Genes, lipids and aging: is it all accounted for by cardiovascular disease risk? *Curr Opin Lipidol* 16: 121–26.

Orlandi A, Bochaton-Piallat M-L, Gabbiani G, Spagnoli LG. 2006. Aging, smooth muscle cells and vascular pathobiology: Implications for atherosclerosis. *Atherosclerosis* 188: 221–30.

Orr R, de Vos NJ, Singh NA, Ross DA, Stavrinos TM, Fiatarone-Singh MA. 2006. Power training improves balance in healthy older adults. *J Gerontol A Biol Sci Med Sci* 61: 78–85.

Ott SM. 2004. Diet for the heart or the bone: a biological tradeoff. *Am J Clin Nutr* 79: 4–5.

Ott SM. 2005. Long-term safety of bisphosphonates. *J Clin Endocrinol Metab* 90: 1897–99.

Ouchi N, Kihara S, Funahashi T, Matsuzawa Y, Walsh K. 2003. Obesity, adiponectin and vascular inflammatory disease. *Curr Opin Lipidol* 14: 561–66.

Packer L. 2005. Oxidants and antioxidants: mechanisms of action and regulation of gene expression by bioflavinoids. In: Surh Y-J, Packer L (eds.), *Oxidative stress, inflammation, and health.* New York: Taylor and Francis, pp. 1–20.

Paffen E, DeMaat MP. 2006. C-reactive protein in atherosclerosis: a casual factor? *Cardiovasc Res* 71: 30–39.

Pagotto U, Marsicano G, Cota D, Lutz B, Pasquali R. 2006. The emerging role of the endocannabinoid system in endocrine regulation and energy balance. *Endocr Rev* 27: 73–100.

Papanicolaou DA, Wilder RL, Manolagas SC, Chrousos GP. 1998. The pathophysiologic roles of interleukin-6 in human disease. *Ann Intern Med* 128: 127–37.

Parfitt AM. 1988. Bone remodelling: relationship to the amount and structure of bone, and the pathogenesis and prevention of fractures. In: Riggs BL, Melton LJ (eds.), *Osteoporosis. Etiology, diagnosis, management.* New York: Raven Press, pp. 45–93.

Park IK, Morrison SJ, Clarke MF. 2004. Bmi1, stem cells, and senescence regulation. *J Clin Invest* 113: 175–79.

Park SC, Huang ES, Kim HS, Park WY (eds.). 2001. *Healthy aging for functional longevity. Molecular and cellular interaction in senescence.* New York: New York Academy of Sciences.

Partridge L, Brand MD. 2005. Special issue on dietary restriction: dietary restriction, longevity and ageing-the current state of our knowledge and ignorance. *Mech Ageing Dev* 126: 911–12.

Partridge L, Gems D. 2002. Mechanisms of ageing: public or private? *Nat Rev Genet* 3: 165–75.

Pasceri V, Yeh ETH. 1999. A tale of two diseases: atherosclerosis and rheumatoid arthritis. *Circulation* 100: 2124–26.

Pasco JA, Henry MJ, Sanders KM, Kotowicz MA, Seeman E, Nicholson GC. 2004. β-adrenergic blockers reduce the risk of fracture partly by increasing bone mineral density: Geelong Osteoporosis Study. *J Bone Miner Res* 19: 19–24.

Patel MB. Arden NK, Masterson LM, Phillips DI, Swaminathan R, Syddall HE, Byrne CD, Wood PJ, Cooper C, Holt RI; Hertfordshire Cohort Study Group. 2005. Investigating the role of the growth hormone insulin-like growth factor (GH-IGF) axis as a determinant of male bone mineral density (BMD). *Bone* 37: 833–41.

Patel MS, Karsenty G. 2002. Regulation of bone formation and vision by LRP5. *N Engl J Med* 246: 1572–73.

Patrignani P, Tacconelli S, Sciulli MG, Capone ML. 2005. New insights into COX-2 biology and inhibition. *Brain Res Brain Res Rev* 48: 352–59.

Peat G, McCarney R, Croft P. 2001. Knee pain and osteoarthritis in older adults: a review of community burden and current use of primary health care. *Ann Rheum Dis* 60: 91–97.

Pei L, Tontonoz P. 2004. Fat's loss is bone's gain. *J Clin Invest* 113: 805–6.

Pelicci PG. 2004. Do tumor suppressive mechanisms contribute to organism aging by inducing stem cell senescence? *J Clin Invest* 113: 4–7.

Pelletier JP, Martel-Pelletier J, Abramson SB. 2001. Osteoarthritis, an inflammatory disease. *Arthritis Rheum* 44: 1237–47.

Perazella MA, Reilly RF. 2003. Chronic kidney disease: A new classification and staging system. *Hosp Physician* 45: 18–22.

Pereg D, Lishner M. 2005. Non-steroidal anti-inflammatory drugs for the prevention and treatment of cancer. *J Intern Med* 258: 115–23.

Perez-Tilve D, Noguerias R, Mallo F, Benoit S, Tschoep M. 2006. Gut hormones: Ghrelin, PYY, and GLP-1 in the regulation of energy balance and metabolism. *Endocrine* 29: 61–71.

Perlin JB, Pogach LM. 2006. Improving the outcomes of metabolic conditions: managing momentum to overcome clinical inertia. *Ann Intern Med* 144: 525–27.

Perls T. 2006. The different paths to 100. *Am J Clin Nutr* 83: 484–87S.

Perls TT, Wilmoth J, Levenson R, Drinkwater M, Cohen M, Bogan H, Joyce E, Brewster S, Kunkel L, Puca A. 2002. Life-long sustained mortality advantage of siblings of centenarians. *Proc Natl Acad Sci USA* 99: 8442–47.

Perrini S, Laviola L, Natalicchio A, Giorgino F. 2005. Associated hormonal declines in aging: DHEAS. *J Endocrinol Invest* 28: 85–93.

Petersen KF, Shulman GI. 2006. Etiology of insulin resistance. *Am J Med* 119(5A): 10–16S.

Phelan EA, Larson EB. 2002. "Successful Aging": Where next? *J Am Geriatr Soc* 1306–8.

Phimphilai M, Zhao Z, Boules H, Roca H, Franceschi RT. 2006. BMP signaling is required for RUNX2-dependent induction of the osteoblast phenotype. *J Bone Miner Res* 21: 637–46.

Picard F, Guarente L. 2005. Molecular links between aging and adipose tissue. *Int J Obes* 29: S36–39.

Pischon T, Girman CJ, Hotamisligil GS, Rifai N, Hu FB, Rimm ER. 2004. Plasma adiponectin levels and risk of myocardial infarction in men. *JAMA* 291: 1730–37.

Plebani M, Laposata M. 2006. Translational research involving new biomarkers of disease. *Am J Clin Pathol* 126: 169–71.

Plum L, Belgardt BF, Brüning JC. 2006. Central insulin action in energy and glucose homeostasis. *J Clin Invest* 116: 1761–66.

Pocai A, Lam TKT, Obici S, Gutierrez-Juarez R, Muse ED, Arduini A, Rossetti L. 2006. Restoration of hypothalamic lipid sensing normalizes energy and glucose homeostasis in overfed rats. *J Clin Invest* 116: 1081–91.

Pocai A, Obici S, Schwartz G, Rossetti L. 2005. A brain-liver circuit regulates glucose homeostasis. *Cell Metab* 1: 53–61.

Pollard JW. 1997. Role of colony-stimulating factor-1 in reproduction and development. *Mol Reprod Dev* 46: 54–61.

Pollard JW, Stanley ER. 1996. Pleiotropic roles for CSF-1 in development defined by the mouse mutation osteopetrotic (op). *Adv Dev Biochem* 4: 153–93.

Porte D Jr. 2006. Central regulation of energy homeostasis. The key role of insulin. Diabetes 55: S155–60.

Prentki M, Nolan CJ. 2006. Islet β cell failure in type 2 diabetes. *J Clin Invest* 116: 1802–12.

Price JS, Waters JG, Darrah C, Pennington C, Edwards DR, Donell ST, Clark IM. 2002. The role of chondrocyte senescence in osteoarthritis. *Aging Cell* 1: 57–65.

Prins JB. 2002. Adipose tissue as an endocrine organ. *Best Pract Res Clin Endocrinol Metab* 16: 639–51.

Pritzker KPH, Gay S, Jiminez SA, Ostergaard K, Pelletier JP, Revell PA, Salter D, van den Berg WB. 2005. Osteoarthritis cartilage histopathology: grading and staging. *Osteoarthritis Cartilage* 14: 13–29.

Provot S, Schipani E. 2005. Molecular mechanisms of endochondral bone development. *Biochem Biophys Res Comm* 328: 658–65.

Quehenberger O. 2005. Molecular mechanisms regulating monocyte recruitment in atherosclerosis. *J Lipid Res* 46: 1582–90.

Rader DJ, Puré E. 2005. Lipoproteins, macrophage function, and atherosclerosis: beyond the foam cell? *Cell Metab* 1: 223–30.

Radin EL. 2004. Who gets osteoarthritis and why? *J Rheumatol* Suppl. 70: 10–15.

Raisz LG. 2005. Pathogenesis of osteoporosis: concepts, conflicts, and prospects. *J Clin Invest* 115: 3318–25.

Rajala MW, Scherer PE. 2003. Mini-review: the adipocyte—at the crossroads of energy homeostasis, inflammation, and atherosclerosis. *Endocrinology* 144: 3765–73.

Ranke MB. 2005. Insulin-like growth factor-1 treatment of growth disorders, diabetes mellitus and insulin resistance. *Trends Endocrinol Metab* 16: 190–97.

Rawlings JS, Rosler KM, Harrison DA. 2004. The JAK/STAT signaling pathway. *J Cell Sci* 117: 1281–83.

Reaven GM. 2005a. Why Syndrome X? From Harold Himsworth to the insulin resistance syndrome. *Cell Metab* 1: 9–14.

Reaven GM. 2005b. The metabolic syndrome: requiescat in pace. *Clin Chem* 51: 931–38.

Reaven GM. 2006. The metabolic syndrome: is this diagnosis necessary? *Am J Clin Nutr* 83: 1237–47.

Reckelhoff JF, Yanes LL, Iliescu R, Fortepiani LA, Granger JP. 2005. Testosterone supplementation in aging men and women: possible impact on cardiovascular-renal disease. *Am J Physiol Renal Physiol* 289: F941–48.

Reeves MJ, Rafferty AP. 2005. Healthy lifestyle characteristics among adults in the United States, 2000. *Arch Intern Med* 165: 854–57.

Reginster JY, Burlet N. 2006. Osteoporosis: a still increasing prevalence. *Bone* 38 (2 Suppl. 1): 4–9.

Reichlin S. 1999. Neuroendocrinology of infection and the innate immune system. *Rec Prog Horm Res* 54: 133–83.

Reinstein E, Ciechanover A. 2006. Narrative review: protein degradation and human diseases: the ubiquitin connection. *Ann Intern Med* 145: 676–84.

Reid IR, Cornish J, Baldock PA. 2006. Nutrition-related peptides and bone homeostasis. *J Bone Miner Res* 21: 495–500.

Reid IR, Tonkin A, Cannon CP. 2005. Comparison of the effects of pravastatin and atorvastatin on fracture incidence in the PROVE IT-TIMI 22 trial-secondary analysis of a randomized controlled trial. *Bone* 37: 190–91.

Reisin E, Alpert MA. 2005. The metabolic syndrome: An overview. *Am J Med Sci* 330: 263.

Reiss AB, Glass AD. 2006. Atherosclerosis: immune and inflammatory aspects. *J Investig Med* 54: 123–31.

Rejnmark L, Vestergaard P, Mosekilde L. 2006. Statin but not non-statin lipid lowering drugs decrease fracture risk: a nation-wide case-control study. *Calcif Tissue Int* 79: 27–36.

Resnick NM, Marcantonio ER. 1997. How should clinical care of the aged differ? *Lancet* 350: 1157–58.

Rhodes CJ. 2005. Type 2 diabetes: A matter of β-cell life and death? *Science* 307: 380–84.

Ribiero S, Horuk R. 2005. The clinical potential of chemokine receptor antagonists. *Pharmacol Ther* 107: 44–58.

Richardson A, Liu F, Adamo ML, Van Remmen H, Nelson JF. 2004. The role of insulin and insulin-like growth factor-1 in mammalian ageing. *Best Prac Res Clin Endocrinol Metab* 18: 393–406.

Rigas B, Kashfi K. 2005. Cancer prevention: a new era beyond cyclooxygenase-2. *J Pharmacol Exp Ther* 314: 1–8.

Rincon M, Rudin E, Barzilai N. 2005. The insulin/IGF-1 signaling in mammals and its relevance to human longevity. *Exp Gerontol* 40: 873–77.

Ristimaki A. 2004. Cyclooxygenase-2: from inflammation to carcinogenesis. *Novartis Found Symp* 256: 215–21.

Ritz E. 2006. Heart and kidney: Fatal twins? *Am J Med* 119(5A): 31–39S.

Ritz E, Fujita T, Shulman GI. 2006. A new dawn in cardiovascular protection III: The challenge of end-organ protection in high-risk patients. *Am J Med* 119(5A): 1–2S.

Ritz P, Berrut G. 2005. Mitochondrial function, energy expenditure, aging and insulin resistance. *Diabetes Metab* 21: 5S65–71.

Roberts AW, Thomas A, Rees A, Evans M. 2003. Peroxisome proliferator-activated receptor-gamma agonists in atherosclerosis: current evidence and future directions. *Curr Opin Lipidol* 14: 567–73.

Roberts DCK. 2006. Metabolic syndrome: an outmoded concept? *Curr Opin Lipidol* 17: 192–94.

Roberts WC. 2006. Atherosclerosis: its cause and its prevention. Am J Cardiol 98: 1550–5.

Roche HM, Phillips C, Gibney MJ. 2005. The metabolic syndrome: the crossroads of diet and genetics. *Proc Nutr Soc* 64: 371–77.

Rockwood K. 2005. What would make a definition of frailty successful? *Age Ageing* 34: 432–34.

Roden DM, Altman RB, Benowitz NL, Flockhart DA, Giacomini KM, Johnson JA, Krauss RM, McLeod HL, Ratain MJ, Relling MV, Ring HZ, Shuldiner AR, Weinshilboum RM, Weiss ST. 2006. Pharmacogenomics: challenges and opportunities. *Ann Intern Med* 145: 749–57.

Rodriguez-Moranta F, Castells A. 2005. Mechanisms of colon cancer prevention with and beyond COX-2 inhibition. *Curr Top Med Chem* 5: 505–16.

Rogers MJ, Frith JC, Luckman P, Coxon FP, Benford HL, Mönkkönen J, Auriola S, Chilton KM, Russell RGG. 1999. Molecular mechanisms of action of bisphosphonates. *Bone* 5 (Suppl.): 73–79S.

Roman-Blas JA, Jimenez SA. 2006. NF-kappaB as a potential therapeutic target in osteoarthritis and rheumatoid arthritis. *Osteoarthritis Cartilage* 14: 839–48.

Rondinone CM. 2006. Adipocyte-derived hormones, cytokines, and mediators. *Endocrine* 29: 81–90.

Roodman GD. 2004. Mechanisms of bone metastases. *N Engl J Med* 350: 1655–64.

Rosen CJ, Brown SA. 2005. A rational approach to evidence gaps in the management of osteoporosis. *Am J Med* 118: 1183–89.

Rosen CJ, Wüster C. 2003. Growth hormone rising: Did we quit too quickly? *J Bone Miner Res* 18: 406–9.

Rosen ED, MacDougald OA. 2006. Adipocyte differentiation from the inside out. Nat Rev Mol Cell Biol 7: 885–96.

Rosen ED, Spiegelman BM. 2006. Adipocytes as regulators of energy balance and glucose homeostasis. Nature 444: 847–53.

Rosenberg IH. 1989. Epidemiologic and methodologic problems in determining nutritional status of older persons (Summary comments). *Am J Clin Nutr* 50 (Suppl.): 1231–33.

Rosenfeld RG. 2006. Introduction. *Horm Res* 65 (Suppl. 1): 1–2.

Roses AD. 1998. Apolipoprotein E and Alzheimer's disease: the tip of the susceptibility iceberg. *Ann N Y Acad Sci* 855: 738–43.

Ross FP. 2000. RANKing the importance of measles virus in Paget's disease. *J Clin Invest* 105: 555–58.

Ross FP, Christiano AM. 2006. Nothing but skin and bone. *J Clin Invest* 116: 1140–49.

Ross PD. 1997. Clinical consequences of vertebral fractures. *Am J Med* 103: 30–42S.

Ross R. 1999. Atherosclerosis: an inflammatory disease. *N Engl J Med* 340: 115–26.

Roubenoff R. 1997. Inflammatory and hormonal mediators of cachexia. *J Nutr* 127: 1014–16S

Roubenoff R. 2003a. Catabolism of aging: is it an inflammatory process? *Curr Opin Clin Nutr Metab Care* 6: 295–99.

Roubenoff R. 2003b. Sarcopenia: Effects on body composition and function. *J Gerontol Med Sci* 58A: 1012–17.

Rovelet-Lecrux A, Hannequin D, Raux G, Le Meur N, Laquerriere A, Vital A, Dumanchin C, Feuillette S, Brice A, Vercelletto M, Dubas F, Frebourg T, Campion D. 2006. APP locus duplication causes autosomal dominant early-onset Alzheimer disease with cerebral amyloid angiopathy. *Nat Genet* 38(1): 24–26.

Rowe JW, Kahn RL. 1987. Human aging: usual and successful. *Science* 237: 143–49.

Rowe JW, Kahn RL. 1997. Successful aging. *The Gerontologist* 37: 433–40.

Ruano-Ravina A, Jato Diaz M. 2006. Autologous chondrocyte implantation: a systematic review. *Osteoarthritis Cartilage* 14: 47–51.

Rubenstein WS, Roy HK. 2005. Practicing medicine at the front line of the genomic revolution. *Arch Intern Med* 165: 1815–17.

Rudman D, Feller AG, Nagraj HS, Gergans GA, Lalitha PY, Goldberg AF, Schlenker RA, Cohn L, Rudman IW, Mattson DE. 1990. Effects of human growth hormone in men over 60 years. *N Engl J Med* 323: 1–6.

Rudolph KL, Chang S, Lee H-W, Blases M, Gottlieb GJ, Greider C, DePinho R. 1999. Longevity, stress response, and cancer in aging telomerase-deficient mice. *Cell* 96: 701–12.

Rule AD, Bergstralh EJ, Slezak JM, Bergert J. Larson TS. 2006. Glomerular filtration rate estimated by cystatin C among different clinical presentations. *Kidney Int* 69: 399–405.

Ruocco MG, Karin M. 2005. IKK{beta} as a target for treatment of inflammation induced bone loss. *Ann Rheum Dis* 64 (Suppl. 4): iv81–85.

Russell M, Berardi P, Gong W, Riabowol K. 2006. Grow-ING, Age-ING and Die-ING: ING proteins link cancer, senescence and apoptosis. *Exp Cell Res* 312: 951–61.

Russell RG, Rogers MJ. 1999. Bisphosphonates: from the laboratory to the clinic and back again. *Bone* 25: 97–106.

Rzepecki P, Nagel-Steger L, Feuerstein S, Linne U, Molt O, Zadmard R, Aschermann K, Wehner M, Schrader T, Riesner D. 2004. Prevention of Alzheimer's disease-associated Abeta aggregation by rationally designed nonpeptidic beta-sheet ligands. *J Biol Chem* 279(46): 47497–505.

Rzonca SO, Suva LJ, Gaddy D, Montague DC, Lecka-Czernik B. 2003. Bone is a target for the antidiabetic compound rosiglitazone. *Endocrinology* 145: 401–6.

Sambrook P, Cooper C. 2006. Osteoporosis. *Lancet* 367: 2010–18.

Sambrook PN, Chen CJS, March L, Cameron ID, Cumming RG, Lord SR, Simpson JM, Seibel MJ. 2006. High bone turnover is an independent predictor of mortality in the frail elderly. *J Bone Miner Res* 21: 549–55.

Sampietro T, Bigazzi F, Dal Pino B, Puntoni M, Bionda A. 2006. HDL: the 'new' target of cardiovascular medicine. *Int J Cardiol* 108: 143–54.

Sandell LJ, Aigner T. 2001. Articular cartilage and changes in arthritis: Cell biology of osteoarthritis. *Arthritis Res* 3: 107–13.

Sandhofer A, Kaser S, Ritsch A, Laimer M, Engl J, Paulweber B, Patsch JR, Ebenbichler CF. 2006. Cholesteryl ester transfer protein in metabolic syndrome. *Obesity* 14: 812–18.

Sapi E. 2004. The role of CSF-1 in normal physiology of mammary gland and breast cancer: an update. *Exp Biol Med* 229: 1–11.

Sapi E, Kacinski BM. 1999. The role of CSF-1 in normal and neoplastic breast physiology. *Proc Soc Exp Biol Med* 220: 1–8.

Sarafidis PA, Nilsson PM. 2006. The metabolic syndrome: a glance at its history. *J Hypertens* 24: 621–26.

Sato T, Dohmae N, Qi Y, Kakuda N, Misonou H, Mitsumori R, Maruyama H, Koo EH, Haass C, Takio K, Morishima-Kawashima M, Ishiura S, Ihara Y. 2003. Potential link between amyloid beta-protein 42 and C-terminal fragment gamma 49–99 of beta-amyloid precursor protein. *J Biol Chem* 278(27): 24294–301.

Satyanarayana A, Greenberg RA, Schaetzlein S, Buer J, Masutomi K, Hahn WC, Zimmermann S, Martens U, Manns MP, Rudolph KL. 2004. Mitogen stimulation cooperates with telomere shortening to activate DNA damage responses and senescence signaling. Mol Cell Biol 24: 5459–74.

Savage R. 2005. Cyclo-oxygenase-2 inhibitors: when should they be used in the elderly? *Drugs Aging* 22: 185–200.

Sayer AA, Syddall HE, Gilbody HJ, Dennison EM, Cooper C. 2004. Does sarcopenia originate in early life? Findings from the Hertfordshire cohort study. *J Gerontol A Biol Sci Med Sci* 59: M930–34.

Schaap LA, Pluijm SM, Smit JH, van Schoor NM, Visser M, Goren LJ, Lips P. 2005. The association of sex hormone levels with poor mobility, low muscle strength and incidence of falls among older men and women. *Clin Endocrinol* 63: 152–60.

Schaffler A, Muller-Ladner U, Scholmerich J, Buchler C. Role of adipose tissue as an inflammatory organ in human diseases. *Endocr Rev* 27: 449–67.

Scherer PE. 2006. Lilly Lecture 2005. Adipose tissue. From lipid storage compartment to endocrine organ. *Diabetes* 55: 1537–45.

Schlienger RG, Kraenzlin ME, Jick SS, Meier Cr. 2004. Use of β-blockers and risk of fracture. *JAMA* 292: 1326–32.

Schneider JG, Semenkovich CF. 2005. PPARα: savior or savage. *Cell Metab* 2: 341–42.

Schoelson SE, Lee J, Yuan M. 2003. Inflammation and the IKKβ/IκB/NF-κβ axis in obesity: and diet-induced insulin resistance. *Int J Obes Relat Metab Disord* 27 (Suppl. 3): S49–52.

Schoppet M, Preissner KT, Hofbauer LC. 2002. RANK ligand and osteoprotegerin: Paracrine regulators of bone metabolism and vascular function. *Arterioscler Thromb Vasc Biol* 22: 549–53.

Schubert M. Jockenhovel F. 2005. Late-onset hypogonadism in the aging male (LOH): definition, diagnostic and clinical aspects. *J Endocrinol Invest* 28: 23–27.

Schuurmans H, Steverink N, Lindenberg S, Frieswijk N, Slaets JP. 2004. Old or frail: what tells us more? *J Gerontol A Biol Sci Med Sci* 59: M962–65.

Schwartz AV, Nevitt MC, Brown BW Jr., Kelsey JL. 2005. Increased falling as a risk factor among older women: the study of osteoporotic fractures. *Am J Epidemiol* 161: 180–85.

Schwartz MW. 2006. Central nervous system regulation of food intake. *Obesity* 14 (Suppl.): 1–8S.

Schwartz MW, Niswender KD. 2004. Adiposity signaling and biological defense against weight gain: absence of protection or central hormone resistance? *J Clin Endocrinol Metab* 89: 5889–97.

Schwartz MW, Porte D Jr. 2005. Diabetes, obesity, and the brain. *Science* 307: 375–79.

Scirica BM, Morrow DA. 2006. Is C-reactive protein an innocent bystander or proatherogenic culprit? *Circulation* 113: 2128–34.

Scott DL, Kingsley GH. 2006. Tumor necrosis factor inhibitors for rheumatoid arthritis. *N Engl J Med* 355: 704–11.

Scranton RE, Young M, Lawler E, Solomon D, Gagnon D, Gaziano JM. 2005. Statin use and fracture risk: study of a US veterans population. *Arch Intern Med* 165: 2007–12.

Seeman E, Bianchi G, Adami S, Kanis J, Khosla S, Orwoll E. 2004. Osteoporosis in men—consensus is premature. *Calcif Tissue Int* 75: 120–22.

Seeman E. Delmas PD. 2006. Bone quality: the material and structural basis of bone strength and fragility. *N Engl J Med* 354: 2250–61.

Seeman TE, Singer BH, Rowe JW, Horwitz RI, McEwen BS. 1997. Price of adaptation: allostatic load and its health consequences. *Arch Intern Med* 157: 2259–68.

Seibel MJ, Robins SP, Bilezikian JP. 1997. Editorial: serum undercarboxylated osteocalcin and the risk of hip fracture. *J Clin Endocrinol Metab* 82: 717–18.

Seidenberg AB, An YH. 2004. Is there an inhibitory effect of COX-2 inhibitors on bone healing? *Pharmacol Res* 50: 151–56.

Selkoe D, Kopan R. 2003. Notch and Presenilin: regulated intramembrane proteolysis links development and degeneration. *Annu Rev Neurosci* 26: 565–97.

Selkoe DJ. 2001. Presenilin, Notch, and the genesis and treatment of Alzheimer's disease. *Proc Natl Acad Sci USA* 98(20): 11039–41.

Selkoe DJ. 2002. Alzheimer's disease is a synaptic failure. *Science* 298(5594): 789–91.

Selkoe DJ. 2004. Alzheimer disease: mechanistic understanding predicts novel therapies. *Ann Intern Med* 140: 627–38.

Semenkovich CF. 2006. Insulin resistance and atherosclerosis. *J Clin Invest* 116: 1813–22.

Semple RK, Chatterjee VKK, O'Rahilly S. 2006. PPARγ and human metabolic diseases. *J Clin Invest* 116: 581–89.

Sepulveda JL, Mehta JL. 2005. C-reactive protein and cardiovascular disease: a critical appraisal. *Curr Opin Cardiol* 20: 407–16.

Serluca FC, Fishman MC. 2006. Big, bad hearts: from flies to man. *Proc Natl Acad Sci USA* 103: 3947–48.

Serrano M, Lin AW, McCurrach ME, Beach D, Lowe SW. 1997. Oncogenic ras provokes premature cell senescence associated with accumulation of p53 and p16[INK4a]. *Cell* 88: 593–602.

Settersten RA Jr. 2005. Linking the two ends of life: what gerontology can learn from childhood studies. *J Gerontol B Psychol Sci Soc Sci* 60: S173–80.

Sharma AM. 2006. The obese patient with diabetes mellitus: from research targets to treatment options. *Am J Med* 119(5A): 17S–3S.

Sharpless NE, DePinho RA. 2004. Telomeres, stem cells, senescence, and cancer. *J Clin Invest* 113: 160–68.

Sharpless NE, DePinho RA. 2006. The mighty mouse: genetically engineered mouse models in cancer drug development. *Nat Rev Drug Discov* 5: 741–54.

Sheikine Y, Hansson GK. 2004. Chemokines and atherosclerosis. *Ann Med* 36: 98–118.

Shlipak MG, Katz R, Sarnak MJ, Fried LJ, Newman AB, Stehman-Breen C, Seliger SL, Kestenbaum B, Psaty B, Tracy RP, Siscovick DS. 2006. Cystatin C and prognosis for cardiovascular and kidney outcomes in elderly persons without chronic kidney disease. *Ann Intern Med* 145: 237–46.

Shoelson SE, Lee J, Goldfine AB. 2006. Inflammation and insulin resistance. *J Clin Invest* 116: 1793–1801.

Siebel MJ, Robins SP, Bilezikian JP. 1997. Editorial: serum undercarboxylated osteocalcin and the risk of hip fracture. *J Clin Endocrinol Metab* 82: 717–18.

Siiteri PK. 2005. The continuing saga of dehydroepiandrosterone (DHEA). *J Clin Endocrinol Metab* 90: 3795–96.

Simmons R. 2006. Developmental origins of adult metabolic disease. *Endocrinol Metab Clin North Am* 35: 193–204.

Simon AM, Manigrasso MB, O'Connor JP. 2002. Cyclo-oxygenase 2 function is essential for bone fracture healing. *J Bone Miner Res* 17: 963–76.

Simon SR. Fabiny AR, Kotch J. 2003. Geriatrics training in general internal medicine fellowship programs: current practice, barriers, and strategies for improvement. *Ann Intern Med* 139: 621–27.

Simonet WS, Lacey DL, Dunstan CR, Kelley M, Chang MS, Luthy R, Nguyen HQ, Wooden S, Bennett L, Boone T, Shimamoto G, DeRose M, Elliott R, Colombero A, Tan HL, Trail G, Sullivan J, Davy E, Bucay N, Renshaw-Gegg L, Hughes TM, Hill D, Pattison W, Campbell P, Sander S, Van G, Tarpley J, Derby P, Lee R, Boyle WJ. 1997. Osteoprotegerin: a novel secreted protein involved in the regulation of bone density. *Cell* 89: 309–19.

Sinclair DA. 2005. Toward a unified theory of caloric restriction and longevity regulation. *Mech Ageing Dev* 126: 987–1002.

Singh MF. 2003. Letters to the Editor: commentary on Dr. Kane's article, "The future of geriatrics: geriatrics at the crossroads." *J Gerontol A: Biol Sci Med Sci* 58: M92–93.

Singh NA, Stavrinos TM, Scarbek Y, Galambos A, Liber C, Fiatarone Singh MA. 2005. A randomized controlled trial of high versus low intensity weight training versus general practitioner care for clinical depression in older adults. *J Gerontol A Biol Sci Med Sci* 60: 765–76.

Siris ES, Miller PD, Barrett-Connor E, Faulkner Kg, Wehren LE, Abbott TA, Berger ML, Santora AC, Sherwood LM. 2001. Identification and fracture outcomes of undiagnosed low bone mineral density in postmenopausal women. Results from the National Osteoporosis Risk Assessment. *JAMA* 286: 2815–22.

Sjoholm Å, Nystrom T. 2006. Inflammation and the etiology of type 2 diabetes. *Diabetes Metab Res Rev* 22: 4–10.

Skurk T, Hauner H. 2004. Obesity and impaired fibrinolysis: role of adipose production of plasminogen activator inhibitor-1. *Int J Obes Relat Metab Disord* 28: 1357–64.

Slaets JPJ. 2006. Vulnerability in the elderly: Frailty. *Med Clin North Am* 90: 593–601.

Smith G, Clarke D. 2006. Assessing the effectiveness of integrated interventions: terminology and approach. *Med Clin North Am* 90: 533–48.

Smith GD, Timpson N, Lawlor DA. 2006. Editorial. C-reactive protein and cardiovascular risk: still an unknown quantity? *Ann Intern Med* 145: 70–72.

Smith RG, Jiang H, Sun Y. 2005. Developments in ghrelin biology and potential clinical relevance. *Trends Endocrinol Metab* 16: 436–42.

Smith SC. 2006. Current and future directions of cardiovascular risk prediction. *Am J Cardiol* 97: 28–32A.

Smits H, Draper P. 1974. Care of the aged: an English lesson? *Ann Intern Med* 80: 747–53.

Smits P, Dy P, Mitra S, Lefebvre V. 2004. Sox5 and Sox 6 are needed to develop and maintain source, columnar, and hypertrophic chondrocytes in the cartilage growth plate. *J Cell Biol* 164: 747–58.

Snijder MB, van Dam RM, Visser M, Seidell JC. 2006. What aspects of body fat are particularly hazardous and how do we measure them? *Int J Epidemiol* 25: 83–92.

Snow MH, Mikuls TR. 2005. Rheumatoid arthritis and cardiovascular disease: the role of systemic inflammation and evolving strategies of prevention. *Curr Opin Rheumatol* 17: 234–41.

Snyder PJ, Peachey H, Berlin JA, Hannoush P, Haddad G, Dlewati A, Santanna J, Loh L, Lenrow DA, Holmes JH, Kapoor SC, Atkinson LE, Strom BL. 2000. Effects of testosterone replacement in hypogonadal men. *J Clin Endocrinol Metab* 85: 2670–77.

Sokoloff L. 1969. *The biology of degenerative joint disease.* Chicago: University of Chicago Press.

Solomon DH. 2005. Selective cyclooxygenase 2 inhibitors and cardiovascular events. *Arthritis Rheum* 52: 1968–78.

Solomon DH, Avorn J, Katz JN, Finkelstein JS, Arnold M, Polinski JM, Brookhart MA. 2005a. Compliance with osteoporosis medications. *Arch Intern Med* 165: 2414–19.

Solomon DH, Finkelstein JS, Wang PS, Avorn J. 2005b. Statin lipid-loweing drugs and bone mineral density. *Pharmacoepidemiol Drug Saf* 14: 219–26.

Sowers JR. 2003. Effects of statins on the vasculature: implications for aggressive lipid management in the cardiovascular metabolic syndrome. *Am J Cardiol* 91 (Suppl.): 14–22B.

Sowers MF, Hayes C, Jamadar D, Capul D, Lachance L, Jannausch M, Welch G. 2003. Magnetic resonance-detected subchondral bone marrow and cartilage defect characteristics associated with pain and X-ray-defined knee osteoarthritis. *Osteoarthritis Cartilage* 11: 387–93.

Spector TD, Cooper C. 1993. Radiographic assessment of osteoarthritis in population studies: Whither Kellgren and Lawrence? *Osteoarthritis Cartilage* 1: 203–6.

Spector TD, Hart DJ, Nandra D, Doyle DV, Mackillop N, Gallimore JR, Pepys MB. 1997. Low-level increases in serum C-reactive protein are present in early osteoarthritis of the knee and predict progressive disease. *Arthritis Rheum* 40: 723–27.

Spillantini MG, Murrell JR, Goedert M, Farlow MR, Klug A, Ghetti B. 1998. Mutation in the tau gene in familial multiple system tauopathy with presenile dementia. *Proc Natl Acad Sci USA* 95(13): 7737–41.

Sporn MB. 2006. The early history of TGF-β, and a brief glimpse of its future. *Cytokine Growth Factor Rev* 17: 3–7.

Stamm JA, Ornstein DL. 2005. The role of statins in cancer prevention and treatment. *Oncology* (Williston Park) 19: 739–50.

Steel K. 2004. Geriatrics: A profession in the making. *J Gerontol A: Biol Sci Med Sci* 59: 1168–69.

Steeve KT, Marc P, Sandrine T, Dominique H, Yannick F. 2004. IL-6, RANKL, TNF-alpha/IL-1: interrelations in bone resorption pathophysiology. Cytokine Growth Factor Rev 15: 49–60.

Steinberg D. 2002. Atherogenesis in perspective: hypercholesterolemia and inflammation as partners in crime. *Nat Med* 8: 16–22.

Steiner DF, Boitard C, Cerasi E, Efendic S, Henquin JC, Ferrannini E. 2006. Toward a systems biology of insulin secretion and type 2 diabetes. *Diabetes* 55: S1–S4.

Stenger S. 2005. Immunological control of tuberculosis: role of tumor necrosis factor and more. *Ann Rheum Dis* 64 (Suppl. 4): iv24–28.

Stephensen CB, Kelley DS. 2006. The innate immune system: friend or foe? *Am J Clin Nutr* 83: 187–88.

Sternberg EM. 1997. Neural-immune interactions in health and disease. *J Clin Invest* 100: 2641–47.

St George-Hyslop PH. 2000. Molecular genetics of Alzheimer's disease. *Biol Psychiatry* 47(3): 183–99.

Stone NJ. 2004. Stopping statins. *Circulation* 110: 2280–82.

Stone NJ, Saxon D. 2005. Approach to treatment of the patient with metabolic syndrome: lifestyle therapy. *Am J Cardiol* 96: 15–21E.

Stopeck AT. 2005. Women's health: the struggle to restore hormonal balance. *Am J Med* 118: 1181–82.

Strain JJ, Hamerman D. 1978. Ombudsmen (medical-psychiatry) rounds: an approach to meeting patient-staff needs. *Ann Intern Med* 88: 550–53.

Strittmatter WJ, Saunders AM, Schmechel D, Pericak-Vance M, Enghild J, Salvesen GS, Roses AD. 1993. Apolipoprotein E: high-avidity binding to beta-amyloid and increased frequency of type 4 allele in late-onset familial Alzheimer disease. *Proc Natl Acad Sci USA* 90(5): 1977–81.

Struglics A, Larsson S, Pratta MA, Kumar S, Lark MW, Lohmander LS. 2006. Human osteoarthritis synovial fluid and joint cartilage contain both aggrecanase-and matrix metalloproteinase-generated aggrecan fragments. *Osteoarthritis Cartilage* 14: 101–13.

Stuchbury G, Munch G. 2005. Alzheimer's associated inflammation, potential drug targets and future therapies. *J Neural Transm* 112: 429–53.

Suda T, Kobayashi K, Jimi E, Udagawa N, Takahashi N. 2001. The molecular basis of osteoblast differentiation and activation. *Novartis Found Symp* 232: 235–47.

Suda T, Takahashi N, Udagawa N, Jimi E, Gillespie M, Martin TJ. 1999. Modulation of osteoclast differentiation and function by the new members of the tumor necrosis factor receptor and ligand families. *Endocr Rev* 20: 345–57.

Sudore RL, Yaffe K, Satterfield S, Harris TB, Mehta KM, Simonsick EM, Newman AB, Rosano C, Rooks R, Rubin SM, Ayonayon HN, Schillinger D. 2006. Limited literacy and mortality in the elderly. The Health, Aging, and Body Composition Study. *J Gen Intern Med* 21: 806–12.

Suganami T, Nishida J, Ogawa Y. 2005. A paracrine loop between adipocytes and macrophages aggravates inflammatory changes: role of free fatty acids and tumor necrosis factor alpha. *Arterioscler Thromb Vasc Biol* 25: 2062–68.

Sullivan PW, Morrato EH, Ghushchyan V, Wyatt HR, Hill JO. 2005. Obesity, inactivity, and the prevalence of diabetes and diabetes-related cardiovascular comorbidities in the US, 2000–2002. *Diabetes Care* 28: 1599–1603.

Surh Y-J. 2005. Transcriptional regulation of cellular antioxidant defense mechanisms. In: Surh Y-J, Packer L (eds.), *Oxidative stress, inflammation, and health*. New York: Taylor and Francis, pp. 21–40.

Suzman RM, Willis DP, Manton KG. 1992. *The oldest old*. New York: Oxford University Press.

Suzuki T, Ando K, Isohara T, Oishi M, Lim GS, Satoh Y, Wasco W, Tanzi RE, Nairn AC, Greengard P, Gandy SE, Kirino Y. 1997. Phosphorylation of Alzheimer beta-amyloid precursor-like proteins. *Biochemistry* 36(15): 4643–49.

Syed AA, Weaver JU. 2005. Glucocorticoid sensitivity: the hypothalamic-pituitary-adrenal-tissue axis. *Obesity Res* 13: 1131–33.

Szmitko PE, Wang CH, Weisel RD, de Almeida JR, Anderson TJ, Verma S. 2003. New markers of inflammation and endothelial cell activation. *Circulation* 108: 1917–23.

Szulc P, Beck TJ, Marchand F, Delmas PD. 2005. Low skeletal muscle mass is associated with poor structural parameters of bone and impaired balance in elderly men: the MINOS study. *J Bone Miner Res* 20: 721–29.

Szulc P, Duboeuf F, Marchand F, Delmas PD. 2004. Hormonal and lifestyle determinants of appendicular skeletal muscle mass in men: the MINOS study. *Am J Clin Nutr* 80: 496–503.

Taniguchi CM, Emanuelli B, Kahn CR. 2006. Critical nodes in signalling pathways: insights into insulin action. *Nat Rev Mol Cell Biol* 7: 85–96.

Tanko LB, Christiansen C, Cox DA, Geiger MJ, McNabb MA, Cummings SR. 2005a. Relationship between osteoporosis and cardiovascular disease in postmenopausal women. *J Bone Miner Res* 20: 1912–20.

Tanko LB, Qin G, Alexandersen P, Bagger YZ, Christiansen C. 2005b. Effective doses of ibandronate do not influence the 3-year progression of aortic calcification in elderly osteoporotic women. *Osteoporos Int* 16: 184–90.

Tanriverdi HA, Barut A, Sarikaya S. 2005. Statins have additive effects to vertebral bone mineral density in combination with risedronate in hypercholesterolemic postmenopausal women. *Eur J Obstet Gynecol Reprod Biol* 120: 63–68.

Tanzi RE, Kovacs DM, Kim TW, Moir RD, Guenette SY, Wasco W. 1996. The gene defects responsible for familial Alzheimer's disease. *Neurobiol Dis* 3(3): 159–68.

Tashjian AH, Gagel RF. 2006. Teriparatide [Human PTH (1–34)]: 2.5 years of experience on the use and safety of the drug for the treatment of osteoporosis. *J Bone Miner Res* 21: 354–65.

Tatar M, Bartke A, Antebi A. 2003. The endocrine regulation of aging by insulin-like signals. *Science* 299: 1346–51.

Teitelbaum SL. 2004a. RANKing c-Jun in osteoclast development. *J Clin Invest* 114: 463–65.

Teitelbaum SL. 2004b. Postmenopausal osteoporosis, T cells, and immune dysfunction. *Proc Natl Acad Sci USA* 101: 16711–12.

Thaper A, Zhang W, Wright G, Doherty M. 2005. Relationship between Heberden's nodes and underlying radiographic changes of osteoarthritis. *Ann Rheum Dis* 64: 1214–16.

Theander-Carrillo C, Wiedmer P, Cettour-Rose P, Nogueiras R, Perez-Tilve D, Pfluger P, Castaneda TR, Muzzin P, Schürmann A, Szanto I, Tschöp MH, Rohner-Jeanrenaud F.

2006. Ghrelin action in the brain controls adipocyte metabolism. *J Clin Invest* 116: 1983–93.

Thomas GS. 2005. Should we screen asymptomatic individuals for coronary artery disease or implement universal lipid-lowering therapy? *Cardiol Rev* 13: 40–45.

Thompson RB. 2001. Foundations for blockbuster drugs in federally sponsored research. *FASEB J* 15: 1671–76.

Thornberry NA, Lazebnik Y. 1999. Caspases: enemies within. *Science* 281: 1312–17.

Thun MJ, Henley SJ, Gansler T. 2004. Inflammation and cancer: an epidemiological perspective. *Novartis Found Symp* 256: 6–21.

Tikellis C, Cooper ME, Thomas MC. 2006. Role of the rennin-angiotensin system in the endocrine pancreas: Implications for the development of diabetes. *Int J Biochem Cell Biol* 38: 737–51.

Tilj H, Moschen AR. 2006. Adipocytokines: mediators linking adipose tissue, inflammation and immunity. *Nat Rev Immunol* 6: 772–83.

Timpson NJ, Lawlor DA, Harbord RM, Gaunt TR, Day IN, Palmer LJ, Hattersley AT, Ebrahim S, Lowe GD, Rumley A, Davey Smith G. 2005. C-reactive protein and its role in metabolic syndrome: Mendelian randomisation study. *Lancet* 366: 1954–59.

Tinetti ME, Fried T. 2004. The end of the disease era. *Am J Med* 116: 179–85.

Tinetti ME, Inouye SK, Gill TM, Doucette JT. 1995. Shared risk factors for falls, incontinence, and functional dependence: unifying the approach to geriatric syndromes. *JAMA* 273: 1348–53.

Togo T, Utani A, Naitoh M, Ohta M, Tsuji Y, Morikawa N, Nakamura M, Suzuki S. 2006. Identification of cartilage progenitor cells in the adult ear perichondrium: utilization for cartilage reconstruction. *Lab Invest* 86: 445–57.

Tomidokoro Y, Ishiguro K, Harigaya Y, Matsubara E, Ikeda M, Park JM, Yasutake K, Kawarabayashi T, Okamoto K, Shoji M. 2001. Abeta amyloidosis induces the initial stage of tau accumulation in APP(Sw) mice. *Neurosci Lett* 299(3): 169–72.

Topol EJ. 2004. Intensive statin therapy: a sea change in cardiovascular prevention. *N Engl J Med* 350: 1562–64.

Topol EJ. 2005. Arthritis medicines and cardiovascular events: "House of Coxibs." *JAMA* 293: 366–68.

Toussaint O, Osiewacz HD, Lithgow GJ, Brack C (eds.). 2000. Healthy aging for functional longevity. Molecular and cellular gerontology. *Ann N Y Acad Sci* 908.

Tracy RD. 2003. Emerging relationships of inflammation, cardiovascular disease and chronic diseases of aging. *Int J Obes* 27 (Suppl. 3): S29–34.

Trujillo ME, Pajvani UB, Scherer PE. 2005. Apotosis through targeted activation of caspase 8 ("ATTAC-mice"). Novel mouse models of inducible and reversible tissue ablation. *Cell Cycle* 4: 1141–45.

Trujillo ME, Scherer PE. 2005. Adiponectin: journey from an adipocyte secretory protein to biomarker of the metabolic syndrome. *J Intern Med* 257: 167–75.

Tsimikas S, Willerson JT, Ridker PM. 2006. C-reactive protein and other emerging blood biomarkers to optimize risk stratification of vulnerable patients. *J Am Coll Cardiol* 47: C19–31.

Tsokos GC, Tsokos M. 2003. The TRAIL to arthritis. *J Clin Invest* 112: 1315–17.

Tyson KL, Reynolds JL, McNair R, Zhang Q, Weissberg PL, Shanahan CM. 2003. Osteo/ chondrocyte transcription factors and their target genes exhibit distinct patterns of expression in human arterial calcification. *Arterioscler Thromb Vasc Biol* 23: 489–94.

Uematsu S, Akira S. 2006. Toll-like receptors and innate immunity. *J Mol Med* 84: 712–25.

Unger RH. 2005. Longevity, lipotoxicity, and leptin: the adipocyte defense against feasting and famine. *Biochimie* 87: 57–64.

Valdes AM, Andrew T, Gerdner JP, Kimura M, Oelsner E, Cherkas LF, Aviv A, Spector TD. 2005. Obesity, cigarette smoking, and telomere length in women. *Lancet* 366: 662–64.

Vamecq J, Latruffe N. 1999. Medical significance of peroxisome proliferator-activated receptors. *Lancet* 354: 141–48.

Vance ML. 2003. Can growth hormone prevent aging? *N Engl J Med* 348: 779–80.

Van Damme J, Mantovani A. 2005. From cytokines to chemokines. *Cytokine Growth Factor Rev* 16: 549–51.

van Wijk JPH, Rabelink TJ. 2004. PPAR-γ agonists: shifting attention from the belly to the heart? *Arterioscler Thromb Vasc Biol* 24: 798–800.

Vasan RS. 2006. Biomarkers of cardiovascular disease. Molecular basis and practical considerations. *Circulation* 113: 2335–62.

Vassar R, Citron M. 2000. Abeta-generating enzymes: recent advances in beta-and gamma-secretase research. *Neuron* 27(3): 419–22.

Veillette CJ, von Schroeder HP. 2004. Endothelin-1 down-regulates the expression of vascular endothelial growth factor-A associated with osteoprogenitor proliferation and differentiation. *Bone* 34: 288–96.

Veldhuis JD, Erickson D, Iranmanesh A, Miles JM, Bowers CY. 2005a. Sex-steroid control of the aging somatotropic axis. *Endocrinol Metab Clin North Am* 34: 877–93.

Veldhuis JD, Iranmanesh A, Weltman A. 1997. Elements in the pathophysiology of diminished growth hormone (GH) secretion in aging humans. *Endocrine* 7: 41–48.

Veldhuis JD, Keenan DM, Iranmanesh A. 2005b. Mechanisms of ensemble failure of the male gonadal axis in aging. *J Endocrinol Invest* 28: 8–13.

Verbrugge LM, Patrick DL. 1995. Seven chronic conditions: their impact on US adults' activity levels and use of medical services. *Am J Public Health* 85: 173–82.

Verma S, Devaraj S, Jinlal I. 2006. C-reactive protein promotes atherothrombosis. *Circulation* 113: 2135–50.

Vermeulen A. 2000. Andropause. *Maturitas* 34: 5–15.

Vignon E. 2004. Radiographic issues in imaging the progression of hip and knee osteoarthritis. *J Rheumatol Suppl.* 70: 36–44.

Vijg J, Suh Y. 2005. Genetics of longevity and aging. *Annu Rev Med* 56: 192–212.

Vila L. 2004. Cyclooxygenase and 5-lipoxygenase pathways in the vessel wall: role in atherosclerosis. *Med Res Rev* 24: 399–424.

Vincent I, Jicha G, Rosado M, Dickson DW. 1997. Aberrant expression of mitotic cdc2/ cyclin B1 kinase in degenerating neurons of Alzheimer's disease brain. *J Neurosci* 17: 3588–98.

Vincent I, Rosado M, Davies P. 1996. Mitotic mechanisms in Alzheimer's disease? *J Cell Biol* 132: 413–25.

Visser M, Pahor M, Taaffe DR, Goodpaster BH, Simonsick EM, Newman AB, Nevitt M, Harris TB. 2002. Relationship of interleukin-6 and tumor necrosis factor-alpha with muscle mass and muscle strength in elderly men and women: the Health ABC Study. *J Gerontol A Biol Sci Med Sci* 57: M326–32.

Vita AJ, Terry RB, Hubert HB, Fries FM. 1998. Aging, health risks, and cumulative disability. *N Engl J Med* 338: 1035–41.

Vogt TM, Ross PD, Palermo L, Musliner T, Genant HK, Black D, Thompson DE. 2000. Vertebral fracture prevalence among women screened for the Fracture Intervention Trial and a simple clinical tool to screen for undiagnosed vertebral fractures. Fracture Intervention Trial Research Group. *Mayo Clin Proc* 75: 888–96.

von Werder K. 1999. The somatopause is no indication for growth hormone therapy. *J Endocrinol Invest* 22: 137–41.

Walker K, Olson MF. 2005. Targeting Ras and Rho GTPases as opportunities for cancer therapeutics. *Curr Opin Genet Dev* 5: 62–68.

Wallace DC. 2005. A mitochondrial paradigm of metabolic and degenerative diseases, aging, and cancer: a dawn for evolutionary medicine. *Annu Rev Genet* 39: 359–407.

Walsh MC, Hunter GR, Livingstone MB. 2006b. Sarcopenia in premenopausal and postmenopausal women with osteopenia, osteoporosis and normal bone mineral density. *Osteoporos Int* 17: 61–67.

Walsh MC, Kim N, Kadono Y, Rho J, Lee SY, Lorenzo J, Choi Y. 2006a. Osteoimmunology: interplay between the immune system and bone metabolism. *Annu Rev Immunol* 24: 33–63.

Walston J. 2004. Frailty: the search for underlying causes. *Sci Aging Knowledge Environ* 2004: pe4.

Wang X, Stocco DM. 2005. The decline in testosterone biosynthesis during male aging: a consequence of multiple alterations. *Mol Cell Endocrinol* 238: 1–7.

Warner TD, Mitchell JA. 2004. Cyclooxygenases: new forms, new inhibitors, and lessons from the clinic. *FASEB J* 18: 790–804.

Waters DD. 2006. What the statin trails have taught us. *Am J Cardiol* 98: 129–34.

Waters MJ, Hoang HN, Fairlie DP, Pelakanos RA, Brown RJ. 2006. New insights into growth hormone action. *J Mol Endocrinol* 36: 1–7.

Watts N. 2002. Bisphosphonates, statins, osteoporosis, and atherosclerosis. *South Med J* 85: 578–82.

Watts NB, D'Alessio DA. 2006. Editorial: type 2 diabetes, thiazolidinediones: bad to the bone? *J Clin Endocrinol Metab* 91: 3276–78.

Watts NB, Harris ST, Mcclung MR, Bilezikian JP, Greenspan SL, Luckey MM. 2006. Bisphosphonates and osteonecrosis of the jaw. Ann Intern Med 145: 791–2.

Weber KT. 2005. The proinflammatory heart failure phenotype: a case of integrative physiology. *Am J Med Sci* 330: 219–26.

Weiss LA, Barrett-Connor E, von Muhlen D, Clark P. 2006. Leptin predicts BMD and bone resorption in older women but not in older men: the Rancho Bernardo Study. *J Bone Miner Res* 21: 758–64.

Weitzmann MN, Pacifici R. 2006. Estrogen deficiency and bone loss: an inflammatory tale. *J Clin Invest* 116: 1186–94.

Wellen KE, Hotamisligil GS. 2005. Inflammation, stress, and diabetes. *J Clin Invest* 115: 1111–19.

Wen Y, Yang S, Liu R, Brun-Zinkernagel AM, Koulen P, Simpkins JW. 2004. Transient cerebral ischemia induces aberrant neuronal cell cycle re-entry and Alzheimer's disease-like tauopathy in female rats. *J Biol Chem* 279: 22684–92.

Wencker D, Chandra M, Nguyen K, Miao W, Garantziotis S, Factor SM, Shirani J, Armstrong RC, Kitsis RN. 2003. A mechanistic role for cardiac myocyte apoptosis in heart failure. *J Clin Invest* 111: 1497–1504.

Werner N, Nickenig G, Laufs UN. 2002. Pleiotropic effects of HMG-CoA reductase inhibitors. *Basic Res Cardiol* 97: 105–16.

Westendorp RGJ. 2006. What is healthy aging in the 21st century? *Am J Clin Nutr* 83: 404–9S

Westendorp RGJ, Wimmer C. 2005. Linking early development with ageing: towards a new research agenda. *Mech Ageing Dev* 126: 419–20.

White EB. 1977. *Essays of EB White*. New York: Harper Collins. Foreword.

White MF. 2002. IRS proteins and the common path to diabetes. *Am J Physiol Endocrinol Metab* 283: E413–22.

Whyte MP. 2006. The long and short of bone therapy. *N Engl J Med* 354: 860–63.

Wick G, Knoflach M, Xu Q. 2004. Autoimmune and inflammatory mechanisms in atherosclerosis. *Annu Rev Immunol* 22: 361–403.

Wick G, Xu Q. 1999. Editorial. Atherosclerosis is a paradigmatic disease of the elderly, the roots of which are laid in youth, whereas the clinically manifested consequences become evident in old age. *Exp Gerontol* 34: 481–82.

Wijchers PJEC, Burbach JPH, Smidt MP. 2006. In control of biology: of mice, men and foxes. *Biochem J* 397: 233–46.

Wildavsky A. 1977. Doing better and feeling worse: the political pathology of health policy. *Daedalus* 106: 105–22.

Wilkins CH, Birge SJ. 2005. Prevention of osteoporotic fractures in the elderly. *Am J Med* 118: 1190–95.

Willerson JT, Ridker PM. 2004. Inflammation as a cardiovascular risk factor. *Circulation* 109 (Suppl. II): II-2-10.

Williams TF. 2004. Commentary. *J Gerontol A: Biol Sci Med Sci* 59: 1170–71.

Wilson JF. 2004. Frailty—and its dangerous effects—might be preventable. *Ann Intern Med* 141: 489–92.

Wilson MMG. 2001. Bitter-sweet memories: truth and fiction. *J Gerontol Med Sci* 56A: M196–99.

Winthrop KL. 2005. Update on tuberculosis and other opportunistic infections associated with drugs blocking tumour necrosis factor α. *Ann Rheum Dis* 64: iv29–30.

Wolf MS, Gazmararian JA, Baker DW. 2005. Health literacy and functional health status among older adults. *Arch Intern Med* 165: 1946–52.

Wolfe MS, Haass C. 2001. The Role of presenilins in gamma-secretase activity. *J Biol Chem* 276: 5413–16.

Wolozin B, Brown J, 3rd, Theisler C, Silberman S. 2004. The cellular biochemistry of cholesterol and statins: insights into the pathophysiology and therapy of Alzheimer's disease. *CNS Drug Rev* 10: 127–46.

Wong JMY, Collins K. 2004. Telomere maintenance and disease. *Lancet* 362: 983–88.

Woo B. 2006. Primary care: the best job in medicine? *N Engl J Med* 355: 864–66.

Woo SB, Hellstein JW, Kalmar JR. 2006. Systematic review: bisphosphonates and osteonecrosis of the jaws. *Ann Intern Med* 144: 753–61.

Woods SC, Benoit SC, Clegg DJ. 2006. The brain-gut-islet connection. Diabetes 55: S114-S121.

Wyss-Coray T. 2006. Inflammation in Alzheimer disease: driving force, bystander or beneficial response? *Nat Med* 12: 1005–15.

Xing L, Boyce BF. 2005. Regulation of apoptosis in osteoclasts and osteoblastic cells. *Biochem Biophys Res Comm* 328: 709–20.

Yach D, Hawkes C, Gould CL, Hofman KJ. 2004. The global burden of chronic diseases. Overcoming impediments to prevention and control. *JAMA* 291: 2616–22.

Yamaguchi T, Kanatani M, Yamauchi M, Kaji H, Sugishita T, Baylink DJ, Mohan S, Chihara K, Sugimoto T. 2006. Serum levels of insulin-like growth factor (IGF); IGF-binding proteins-3,-4, and-5; and their relationships to bone mineral density and the risk of vertebral fractures in postmenopausal women. *Calcif Tissue Int* 78: 18–24.

Yancik R, Ries LAG. 2004. Cancer in older persons: magnitude of the problem and efforts to advance the aging/cancer research interface. In: Balducci L, Lyman GH, Ershler WB, Extermann M (eds.), *Comprehensive geriatric oncology,* 2nd ed. UK: Taylor & Francis Group, pp. 38–46.

Yang J, Anzo M, Cohen P. 2005. Control of aging and longevity by IGF-1 signaling. *Exp Gerontol* 40: 867–72.

Yang Y, Geldmacher DS, Herrup K. 2001. DNA replication precedes neuronal cell death in Alzheimer's disease. *J Neurosci* 21: 2661–68.

Yang Y, Mufson EJ, Herrup K. 2003. Neuronal cell death is preceded by cell cycle events at all stages of Alzheimer's disease. *J Neurosci* 23: 2557–63.

Ye P, Zhang XJ, Wang ZJ, Zhang C. 2006. Effect of aging on the expression of peroxisome proliferator-activated receptor gamma and the possible relation to insulin resistance. *Gerontology* 52: 69–75.

Yeh ETH. 2004. CRP as a mediator of disease. *Circulation* 109 (Suppl. II): II-11-14.

Yi-Hao Y, Ginsberg HN. 2005. Adipocyte signaling and lipid homeostasis: sequelae of insulin-resistant adipose tissue. *Circ Res* 96: 1042–52.

Yin T, Li L. 2006. The stem cell niches in bone. *J Clin Invest* 116: 1195–1201.

Yki-Järvinen H. 2004. Thiazolidinediones. *N Engl J Med* 351: 1106–18.

Ylitalo R. 2000. Bisphosphonates and atherosclerosis. *Gen Pharmacol* 35: 287–96.

Yoneda T, Hiraga T. 2005. Crosstalk between cancer cells and bone microenvironment. *Biochem Biophys Res Comm* 328: 679–87.

Yura S, Itoh H, Sagawa N, Yamamoto H, Masuzaki H, Nakao K, Kawamura M, Takemura M, Kakui K, Ogawa Y, Fujii S. 2005. Role of premature leptin surge in obesity resulting from intrauterine undernutrition. *Cell Metab* 1: 371–78.

Zaidi M, Blair HC, Moonga BS, Abe E, Huang CL-H. 2003. Osteoclastogenesis, bone resorption, and osteoclast-based therapeutics. *J Bone Miner Res* 18: 599–609.

Zamboni M, Mazzali G, Zoico E, Harris TB, Meigs JB, Di Francesco V, Fantin F, Bissoli L, Bosello O. 2005. Health consequences of obesity in the elderly: a review of four unresolved questions. *Int J Obes Relat Metab Disord* 29: 1011–29.

Zerhouni EA. 2005. Translational and clinical science: time for a new vision. *N Engl J Med* 353: 1621–23.

Zhang L, Chawla A. 2004. Role of PPARγ in macrophage biology and atherosclerosis. *Trends Endocrinol Metab* 15: 500–505.

Zhang W, Doherty M. 2005. How important are genetic factors in osteoarthritis? Contributions from family studies. *J Rheumatol* 32: 1139–42.

Zhang X, Schwarz EM, Young DA, Puzas JE, Rosier RN, O'Keefe RJ. 2002. Cyclooxygenase-2 regulates mesenchymal cell differentiation into the osteoblast lineage and is critically involved in bone repair. *J Clin Invest* 109: 1405–15.

Zhou Y, Eppenberger-Castori S, Marx C, Yan C, Scott GK, Eppenberger U, Benz CC. 2005. Activation of nuclear factor-kB (NF-kB) identifies a high risk subset of hormone-dependent breast cancers. *Int J Biochem Cell Biol* 37: 1130–44.

Zhu X, Raina AK, Rottkamp CA, Aliev G, Perry G, Boux H, Smith MA. 2001. Activation and redistribution of c-jun N-terminal kinase/stress activated protein kinase in degenerating neurons in Alzheimer's disease. *J Neurochem* 76: 435–41.

Zick Y. 2003. Role of Ser/Thr kinases in the uncoupling of insulin signaling. *Int J Obes Relat Metab Disord* 27 (Suppl. 3): S56–60.

Zimmet P, Thomas CR. 2003. Genotype, obesity and cardiovascular disease: has technical and social advancement outstripped evolution? *J Intern Med* 254: 114–25.

Index